Plant Taxonomy: Classical and Modern Methods

Plant Taxonomy: Classical and Modern Methods

Edited by Freddie Casey

New York

Published by Syrawood Publishing House,
750 Third Avenue, 9th Floor,
New York, NY 10017, USA
www.syrawoodpublishinghouse.com

Plant Taxonomy: Classical and Modern Methods
Edited by Freddie Casey

© 2020 Syrawood Publishing House

International Standard Book Number: 978-1-68286-856-0 (Hardback)

This book contains information obtained from authentic and highly regarded sources. Copyright for all individual chapters remain with the respective authors as indicated. All chapters are published with permission under the Creative Commons Attribution License or equivalent. A wide variety of references are listed. Permission and sources are indicated; for detailed attributions, please refer to the permissions page and list of contributors. Reasonable efforts have been made to publish reliable data and information, but the authors, editors and publisher cannot assume any responsibility for the validity of all materials or the consequences of their use.

Trademark Notice: Registered trademark of products or corporate names are used only for explanation and identification without intent to infringe.

Cataloging-in-Publication Data

Plant taxonomy : classical and modern methods / edited by Freddie Casey.
  p. cm.
Includes bibliographical references and index.
ISBN 978-1-68286-856-0
1. Plants--Classification. 2. Plants--Classification--Methodology. I. Casey, Freddie.
QK95 .P53 2020
581.012--dc23

TABLE OF CONTENTS

Preface VII

Chapter 1 Nutlet Morphology and its Taxonomic Significance in the Genus
Mentha L. (Lamiaceae) 1
Gül Tarimcilar, Özer Yilmaz, Ruziye Daşkin and Gönül Kaynak

Chapter 2 Taxonomic Variation among *Schinus Molle* L. Plants Associated
with a Slight Change in Elevation 11
Abeer Al-Andal, Mahmoud Moustafa and Suliman Alruman

Chapter 3 Numerical Taxonomic Analysis in Leaf Architectural Traits of
some *Hoya* R. Br. Species (Apocynaceae) 21
Jess H. Jumawan and Inocencio E. Buot, Jr

Chapter 4 *Camellia* (Theaceae) Classification with Support Vector Machines
based on Fractal Parameters and Red, Green, and Blue Intensity
of Leaves 30
W. Jiang, Z. M. Tao, Z. G. Wu, N. Mantri, H. F. Lu and Z. S. Liang

Chapter 5 Pollen Characters as Taxonomic Evidence in some Species of
Dipsacaceae 46
Ebadi-Nahari Mostafa, Nikzat-Siahkolaee Sedigheh and
Eftekharian Rosa

Chapter 6 Checklist of Mosses (Bryophyta) of Gangetic Plains 54
Krishna Kumar Rawat, Afroz Alam and Praveen Kumar Verma

Chapter 7 Taxonomic Revision of Saudi Arabian *Tetraena* Maxim. and
Zygophyllum L. (Zygophyllaceae) with one New Variety and
Four New Combinations 64
Dhafer Ahmed Alzahrani and Enas Jameel Albokhari

Chapter 8 Three Lichen Taxa new for Turkey 89
Kenan Yazici and André Aptroot

Chapter 9 Notes on the Genus *Tylophora* R. Br. (Asclepiadaceae) 96
L. Rasingam, J. Swamy and S. Nagaraju

Chapter 10 Phylogeny of *Galium* L. (Rubiaceae) from Korea and Japan
based on Chloroplast DNA Sequence 102
Keum Seon Jeong, Jae Kwon Shin, Masayuki Maki and
Jae-Hong Pak

Chapter 11 Typification of Fourteen Names of Twelve Recognized Taxa in
            Leucas R. Br. (Lamiaceae) and one New Combination 112
            Rajeev Kumar Singh

Chapter 12 Taxonomy and Reproductive Biology of the Genus *Zephyranthes*
            Herb. (Liliaceae).. 123
            Sumona Afroz, M. Oliur Rahman and Md. Abul Hassan

Chapter 13 Type Specimens of Names in *Bauhinia* and *Phanera*
            (Fabaceae: Caesalpinioideae) at Central National Herbarium,
            Howrah (Cal) 136
            S. Bandyopadhyay and P. P. Ghoshal

Chapter 14 Indian Cheilanthoid Fern - A Numerical Taxonomic Approach 144
            Kakali Sen and Radhanath Mukhopadhyay

Chapter 15 New Angiospermic Taxa for the Flora..154
            M. Oliur Rahman and Md. Abul Hassan

Chapter 16 Updated Nomenclature and Taxonomic Status of the Plants of
            Bangladesh Included in Hook. F. The *Flora of British India*....................................161
            M. Enamur Rashid and M. Atiqur Rahman

Chapter 17 An Annotated Checklist of the Angiospermic Flora of Rajkandi
            Reserve Forest of Moulvibazar..179
            A. K. M. Kamrul Haque, Saleh Ahammad Khan,
            Sarder Nasir Uddin and Shayla Sharmin Shetu

Chapter 18 Species Delineation of the Genus *Diplazium* Swartz (Athyriaceae)
            using Leaf Architecture Characters... 200
            Jennifer M. Conda and Inocencio E. Buot, Jr

Chapter 19 Taxonomic Revision of the Genus *Crinum* L. (Liliaceae) of
            Bangladesh.. 211
            Sumona Afroz, M. Oliur Rahman and Md. Abul Hassan

            Permissions

            List of Contributors

            Index

PREFACE

The science that finds, identifies, classifies, describes and names plants is called plant taxonomy. It is closely associated with plant systematics. Plant taxonomy facilitates an organized system for the cataloging and naming of specimens. Identification, classification and description are the main goals of plant taxonomy. Plant identification is a process of identifying an unknown plant by comparing it with previously collected specimens or through an identification manual. Plant classification is the practice of placing known plants into categories or groups to show some relationship. Giving a formal description of a newly discovered species usually in the form of a scientific paper using ICN guidelines is called plant description. This book provides significant information about this discipline to help develop a good understanding of plant taxonomy and related fields. Coherent flow of topics, student-friendly language and extensive use of examples make it an invaluable source of knowledge. This book will prove to be immensely beneficial to students and researchers in this field of study.

Various studies have approached the subject by analyzing it with a single perspective, but the present book provides diverse methodologies and techniques to address this field. This book contains theories and applications needed for understanding the subject from different perspectives. The aim is to keep the readers informed about the progresses in the field; therefore, the contributions were carefully examined to compile novel researches by specialists from across the globe.

Indeed, the job of the editor is the most crucial and challenging in compiling all chapters into a single book. In the end, I would extend my sincere thanks to the chapter authors for their profound work. I am also thankful for the support provided by my family and colleagues during the compilation of this book.

Editor

NUTLET MORPHOLOGY AND ITS TAXONOMIC SIGNIFICANCE IN THE GENUS *MENTHA* L. (LAMIACEAE) FROM TURKEY

GÜL TARIMCILAR, ÖZER YILMAZ, RUZİYE DAŞKIN[1] AND GÖNÜL KAYNAK

Department of Biology, Faculty of Arts and Science, Uludag University, 16059 Görükle Bursa, Turkey

Keywords: Nutlet morphology; Taxonomy; SEM; *Mentha*; Lamiaceae; Turkey.

Abstract

The nutlet morphology of 11 taxa of *Mentha* L. (*M. pulegium, M. aquatica, M. × piperita, M. x dumetorum, M. spicata* subsp. *spicata, M. spicata* subsp. *tomentosa, M. × villoso-nervata, M. longifolia* subsp. *longifolia, M. longifolia* subsp. *typhoides, M. × rotundifolia* and *M. suaveolens*) distributed throughout Turkey was investigated by scanning electron microscopy (SEM). The shape of all studied nutlets was broadly oblong or ovoid. Nutlet size ranged from 0.54 to 0.97 mm in length and from 0.37 to 0.66 mm in width. The smallest and biggest nutlets were found in *M. × villoso-nervata* and *M. aquatica*, respectively. The *Mentha* taxa studied can be divided into three groups, based on nut sculpturing type such as distinctly bireticulate, inconspicuously bireticulate and reticulate. This study has shown that some nutlet morphological characteristics can be utilised as additional diagnostic characters in delimitations of *Mentha* at the species and infraspecific levels.

Introduction

Mentha L., one of the most important genera of the family Lamiaceae, has worldwide distribution and it consists of perennial aromatic herbs. Some *Mentha* species, such as *M. pulegium* L., *M. longifolia* (L.) Huds., *M. spicata* L., *M. × piperita* L. and *M. × villoso-nervata* Opiz, are traditionally used in folk medicine (Baytop, 1999). Mint oil and their constituents obtained from different species of *Mentha* are also used in perfumery, cosmetics and food industries (Kokkini, 1994).

Mentha is a taxonomically difficult genus because of extensive hybridization, vegetative propagation, polyploidisation and cultivation (Harley, 1972; Harley and Brighton, 1977; Tucker *et al.*, 1980). The genus comprises 18 species and 11 hybrids placed into four sections, namely *Pulegium, Tubulosae, Eriodontes* and *Mentha* according to the latest taxonomic treatment (Tucker and Naczi, 2007). Harley (1982) recognized 11 *Mentha* taxa belonging to two sections (*Pulegium* and *Mentha*) from Turkey and then two hybrids have been added to Flora of Turkey (Tarimcilar and Kaynak, 1997a, b). In this study, the treatment of Harley (1982) has been followed for the nomenlature of *Mentha*.

There are some studies about monophyly of *Mentha* and phylogenetic relationships within the genus (Gobert *et al.*, 2002; Bunsawat *et al.*, 2004; Shasany *et al.*, 2005). Saric-Kundelic *et al.* (2009) investigated the utility of morphological, anatomical and phytochemical characters for the identification of *Mentha* species, hybrids, varieties and cultivars in Bosnia-Herzegovina and Slovakia. In various genera of family Lamiaceae, the nutlet morphology, anatomy, pericarp structure and their taxonomic significance have been reported by some studies (Husain *et al.*, 1990; Marin *et al.*, 1994; Ryding, 2010). However, accounts on the mericarp morphology of some

---

[1]Corresponding author. E-mail: ruziyeg@uludag.edu.tr

taxa of *Mentha* examined in this study are rather limited (Duletic-Lausevic and Marin, 1999; Moon *et al.*, 2009). We aim in this study, with the aid of scanning electron microscope (SEM), to provide detailed data on nutlet morphology of 11 *Mentha* taxa found in Turkey and to determine which characteristics of their nutlets may be used for taxonomic purposes.

Materials and Methods

Plant materials:

Nutlets of 11 taxa of *Mentha* collected from different parts of Turkey were investigated. The materials used in this study were composed mainly of herbarium specimens, which were deposited in the herbarium of Uludag University (BULU). The specimens used for SEM micrographs were presented in Table 1.

Nutlet size and SEM analyses:

For nutlet length and width, 50 nutlets were measured per taxon. However, at least 10 nutlets were measured for hybrids. In order to ensure that the nutlets were of normal size and maturity, they were examined using a stereomicroscope. For SEM, nutlets of taxa were transferred directly to a double-sided tape-affixed stub and were coated with gold-palladium, using a BAL–TEC SCD 005 sputter. The micrographs were obtained from a CARL ZEISS Evo 40 SEM using a voltage of 20 kV at the Microscopy Laboratory of Science and Art Faculty of Uludag University. The micrographs were used to describe surface sculpturing type of nutlets. The terminology for nutlet shape and surface sculpturing mainly follows that of Barthlott (1981) and Stearn (1983).

Table 1. List of taxa used for SEM micrograph (GT- Gül Tarımcılar).

No.	Taxon	Collection data	Vouchers
1	*M. aquatica* L.	A2 Bursa: Fadilli village, 9 m, 3.9.2004	GT 30514
2	*M.* × *dumetorum* Schult.	A1 Kirklareli: Babaeski, 60 m, 23.8.2003	GT 30448
3	*M. pulegium* L.	A2 Istanbul: Cavusbasi, 16.8.2005	GT 30533
4	*M. longifolia* (L.) Huds. subsp. *longifolia*	A2 Bursa: Gemlik, Hayriye village, 10 m, 8.9.2006	GT 30592
5	*M. longifolia* (L.) Huds. subsp. *typhoides* (Briq.) Harley	A2 Istanbul: Sile, 15.8.2005	GT30530
6	*M.* × *piperita* L.	A2 Istanbul: Cavusbasi, Kavaklık, 16.8.2005	GT 30535
7	*M.* × *rotundifolia* (L.) Huds.	B1 Balikesir: Bandırma to Erdek, 130 m, 27.8.2004	GT 30508
8	*M. spicata* L. subsp. *spicata*	A1 Tekirdag: 1 km to Hayrabolu, 70 m, 23.8.2003	GT 30452
9	*M. spicata* L. subsp. *tomentosa* (Briq.) Harley	A2 Bilecik: Pazaryeri, Bahcesultan, 1050 m, 6.9.2006	GT 30562
10	*M. suaveolens* Ehrth.	A2 Yalova: Sultaniye, 25 m, 7.6.2006	GT 30570
11	*M.* × *villoso-nervata* Opiz.	B1 Canakkale: Saros, Kocacesme village, 35 m, 25.8.2004	GT 30470

Results and Discussion

The characteristics of nutlet (i.e. size, colour, presence or absence of trichomes and surface sculpturing) are summarized in Table 2. Micrographs of nutlets belonging to all studied taxa are presented in Figures 1-4. We found that the shape of all studied nutlets was broadly oblong or ovoid and that nutlet colour varied from pale to dark brown. The nutlets of *M. pulegium*, *M. aquatica* and *M. dumetorum* were pale brown, while those of *M.* × *piperita*, *M.* × *villoso-nervata* and *M.* × *rotundifolia* were dark brown. However, the colour of the nutlets in *M. spicata* subsp. *spicata*, *M. spicata* subsp. *tomentosa*, *M. longifolia* subsp. *longifolia*, *M. longifolia* subsp. *typhoides* and *M. suaveolens* varied from chestnut brown to dark brown. Moreover, short or long trichomes were observed on the surface of nutlets of *M. aquatica*, *M.* × *dumetorum*, *M. spicata* subsp. *tomentosa* and *M. longifolia* subsp. *longifolia*. Nutlet size ranged from 0.54 to 0.97 mm in length and from 0.37 to 0.66 mm in width. The smallest nutlet was found in *M.* × *villoso-nervata* and the biggest nutlet was found in *M. aquatica* (Table 2).

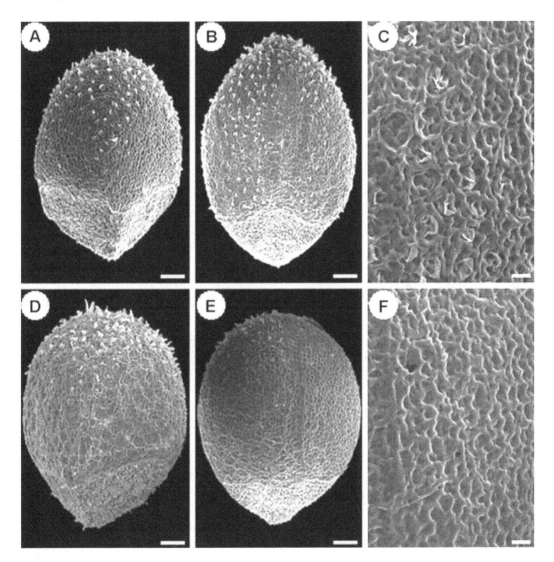

Fig. 1. SEM micrographs of nutlets of *Mentha aquatica* (A-C); *M.* × *dumetorum* (D-F); Ventral view (A, D); dorsal view (B, E); surface sculpturing (C, F). Scale bars: A, B, D, E = 100 µm; C, F = 20 µm.

Under SEM, three types were observed in the *Mentha* taxa based on surface sculpturing pattern:

Type I. Distinctly bireticulate: a surface with penta- or hexagonal-shaped small cells, and the walls of these cells are high, irregular and having depressions. This sculpturing pattern was seen in *M. aquatica* and *M.* × *dumetorum* (Fig. 1C, F).

Table 2. Nutlet characteristics of the studied taxa of *Mentha* L.

Taxon	Length (mm) Mean± SD	Width (mm) Mean± SD	Sculpture	Presence/ absence of trichomes	Colour	Figures
M. aquatica	0.9± 0.07	0.6± 0.05	TYPE I	short hair	pale brown	Fig. 1A-C
M. × *dumetorum*	0.8± 0.15	0.6± 0.06	TYPE I	short hair	pale brown	Fig. 1D-F
M. pulegium	0.7± 0.01	0.5± 0.04	TYPE II	absent	pale brown	Fig. 2A-C
M. × *piperita*	0.7± 0.04	0.5± 0.04	TYPE II	absent	dark brown	Fig. 2D-F
M. spicata subsp. *spicata*	0.8± 0.02	0.6± 0.01	TYPE II	absent	chestnut to dark brown	Fig. 2G-I
M. spicata subsp. *tomentosa*	0.7± 0.01	0.5± 0.01	TYPE III	scarcely hair	chestnut to dark brown	Fig. 3A-C
M. longifolia subsp. *longifolia*	0.6± 0.06	0.5± 0.07	TYPE III	long hair	chestnut to dark brown	Fig. 3D-F
M. longifolia subsp. *typhoides*	0.7± 0.01	0.5± 0.03	TYPE III	absent	chestnut to dark brown	Fig. 3G-I
M. × *villoso-nervata*	0.6± 0.02	0.4± 0.03	TYPE III	absent	dark brown	Fig. 4A-C
M. suaveolens	0.6± 0.02	0.5± 0.02	TYPE III	absent	chestnut to dark brown	Fig. 4D-F
M. × *rotundifolia*	0.6± 0.02	0.4± 0.01	TYPE III	absent	dark brown	Fig. 4G-I

Type II. Inconspicuously bireticulate: a surface covers inconspicuously penta- or hexagonal-shaped small cells, and these cells having various walls. *M. pulegium*, *M.* × *piperita* and *M. spicata* subsp. *spicata* exhibited this type of sculpturing. Only in *M. pulegium*, the nutlets with cells having rigid cell boundary and having star-shaped extensions at their centres (Fig. 2C). The nutlets of *M.* × *piperita* and *M. spicata* subsp. *spicata* with cells having wrinkled or often unclear walls (Fig. 2F, I).

Type III. Reticulate: a surface with penta- or hexagonal-shaped cells having large lumen and smooth, regular walls and forming a net-like appearance on their surface. The nutlets of *M. spicata* subsp. *tomentosa*, *M. longifolia* subsp. *longifolia*, *M. longifolia* subsp. *typhoides*, *M.* × *villoso-nervata*, *M. suaveolens* and *M.* × *rotundifolia* exhibited this type (Figs 3C, F, I; 4C, F, I).

When the nutlet characteristics of the investigated *Mentha* taxa were compared with previous literature (Ball, 1972; Borisova, 1977; Tarimcilar and Kaynak, 2002), our results are more or less similar to their findings. The shape of nutlets examined in this study was broadly oblong or ovoid. Borisova (1977), Harley (1982) and Tarimcilar and Kaynak (2002) have reported that the nutlet shape of the genus *Mentha* varies from globose to ovoid or obovoid.

Duletic-Lausevic and Marin (1999) found nutlet dimensions 0.7 × 0.5 mm in *M. pulegium* and *M. longifolia*, 0.8 × 0.6 mm in *M. aquatica*, 0.6 × 0.4 mm in *M. spicata* and *M.* × *rotundifolia*, and 0.6 × 0.5 mm in *M. suaveolens*. Moon et al. (2009) examined nutlet characteristics (i.e. size, colour, shape and surface sculpturing) of *Mentha aquatica*, *M. longifolia*, and *M. suaveolens* and reported the length and width measurements (mm) as 1±0.05 × 0.7±0.02, 0.6±0.03 × 0.5±0.02 and 0.6±0.03 × 0.4±0.02, respectively. Nutlet shape of these taxa is widely elliptic, surface sculpturing type is reticulate, and colour varies from yellowish brown to reddish dark brown (Moon et al., 2009).

According to our results, the nutlets of *M. aquatica*, *M.* × *dumetorum*, *M. spicata* subsp. *tomentosa* and *M. longifolia* subsp. *longifolia* have trichomes. The presence or absence of trichomes on nutlet is an important character to discriminate *M. longifolia* subsp. *longifolia* and subsp. *typhoides* which have the similar nutlet size, sculpturing and colour (Table 2). On the other hand, Duletic-Lausevic and Marin (1999) stated that the nutlets of *M. spicata*, *M. rotundifolia* and *M. suaveolens* lack trichomes and that *M. aquatica* and *M. longifolia* exhibit nutlets with or without trichomes.

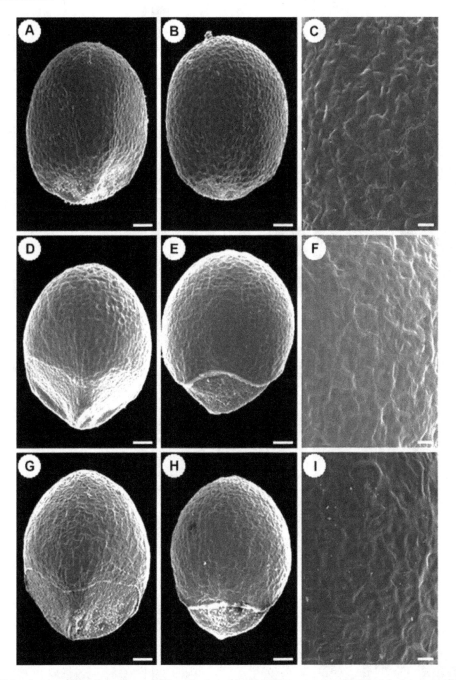

Fig. 2. SEM micrographs of nutlets of *M. pulegium* (A-C); *M.* × *piperita* (D-F); *M. spicata* subsp. *spicata* (G-I). Ventral view (A, D, G); dorsal view (B, E, H); surface sculpturing (C, F, I). Scale bars: A, B, D, E, G, H = 100 μm; C, F, I = 20 μm.

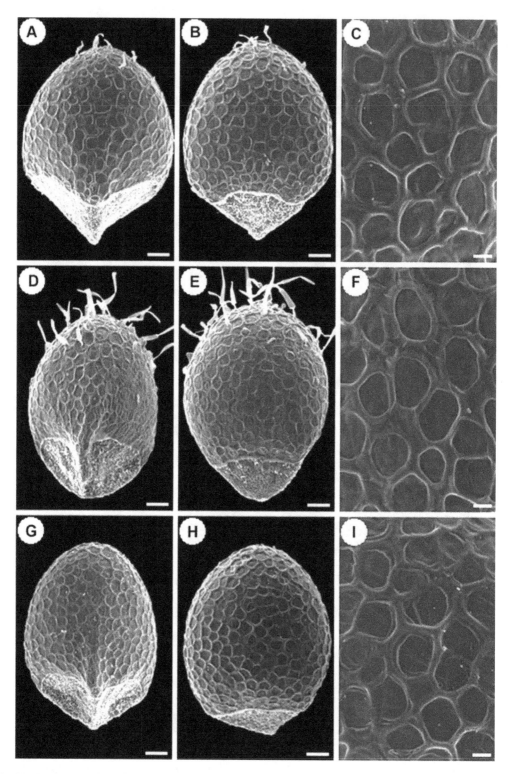

Fig. 3. SEM micrographs of nutlets of *M. spicata* subsp. *tomentosa* (A-C); *M. longifolia* subsp. *longifolia* (D-F); *M. longifolia* subsp. *typhoides* (G-I). Ventral view (A, D, G); dorsal view (B, E, H); surface sculpturing (C, F, I). Scale bars: A, B, D, E, G, H = 100 μm; C, F, I = 20 μm.

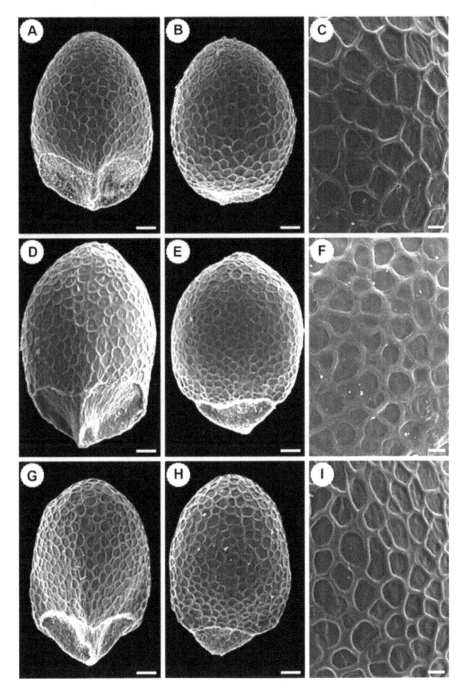

Fig. 4. SEM micrographs of nutlets of *M.* × *villoso-nervata* (A-C); *M. suaveolens* (D-F); *M.* × *rotundifolia* (G-I). Ventral view (A, D, G); dorsal view (B, E, H); surface sculpturing (C, F, I). Scale bars: A, B, D, E, G, H = 100 μm; C, F, I = 20 μm.

Mentha taxa employed in this study can be divided into three informal groups, with regard to nutlet characteristics basically sculpturing patterns. Group I includes *M. aquatica*, *M.* × *dumetorum* (*M. aquatica* × *M. longifolia*) and they are similar to each other both in terms of the morphological features and the nutlet characteristics. However, *M.* × *dumetorum* differs from *M. aquatica* in its more oblong spikes and narrower leaves (Tarimcilar and Kaynak, 1997a, 2002).

Group II consists of *M. pulegium, M. spicata* subsp. *spicata* and *M. × piperita* (*M. aquatica × M. spicata*). Of the studied *Mentha* taxa, only *M. pulegium* is located in sect. *Pulegium*, whereas the others are included in sect. *Mentha*. Sect. *Pulegium* is distinguished from sect. *Mentha* by its bracts similar to leaves, tubular calyx, weakly 2-lipped, with distinctly unequal calyx teeth, hairy within calyx throat, gibbous corolla tube. Sect. *Mentha* have variable bracts, calyx tubular or campanulate, with more or less equal calyx teeth, glabrous calyx throat and straight corolla tube (Harley, 1982). Moreover, the inflorescence of *M. × piperita* is morphologically similar to *M. spicata* in that it forms a terminal spike, but it differs from *M. aquatica* in its more lanceolate leaves that have shorter petioles (3-9 mm or rarely more).

Group III includes *M. spicata* subsp. *tomentosa*, *M. longifolia* subsp. *longifolia*, *M. longifolia* subsp. *typhoides*, *M. × villoso-nervata*, *M. suaveolens* and *M. × rotundifolia*. The nutlet surfaces of this group are covered with penta- or hexagonal-shaped cells that form a particularly net-like appearance. *M. × villoso-nervata* (*M. spicata × M. longifolia*) is morphologically different from the parents in its narrower spikes and smaller leaves and calyx (Tarimcilar and Kaynak, 1997b, 2002). *M. × rotundifolia* (*M. suaveolens × M. longifolia*) resembles *M. suaveolens* in its pale green and strongly rugose leaves, but it differs in that its leaves are more oblong and have an acute apex (Harley, 1982; Tarimcilar and Kaynak, 2002).

Hybrids can be distinguished from their parental species in terms of some nutlet features. As seen in Table 2, *M. × dumetorum* mainly differs from *M. aquatica* and *M. longifolia* with its smaller and distinctly bireticulate sculpturing nutlet. The nutlets of *M. × piperita* are smaller than those of *M. spicata* subsp. *spicata,* but they are more similar to *M. spicata* than *M. aquatica* in terms of nutlet characteristics. They are easily distinguishable from *M. aquatica* due to its inconspicuously bireticulate, glabrous and dark brown nutlet. *M. × villoso-nervata* differs from *M. longifolia* subsp. *longifolia* and *M. spicata* subsp. *spicata* by its glabrous, dark brown and reticulate nutlet, respectively. The nutlet characteristics of *M. × rotundifolia* and *M. suaveolens* display a great similarity with each other.

A key can be established based on nutlet chacteristics for Turkish *Mentha* taxa:

1	Nutlet sculpturing bireticulate	2
-	Nutlet sculpturing reticulate	6
2	Nutlet sculpturing distinctly bireticulate	3
-	Nutlet sculpturing inconspicuously bireticulate	4
3	Nutlets at least 0.83 mm long	*M. aquatica*
-	Nutlets at least 0.65 mm long	*M. × dumetorum*
4	Nutlets 0.78-0.82 mm long	*M. spicata* subsp. *spicata*
-	Nutlets shorter than 0.78 mm	5
5	Nutlet cells with star-shaped extensions at their centres	*M. pulegium*
-	Nutlet cells without star-shaped extensions at their centres	*M. × piperita*
6	Nutlets without hair	7
-	Nutlets with hair	10
7	Nutlets 0.47-0.53 mm wide	8
-	Nutlets 0.37-0.43 mm wide	9
8	Nutlets 0.69-0.71 mm long	*M. longifolia* subsp. *typhoides*
-	Nutlets 0.58-0.62 mm long	*M. suaveolens*

9	Nutlets at least 0.39 mm wide	*M.* × *rotundifolia*
-	Nutlets at least 0.37 mm wide	*M.* × *villoso-nervata*
10	Nutlets 0.69-0.71 mm long	*M. spicata* subsp. *tomentosa*
-	Nutlets 0.54-0.66 mm long	*M. longifolia* subsp. *longifolia*

The utility of nutlet characters, i.e. shape, size, presence or absence of hairs, nature of indumentum, surface sculpturing, exocarp cellular morphology and anatomy of the nutlet has been shown at various taxonomic levels in different genera of Lamiaceae (Husain *et al.*, 1990; Marin *et al.*, 1994; Duletic-Lausevic and Marin, 1999; Moon and Hong, 2006). Our findings also showed that the nutlet size, presence/absence of trichomes, surface sculpturing pattern are valuable diagnostic characteristics for separating closely related taxa of *Mentha*. In conclusion, we can say that nutlet morphological characteristics combined with other morphological characters can be used for delimation of taxa at the species and infraspecific levels in the genus *Mentha*. Furthermore, this study provides the detailed data on the nutlet features of Turkish *Mentha* taxa.

Acknowledgements

We thank Research Foundation of Uludag University (project numbers F-2003/3 and F-2005/4) for financial support.

References

Ball, P.W. 1972. *Mentha* L. *In:* Tutin, T.G., Heywood, V.H., Burges, N.A., Moore, D.M., Valentine, D.H., Walters, S. and Webb, B.A. (Eds), Flora Europaea. Vol. **3**. Cambridge Univ. Press, Cambridge, pp. 183-186.

Barthlott, W. 1981. Epidermal and seed surface characters of plants: systematic applicability and some evolutionary aspects. Nord. J. Bot. **1**: 345-355.

Baytop, T. (Ed.). 1999. Türkiye'de Bitkiler ile Tedavi, Nobel Kitabevleri, pp. 302-304.

Borisova, A.G. 1977. *Mentha* L. *In:* Shishkin, B.K. (Ed.), Flora of the U.S.S.R. Vol. **21**. Translated from Russian Israel Program for Scientific Translations, Jerusalem, pp. 427-449.

Bunsawat, J., Elliott N.E., Hertweck, K.L., Sproles, E. and Alice, L.A. 2004. Phylogenetics of *Mentha* (Lamiaceae): Evidence from chloroplast DNA sequences. Syst. Bot. **29**: 959-964.

Duletic-Lausevic, S. and Marin, P.D. 1999. Pericarp structure and myxocarpy in selected genera of *Nepetoideae* (Lamiaceae). Nord. J. Bot. **19**: 435-446.

Gobert, V., Moja, S., Colson, M. and Taberlet, P. 2002. Hybridization in the section *Mentha* (Lamiaceae) inferred from AFLP markers. Amer. J. Bot. **89**: 2017-2023.

Harley, R.M. 1972. Notes on the genus *Mentha* L. (Labiatae). Bot. J. Linn. Soc. **65**: 250-253.

Harley, R.M. 1982. *Mentha* L. *In:* Davis, P.H. (Ed.), Flora of Turkey and the East Aegean Islands. Vol. **7**. Edinburgh Univ. Press, Edinburgh, pp. 384-394.

Harley, R.M. and Brighton, C.A. 1977. Chromosome numbers in the genus *Mentha* L. Bot. J. Linn. Soc. **74**: 71-96.

Husain, S.Z., Marin, P.D., Silic, C., Qaiser, M. and Petkovic, B. 1990. A micromorphological study of some representative genera in the tribe *Saturejeae* (Lamiaceae). Bot. J. Linn. Soc. **103**: 59-80.

Kokkini, S. 1994. Herbs of the Labiatae. *In:* Macrae, R., Robinson, R.K. and Sadler, M.J. (Eds), Encyclopedia of Food Science, Food Technology and Nutrition, Vols. **1-8**. Academic Press, London, pp. 2342-2348.

Marin, P.D., Petkovic B.P. and Duletic, S. 1994. Nutlet sculpturing of selected *Teucrium* species (Lamiaceae): A character of taxonomic significance. Plant Syst. Evol. **192**: 199-214.

Moon, H. and Hong, S. 2006. Nutlet morphology and anatomy of the genus *Lycopus* (Lamiaceae: Mentheae). J. Pl. Res. **119**: 633-644.

Moon, H., Hong, S., Smets, E. and Huysmans, S. 2009. Micromorphology and character evolution of nutlets in tribe Mentheae (Nepetioideae, Lamiaceae). Syst. Bot. **34**: 760-776.

Ryding, O. 2010. Pericarp structure and phylogeny of tribe Mentheae (Lamiaceae). Plant Syst. Evol. **285**: 165-175.

Saric-Kundelic, B., Fialova, S., Dobes, C., Olzant, S., Tekelova, D., Grancai, D., Reznicek, G. and Saukel, J. 2009. Multivariate numerical taxonomy of *Mentha* species, hybrids, varieties and cultivars. Sci. Pharm. **77**: 851-876.

Shasany, A.K., Darokar, M.P., Dhawan, S., Gupta, A.K., Shukla, A.K., Patra, N.K. and Khanuja, S.P.S. 2005. Use of RAPD and AFLP markers to identify inter- and intrasepecific hybrids of *Mentha*. J. Heredity **96**: 542-549.

Stearn, W. T. (Ed.) 1983. Botanical Latin. 3rd rev., David & Charles Inc., Vermont.

Tarimcilar, G. and Kaynak, G. 1997a. A new record for the Flora of Turkey. Turk. J. Bot. **21**: 247-249.

Tarimcilar, G. and Kaynak, G. 1997b. A new record for the Flora of Turkey. Lagascalia **20**: 113-115.

Tarimcilar, G. and Kaynak, G. 2002. A morphological study on *Mentha* L. (Labiatae) taxa of Black Sea region. Süleyman Demirel Üniv. Fen Bil. Enst. Derg. **5**: 194-229.

Tucker, A.O., Harley, R.M. and Fairbrothers, D.E. 1980. The Linnean Types of *Mentha* (Lamiaceae). Taxon **29**: 233-255.

Tucker, A.O. and Naczi, R.F.C. 2007. *Mentha*: An overview of its classification and relationships. *In:* Lawrence, B.M. (Ed.), Mint: the genus *Mentha*. CRC Press, London, pp. 3-4.

TAXONOMIC VARIATION AMONG *SCHINUS MOLLE* L. PLANTS ASSOCIATED WITH A SLIGHT CHANGE IN ELEVATION

ABEER AL-ANDAL, MAHMOUD MOUSTAFA[1,2] AND SULIMAN ALRUMAN[1]

Department of Biology, College of Science, King Khalid University, Abha, Kingdom of Saudi Arabia

Keywords: RAPD; ISSR; Mixed RAPD; *Schinus molle* L.

Abstract

This study examined the degree of variations in DNA fingerprints associated with slight altitudinal change of *Schinus molle* grown in Abha region, Saudi Arabia. Seven populations from *Schinus molle* plants located at 2193.0, 2246.0, 2197.7, 2441.0, 2372.0, 2250.6 and 2175.0 meters had been investigated. The degree of genetic variability was evaluated using random amplified polymorphic DNA (RAPD), mixed RAPD and inter-simple sequence repeat markers (ISSR). The genetic similarity coefficients from RAPD analysis revealed the maximum similarity value (89.9%) was between population at 2250.6 m and population at 2175.0 m. The genetic similarity coefficients from mixed RAPD primers displayed the highest similarity value (87.6%) between population at 2246.0 m and population at 2197.7 m. Similarity coefficients from ISSR analysis revealed the highest similarity value (86.2%) among populations at 2193.0 m, 2246.0 m, 2441.0 m and at 2250.6 m. Super tree analysis (RAPD + mixed RAPD + ISSR) showed the highest similarity value (85.5%) between population at 2441.0 m and population at 2250.6 m. In conclusion, marker systems including RAPD, mixed RAPD and ISSR, alone or combined can be effectively used in determining the genetic relationship among *Schinus molle* plants even at very close populations.

Introduction

Abha region has a specialized environmental condition among all other areas in the Kingdom of Saudi Arabia which have an indirect effect on the weed plants growth. *S. molle* plants (family, Anacardiaceae) are among the most common weed in Saudi Arabia especially in Tharawat Mountains. The tree of *S. molle* plant is an evergreen, dioecious, grows up to 20 meters in height. Flowers are small, with yellowish white petals and all plant parts especially fruits having strong aroma (Lim, 2012). *S. molle* is common weed in South America and recently into many tropical and subtropical countries (Olafsson *et al.*, 1997). In Abha region, *S. molle* tree has been planted in many areas as in valleys, public gardens and for house decorations. After that, the plant became a common weed in many areas of Abha city growing beside road and next the wall of houses as it is reproduced by seeds. *S. molle* plant showed to be resistant to the harsh environmental condition such as high temperature, cold and increasing soil salinity (Lim, 2012). In addition, *S. molle* plants usually used for the restoration of degraded areas and showed tolerance to heavy metals (Doganlar *et al.*, 2012; Pereira *et al.*, 2016). Toward this approach examine the genetic diversity of *S. molle* plant is highly needed as no reports available.

In recent years, a number of randomly amplified polymorphic DNA–polymerase chain reaction (RAPD-PCR) and inter-simple sequence repeat–polymerase chain reaction (ISSR-PCR) markers had been used to study genetic diversity among plant species. For example, RAPD

---

[1]Corresponding author. Email: mfmostfa@kku.edu.sa
[1]Research Center for Advanced Materials Science (RCAMS), King Khalid University, Abha, Saudi Arabia.
[2]Department of Botany, Faculty of Science, South Valley University, Qena, Egypt.

technique was successfully applied genetically to distinguish among *Ocimum* spp. (Vieria *et al.*, 2003), to study the genetic diversity in *Monodora myristica* (Uyoh *et al.*, 2014) and various population of *Ziziphus spina-christi* L. (Moustafa *et al.*, 2016). ISSR technique was used to study genetic diversity of the *Lens* spp. (Fikiru *et al.*, 2007), and genetic relationships of *Chukrasia* spp. (Wu *et al.*, 2014).

Therefore, the aim of this research is to study genetic diversity of *S. molle* plants growing at close locations in Abha region, KSA. To the best of our knowledge, there are few reports indicating the use of mix primer to estimate the genetic diversity among plant /or to study plant DNA fingerprint. Therefore, this study also aimed to check the status of DAN fingerprints using mixed primers.

Materials and Methods
Plant material

Seven locations at various elevations in Abha region, KSA, include 2193.0, 2246.0, 2197.7, 2441.0, 2372.0, 2250.6 and 2175.0 meters have been selected (Fig. 1). At each site, random samples of young fresh leaves from *S. molle* trees having a height 1500 cm had been collected.

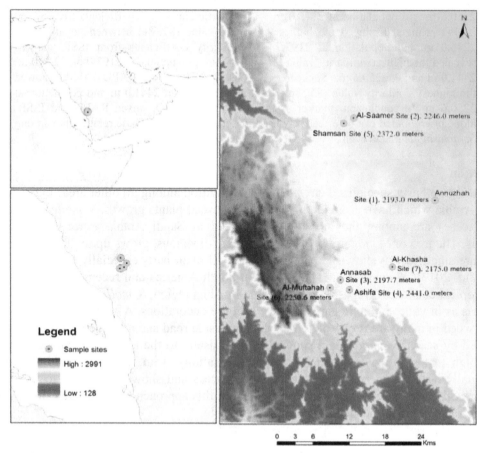

Fig. 1. Sampling sites in Abha region, KSA. Site (1), (2193.0); Site (2), (2246.0); Site (3), (2197.7); Site (4), (2441.0), Site (5), (2372.0), Site (6), (2250.6) and Site (7), (2175.0 m).

Extraction the genomic DNA from leaves of S. molle plants

Genomic DNA was extracted from fresh young leaves of *S. molle* plants by using DNeasy plant mini kit. DNA concentration was estimated by a Thermo Scientific™ BioMate 3S UV-Visible at 260 nm.

PCR amplification

Eight RAPD, nine ISSR and eight mixed RAPD markers were used in this study (Table 1). PCR reaction consists from 1 X GoTaq Green Master Mix, 4 µl from each primer, 20 ng of genomic DNA and nuclease-free water to get a final 25 µl volume. PTC 200 Peltier Thermal Cycler (MJ Research - USA) adjusted as follows: Initial degree at 94°C for 5 minutes followed by forty nine cycles at 92°C for 1 minute, primer annealing temperature at 29°C for 1 minute, extension at 72°C for 2 minutes and final process for primer extension at 72°C for 7 minutes. An equal amount of each amplified product of 20 ul was separated by electrophoresis using 1.3 % agarose gels in 0.5X TBE buffer. Stained gel with ethidium bromide was photographed by gel documentation system using UV transilluminator at 365 nm (Hashemi *et al.*, 2009). Each experiment was repeated three times and molecular weight of RAPD-PCR, mixed RAPD-PCR and ISSR-PCR fragments were estimated using marker 1 kb DNA ladder between 250 to 10,000 bp.

Table 1. RAPD, mixed RAPD and ISSR primers.

RAPD primers	Sequence of primer (5' – 3')
Oligo 342	GAGATCCCTC
Oligo 345	GCGTGACCCG
Oligo 349	GGAGCCCCCT
Oligo 33	CCGGCTGGAA
OPK-8	GAACACTGGG
OPJ-1	CCCGGCATAA
Oligo 214	CATGTGCTTG
Oligo 213	CAGCGAACTA
Mixed RAPD primers	**Sequence of primer (5' – 3')**
Oligo 203+Oligo 342	CACGGCGAGT+GAGATCCCTC
Oligo 203+ Oligo 345	CACGGCGAGT+GCGTGACCCG
Oligo 203+Oligo 42	CACGGCGAGT+TTAACCCGGC
Oligo 203+Oligo 349	CACGGCGAGT+GGAGCCCCCT
Oligo 203+Oligo 214	CACGGCGAGT+CATGTGCTTG
Oligo 203+Oligo 213	CACGGCGAGT+CAGCGAACTA
Oligo 203+Oligo 33	CACGGCGAGT+CCGGCTGGAA
Oligo 203+OPK-8	CACGGCGAGT+GAACACTGGG
ISSR primers	**Sequence of primer (5' – 3')**
Primer (3)	TGGATGGATGGATGGA
Primer (4)	CACACACA CACACA AG
UBC 823	TCT CTC TCT CTC TCC
UBC 824	TCT CTC TCT CTC TCG
UBC 826	ACA CAC ACA CAC ACC
HB 14	CTC CTCCTC GC
Primer (1)	GAGAGAGAGAGAGAGAC
Primer (2)	GAGAGAGAGAGAGAGAGAG
HB 11	GTG TGT GT GTGTCC

Data analysis

All scored fragments gained from RAPD-PCR, mixed RAPD-PCR and ISSR-PCR were manually recorded as present (1) or absent (0). Matrix of similarity based on binary-double zeros-S3, and squared Euclidean distance was used to calculate the distances and to generate dendrogram (Sneath and Sokal, 1973). Polymorphism percentage was estimated by calculating polymorphic bands/total number of bands.

Results

RAPD analysis

RAPD primers produced a total of 109 scorable bands from genotypes of *S. molle*, out of which 23.0 (21.1%) were found to be polymorphic, 1.00 (0.91%) to be monomorphic bands and 85.0 (77.9%) to be unique bands. Primer Oligo 345, yielded the maximum number of bands (20.0 bands) while the lowest number of bands (3.00 bands) obtained from Primer Oligo 214. The percentage of polymorphism ranged from 0.00% (Primer Oligo 33 and Primer Oligo 214) to 57.1% (Primer Oligo 342). The maximum number of unique bands (17.0 bands) was recorded from Primer Oligo 33, while the lowest number of unique bands (3.00 bands) from the Primer Oligo 342 and Primer Oligo 214 (Table 2 and Fig. 2 Panel A).

The genetic similarity coefficients (Table 3) revealed that the maximum similarity value (89.9%) was between population at 2250.6 m and population at 2175.0 m, while the least similarity value (72.5%) between population at 2246.0 m and population at 2372.0 m.

Dendrogram analysis (Fig. 3 Panel A) showed that population at 2193.0, 2197.7 and 2441.0 m found to be forming one cluster whereas population at 2246.0 m separated from them in a single cluster while population at 2372.0, 2250.6 and 2175.0 m found to be forming another one cluster.

Table 2. Polymorphism of eight RAPD primers.

Primer ID	Total no. of bands per primer	No. of polymorphic bands	No. of monomorphic bands	No. of unique bands	Polymorphism %
Oligo342	7.00	4.00	0.00	3.00	57.1
Oligo 345	20.0	7.00	0.00	13.0	35.0
Oligo 349	17.0	5.00	1.00	11.0	29.4
Oligo 33	17.0	0.00	0.00	17.0	0.00
OPK-8	19.0	3.00	0.00	16.0	15.7
OPJ-1	13.0	1.00	0.00	12.0	7.69
Oligo 214	3.00	0.00	0.00	3.00	0.00
Oligo 213	13.0	3.00	0.00	10.0	23.0
Total	109	23.0	1.00	85.0	20.9

Mixed RAPD analysis

Mixed RAPD primers generated a total of 100 reproducible bands of which (19.0%) were polymorphic bands, (81.0%) unique bands, and no any monomorphic bands (Table 4 and Fig. 2 Panel B). Primer OPK-8 produced the highest number of bands (21.0) while primer Oligo 42 gave the minimum number of bands (3.00). Primer Oligo 33 showed the highest percentage value of polymorphism of 50.0% and the zero polymorphism rate gained from the primer Oligo 42 and primer Oligo 214. The maximum number of unique bands were (18.0 bands) gained from primer OPK-8, while the minimum numbers were (3.00) gained from primer Oligo 42.

Table 3. Genetic similarity among *S. molle* plants based on RAPD markers.

	2193.0 m	2246.0 m	2197.7 m	2441.0 m	2372.0 m	2250.6 m	2175.0 m
2193.0 m	1.00						
2246.0 m	0.7614	1.00					
2197.7 m	0.8216	0.7821	1.00				
2441.0 m	0.7956	0.7684	0.828	1.00			
2372.0 m	0.7399	0.7251	0.7753	0.8022	1.00		
2250.6 m	0.8404	0.828	0.8705	0.8705	0.8821	1.00	
2175.0 m	0.7471	0.7326	0.7821	0.7956	0.7821	0.899	1.00

The genetic similarity coefficients displayed the highest similarity value (87.6%) between population at 2246.0 m and population at 2197.7 m, while the least similarity value (72.6%) was recorded between population at 2372.0 m and population at 2175.0 m (Table 5).

Resulted dendrogram showed that populations at 2193.0, 2246.0 and 2197.7 m found to be forming one cluster whereas population at 2441.0 m and population at 2372.0 m clustered together as well as population at 2250.6 m and population at 2175.0 m (Fig. 3 Panel B).

Table 4. Polymorphism of eight mixed RAPD primers.

Primer ID	Total no. of bands per primer	No. of polymorphic bands	No. of monomorphic bands	No. of unique bands	Polymorphism %
Oligo 342	17.0	3.00	0.00	14.0	17.6
Oligo 345	16.0	6.00	0.00	10.0	37.5
Oligo 42	3.00	0.00	0.00	3.00	0.00
Oligo 349	15.0	1.00	0.00	14.0	6.66
Oligo 214	10.0	0.00	0.00	10.0	0.00
Oligo 213	8.00	1.00	0.00	7.00	12.5
Oligo 33	10.0	5.00	0.00	5.00	50.0
OPK 8	21.0	3.00	0.00	18.0	14.2
Total	100	19.0	0.00	81.0	17.3

ISSR analysis

A total of 231 counted bands were generated by using the nine ISSR primers from *S. molle* genetic materials (Table 6 and Fig. 2 Panel C). Sixty-two polymorphic bands (26.8%), 1.00 (0.43%) monomorphic bands, 168 (72.7%) unique bands with polymorphism rate 23.2% were recorded. Primer UBC 826 generated the maximum number of bands (63.0), while primer UBC 824 showed the minimum number of bands (6.00). Primer (1) showed the highest rate of polymorphism (51.4%) and Primer (3) showed the least rate numbers (4.54%). The highest number of unique bands (46.0) was recorded from primer UBC 826, while the least number of unique bands (5.00) resulted from primer UBC 824.

Resulted genetic similarity coefficients exhibited the highest similarity value among populations at 2193.0 m, 2246.0 m, 2441.0 m and population at 2250.6 m recording 86.2%, while the least similarity value between population at 2246.0 m and population at 2175.0 m with value of 69.1% (Table 7).

A dendrogram pattern revealed that population at 2193.0 m and population at 2246.0 m formed one cluster whereas the populations at 2197.7 m, 2441.0 m, 2250.6 m and 2372.0 m found to be in another cluster and population at 2175.0 m formed out-group from the in-group including populations at 2193.0 m, 2246.0 m, 2197.7 m, 2441.0 m, 2250.6 m and 2372.0 m (Fig. 3 Panel C).

Table 5. Genetic similarity among *S. molle* plants based on mixed RAPD markers.

	2193.0-m	2246.0-m	2197.7-m	2441.0-m	2372.0-m	2250.6-m	2175.0-m
2193.0 m	1.00						
2246.0 m	0.8701	1.00					
2197.7 m	0.8439	0.8764	1.00				
2441.0 m	0.8235	0.8439	0.8571	1.00			
2372.0 m	0.7654	0.7879	0.8024	0.8372	1.00		
2250.6 m	0.7952	0.7879	0.8166	0.8235	0.75	1.00	
2175.0 m	0.7578	0.7654	0.8095	0.7879	0.7261	0.8166	1.00

Table 6. Polymorphism of nine ISSR primers.

Primer ID	Total no. of bands per primer	No. of polymorphic bands	No. of monomorphic bands	No. of unique bands	Polymorphism %
Primer (3)	22.0	1.00	0.00	21.0	4.54
Primer (4)	26.0	5.00	0.00	21.0	19.2
UBC 823	18.0	5.00	0.00	13.0	27.7
UBC 824	6.00	1.00	0.00	5.00	16.6
UBC 826	63.0	16.0	1.00	46.0	25.3
HB 14	30.0	12.0	0.00	18.0	40.0
Primer (1)	35.0	18.0	0.00	17.0	51.4
Primer (2)	20.0	3.00	0.00	17.0	15.0
HB 11	11.0	1.00	0.00	10.0	9.09
Total	231	62.0	1.00	168	23.2

Table 7. Genetic similarity among *S. molle* plants based on ISSR markers.

	2193.0 m	2246.0 m	2197.7 m	2441.0 m	2372.0 m	2250.6 m	2175.0 m
2193.0 m	1.00						
2246.0 m	0.8621	1.00					
2197.7 m	0.8123	0.8123	1.00				
2441.0 m	0.8093	0.7906	0.8304	1.00			
2372.0 m	0.7969	0.7713	0.7874	0.8392	1.00		
2250.6 m	0.8093	0.7713	0.8304	0.8621	0.8333	1.00	
2175.0 m	0.7273	0.6912	0.7514	0.7411	0.7131	0.7874	1.00

Super tree analysis (RAPD + mixed RAPD + ISSR)

A combined analysis using pooled RAPD, mixed RAPD and ISSR data showed that there are 20.5 % polymorphism among studied population growing at various height. The highest similarity values (85.5%) was found between populations at 2441.0 m and population at 2250.6 m and the lowest similarity values between population at 2246.0 m and population at 2175.0 m (71.9%) Table (8). Resulted dendrogram revealed that population at 2193.0-m and population at 2246.0-m clustered together whereas populations at 2197.7, 2441.0, 2250.6 and 2372.0 m found to be forming one cluster while population at 2175.0 m separated from them in a single cluster (Fig. 3 Panel D).

Table 8. Genetic similarity among *S. molle* plants based on combined analysis.

	2193 m	2246 m	2197.7 m	2441 m	2372 m	2250.6 m	2175 m
2193 m	1.00						
2246 m	0.8406	1.00					
2197.7 m	0.822	0.8204	1.00				
2441 m	0.8092	0.7978	0.836	1.00			
2372 m	0.7761	0.764	0.7879	0.8298	1.00		
2250.6 m	0.814	0.7895	0.8375	0.8557	0.8282	1.00	
2175 m	0.7393	0.7191	0.7727	0.7658	0.7338	0.8235	1.00

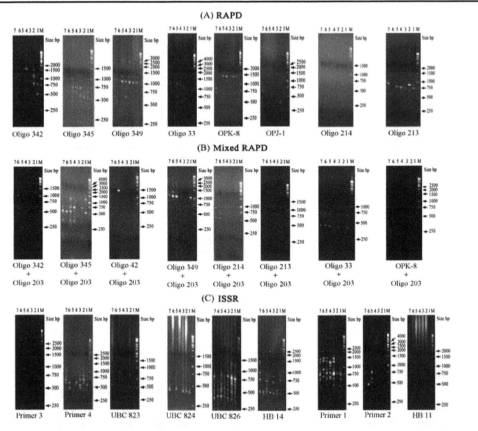

Fig. 2. RAPD, mixed RAPD and ISSR profiles of *S. molle* plants. Lane 1, 2193.0; Lane 2, 2246.0; Lane 3, 2197.7; Lane 4, 2441.0; Lane 5, 2372.0; Lane 6, 2250.6; Lane 7, 2175.0; M-1kb DNA Ladder.

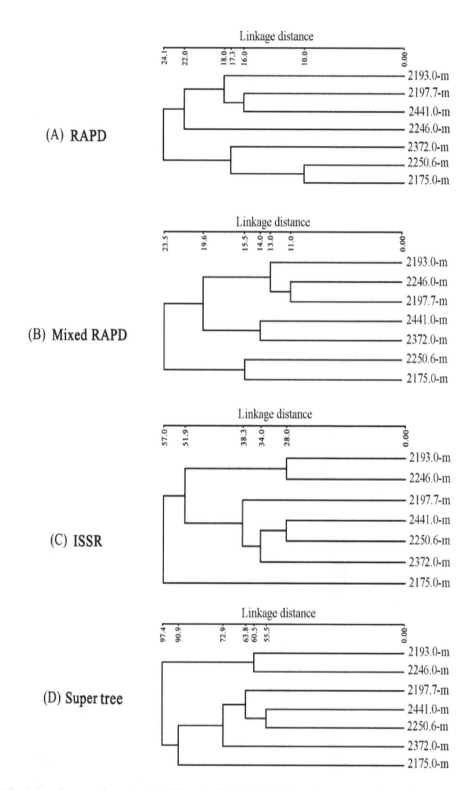

Fig. 3. Dendrogram based on RAPD, mixed RAPD, ISSR and super tree data of *S. molle*.

Discussion

This research article reports the use of the RAPD, mixed RAPD and ISSR makers to the *S. molle* plant and revealed its efficiency to determinate the DNA fingerprints. Also it revealed that RAPD, mixed RAPD and ISSR markers could be used alone or in combination to estimate the genetic diversifications of *S. molle* plants. Polymorphism rate obtained either from RAPD, mixed RAPD, ISSR or from combined analysis all showed that there was high genetic variability among *S. molle* at a very close distance populations. The percentage of polymorphism almost same to that detected in other examined plants that they have a wide genetic variability. For example, Adawy *et al.* (2004) and Hussein *et al.* (2005) found that RAPD polymorphism rate in various Egyptian date palm cultivars (*Phoniex dactylifera* L.) is in the range of 25.2% and for ISSR technique in the range of 28.6%. Among the studied *Pistacia vera* (L.) various cultivars polymorphism rate based on ISSR markers was 46.4% and 100% among *Mangifera indica* (L.) based on ISSR markers (Noroozi *et al.*, 2009; Souza *et al.*, 2011). RADP, mixed RAPD and ISSR showed various degrees in their ability to detect the diversifications among populations of *S. molle* plants. This variation may be due to that the genome *S. molle* plants having a considerable number of alleles per locus/or loci that vary in their distribution. Izzatullayeva *et al.* (2014) reported that such difference between RAPD and ISSR markers due to the fact of abundant nature of microsatellites that results from slippage in DNA replication. This explains why this plant can be found in various habitats vary from salinity soil to alkalinity soil and in different temperature condition ranging from very low to very high (Lim, 2012). In this study, total number of unique bands obtained from ISSR-PCR of *S. molle* plant more than that of RAPD-PCR and mixed RAPD-PCR. The results also showed that ISSR fingerprinting had a high number of scored bands and high polymorphic percentage rate. This in agreement with earlier studies showed that ISSR fingerprinting was more efficient than the RAPD assay in assessing genetic variation in *Arthrocnemum macrostachyum* (Saleh, 2011). Again this variation among RAPD, mixed RAPD and ISSR probably due to that amplified profiles of PCR of RAPD, ISSR, or mixed RAPD originated from different variable numbers of repetitive and non-repetitive sequence on the genomes of *S. molle* plant (Thormann *et al.*, 1994). The presence of monomorphic bands from RAPD-PCR or from ISSR-PCR indication to the sharing characters based on the DNA fragment in genomic *S. molle* plants. Cluster analysis based on RAPD, mixed RAPD and ISSR markers individually or combined showed that the three markers differ from each other in the manner of distributing *S. molle* populations.

In our study, the amount of genetic similarity among various populations of *S. molle* plants based on RAPD markers were in range between 72.5% to 89.9% and for mixed RAPD between 72.6% to 87.6% and for ISSR 69.1% to 86.2% and for the sum of all data between 71.9% to 85.5%. These values to some extent are in accordance with the basis proofed by Weier *et al.* (1982) that operational taxonomic units between 85 to 100% among the same plant species and more than 65% between the same plant genus. In conclusion, our study confirms that there were a wide genetic diversity among *S. molle* plants that can be evaluated by using RAPD, mixed RAPD and ISSR markers.

Acknowledgements

The authors are thankful to King Abdul-Aziz City for Science and Technology (KACST) for providing financial support (No.AT-36-305).

References

Adawy, S.S., Hussein, E.H.A., El-Khishin, D., Saker, M.M., Mohamed, A.A. and El-Itriby, H.A. 2004. Genotyping Egyptian date palm cultivars using RAPD, ISSR, AFLP markers and estimation of genetic stability among tissue culture derived plants. Arab J. Biotech.**8**: 99-114.

Doganlar, Z. B., Doganlar, O., Erdogan, S. and Onal, Y. 2012. Heavy metal pollution and physiological changes in the leaves of some shrub, palm and tree species in urban areas of Adana, Turkey. Chem Spec Bioavailab **24**: 65-78.

Fikiru, E., Tesfaye, K. and Bekele, E. 2007. Genetic diversity and population structure of Ethiopian lentil (*Lens culinaris* Medikus) landraces as revealed by ISSR marker. Afr. J. Biotechnol. **6**: 1460-1468.

Hashemi, S.H., Mirmohammadi-Maibody, S.A.M., Nematzadeh G.A. and Arzani, A. 2009. Identification of rice hybrids using microsatellite and RAPD markers. Afr. J. Biotechnol. **8**: 2094-2101.

Hussein, E.H.A., Adawy, S.S., Ismail, S.E.M.E. and El-Itriby, H.A. 2005. Molecular characterization of some Egyptian date palm germplasm using RAPD and ISSR markers. Arab J. Biotechn. **8**: 83-98.

Izzatullayeva, V., Akparov, Z., Babayeva, S., Ojaghi, J. and Abbasov, M. 2014. Efficiency of using RAPD and ISSR markers in evaluation of genetic diversity in sugar beet. Turk. J. Biol. **38**: 429-438.

Lim, T.K. 2012. Edible medicinal and non-medicinal plants: volume 1, fruits: *Schinus molle*. Springer, Netherlands, 153-159 pp.

Moustafa, M.F., Hesham, A., Quraishi, M.S. and Alrumman S.A. 2016. Variations in genetic and chemical constituents of *Ziziphus spina-christi* L. populations grown at various altitudinal zonation up to 2227 m height.Genet. Eng. Biotechnol. **14**: 349-362.

Noroozi, S., Baghizadeh, A. and Javaran, M.J. 2009. The genetic diversity of Iranian pistachio (*Pistacia vera* L.) cultivars revealed by ISSR markers. Bio Di Con. **2**: 50-56.

Olafsson, K., Jaroszewski, J.W., Smitt, U.W. and Nyman, U. 1997. Isolation of angiotensin converting enzyme (ACE) inhibiting triterpenes from *Schinusmolle*. Planta Med. **63**: 352-355.

Pereira, M. P., Rodrigues, L. C. A., Corrêa, F. F., Castro, E. M., Ribeiro, V. E. and Pereira, F. J. 2016. Cadmium tolerance in *Schinus molle* trees is modulated by enhanced leaf anatomy and photosynthesis. Trees**30**: 807-814.

Saleh, B. 2011. Efficiency of RAPD and ISSR markers in assessing genetic variation in *Arthrocnemum macrostachyum* (Chenopodiaceae). Braz. Arch. Biol. Technol. **54**: 859-866.

Sneath, P.H.A. and Sokal, R.R. 1973. Numerical taxonomy: the principles and practice of numerical classification. W.H. Freeman and Company, San Francisco, California, CA, USA.

Souza, I.G.B., Valente, S.E.S., Britto, F.B., de Souza, V.A.B. and Lima, P.S.C. 2011. RAPD analysis of the genetic diversity of mango (*Mangifera indica*) germplasm in Brazil. Genet. Mol. Res. **10**: 3080-3089.

Thormann, C.E., Ferreira, M.E., Camargo, L.E.A., Tivang, J.G. and Osborn, T.C. 1994. Comparison of RFLP and RAPD Markers to Estimating Genetic Relationships Within and Among Cruciferous Species. Theor Appl Genet **88**: 973–980.

Uyoh, E.A., Umego, C. and Aikpokpodion, P.O. 2014. Genetic diversity in African ntmeg (*Monodora myristica*) a ccessions from South Eastern Nigeria. Afr. J. Biotechol. **13**: 4105-4111.

Vieira, R.F., Goldsbrough, P. and Simon, J.E. 2003. Genetic diversity of basil (*Ocimum* spp.) based on RAPD markers. J. Amer. Soc. Hort. Sci. **128**: 94-99.

Weier, T.E., Stocking, C.R., Barbour, M.G. and Rost, T.L. 1982. Botany: an Introduction to Plant Biology. John Wiley and Sons, New York.

Wu, C., Zhong, C., Zhang, Y., Jiang, Q., Chen, Y., Chen, Z., Pinyopusarerk, K. and Bush, D. 2014. Genetic diversity and genetic relationships of *Chukrasia* spp. (Meliaceae) as revealed by inter simple sequence repeat (ISSR) markers. Trees **28**: 1847-1857.

NUMERICAL TAXONOMIC ANALYSIS IN LEAF ARCHITECTURAL TRAITS OF SOME *HOYA* R. BR. SPECIES (APOCYNACEAE) FROM PHILIPPINES

JESS H. JUMAWAN[1] AND INOCENCIO E. BUOT, JR

Institute of Biological Sciences, University of the Philippines, Los Baños, College, Laguna, Philippines

Keywords: Cluster Analysis; Multivariate Analysis; Numerical Taxonomy; Principal Coordinate Analysis.

Abstract

The present study examines the leaf variations in leaf traits of four *Hoya* R. Br. species from Philippines namely: (1) *H. buotii* Kloppenburg, (2) *H. halconensis* Kloppenburg, (3) *H. mindorensis* Schlechter red bearing flowers; and (4) *H. mindorensis* Schlechter yellow bearing flowers. Leaf samples (n= 30 leaves) were collected from each plant group and measured with nine architectural traits. The results showed variability in the leaves using univariate and multivariate analysis. Data ordination depicted variations in leaf morphology. The two plant groups *H. mindorensis* red bearing flowers and *H. mindorensis* yellow bearing flowers were consistently variable as supported by principal coordinate analysis, cluster analysis and two way Anova (P<0.001). The variability of the two plant groups could be due to developmental instability, plasticity or taxonomic identity, one being the subspecies of the other. Hence, a closer study to investigate the significant variability of the two plant groups was recommended. Distinct separation of *H. buotii* and *H. halconensis* was detected being regularly mistaken as one species. The study demonstrated the applicability of multivariate analysis as effective tool in numerical taxonomy. Multivariate analysis can be employed to demonstrate likelihood of relationship among various *Hoya* species.

Introduction

Most *Hoya* species of Philippines were considered endemic to the country with several new discoveries for the past decade. The genus *Hoya*, is commonly known as wax plant belong to family Apocynaceae, was considered to be taxonomically complex (Wanntorp *et al.*, 2006). The estimated number of *Hoya* species in the country ranges from 80 – 104 (Kloppenburg *et al.*, 2012 and Aurigue *et al.*, 2013). The Philippines was considered as one of the richest and most diverse range of *Hoya* species which are located all throughout the archipelago (Kloppenburg and Siar, 2008).

Identification of *Hoya* species largely depend on traditional taxonomy that put emphasis on reproductive characters. Descriptions on qualitative and quantitative characteristics of inflorescence, corolla, corona and pollinarium were very significant in identification of *Hoya* species (Kleijn and Van Donkelaar, 2001; Omlor, 1996; Forster and Little, 1996). Nomenclature issues were still largely unresolved for various taxa (Rodda and Juhoneweb, 2013). Many species were documented to exhibit phenotypic plasticity in morphological characters (Tungmunnithum *et al.*, 2011). The structure of many *Hoya* species was described to possess complex corona morphology (Kunze, 2008). Many of these problems in *Hoya* taxonomy had risen due to dependence on reproductive parts. Reproductive features are not present all the time and makes difficulty in identification.

---

[1]Corresponding author. Email: jehoju@gmail.com

DNA barcode was the method suggested to properly identify the endemic Philippine *Hoya* species (Maranan and Diaz, 2013). The technique was regarded as an effective tool for species identification but considerably weak attempt to discovery and description of species (Wheeler, 2004). Aside from being an expensive method for species identification, DNA barcoding is insufficient in terms of theoretical basis of traditional taxonomy (Lipscomb *et al.*, 2003).

Leaves of Philippine *Hoya* species are present throughout the year and can be used extensively for detecting variations. Leaf characters were proven to be valuable in taxonomic studies of tropical plants which seldom produce flowers and angiosperm remains as fossils (Hickey and Taylor, 1991; Dilcher, 1974). Leaf morphological characters of Gunneraceae were subjected to multivariate analysis to support genus monophyly (Fuller, 2005). Leaf morphometric data are important and the variation displayed by morphological traits reflects the evolutionary arrangement manifested as morphological changes (Otte and Endler, 1989). Multivariate analysis is a tool in the examination of leaf morphometric traits, an important component in the field of numerical taxonomy. The main objective of the study was to examine the variations of the leaf morphometric traits of the selected *Hoya* species namely: *Hoya buotii, Hoya halconensis,* and *Hoya mindorensis*.

Materials and Methods

Plant materials

The *Hoya* species were acquired from the propagated plant collections of Dr. I. E. Buot Jr., Professor and curator of IBS Herbarium, PBD in UPLB. There were 3 species of *Hoya* included in the study which were *Hoya buotii, Hoya halconensis,* and *Hoya mindorensis*. However, it was noticed that *H. mindorensis* bears two flower types: the red bearing plants and the yellow bearing plant. For the purpose of this examination, the analyses were conducted in four plant groups: (1) *H. buotii*, (2) *H. halconensis* (3) *H. mindorensis* red bearing flowers; and (4) *H. mindorensis* yellow bearing flowers. The selected *Hoya* species usually encountered confusion in proper taxonomic identification.

Leaf character selection and measurements

The selection of leaf morphometric characters were based from manual of leaf architecture with modifications (Leaf Architecture Working Group, 1999). A total of nine morphometric traits were chosen in the study. The description and illustration of the parameters considered for leaf morphometric measurements were shown in Table 1 and Fig. 1.

Table 1. Parameters of leaf morphometric measurements used in the analysis.

Code	Description of characters
LL	Lamina length
LW	Lamina width
PW	Petiole width
WL	Width in left side of lamina
WR	Width in right side of lamina
VL	Number of secondary veins in left side of the lamina
VR	Number of secondary veins in right side of the lamina
LR	Leaf ratio (LL/LW)
LA	Leaf area (LL X LW X 2/3)

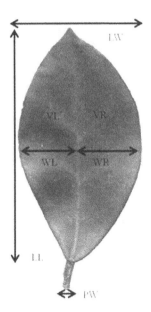

Fig. 1. The illustration of the parameters taken for leaf morphometric measurements.

Data analysis

There were four plant groups, nine leaf morphometric data, and thirty leaf sample replicates which has a total of 1,080 data sets. The data generated from the leaf morphometric traits were subjected to univariate and multivariate statistical analyses. The univariate data comprised the minimum value, maximum value, mean and standard deviation. The univariate data sets were plotted in a box and whisker to evaluate the distribution of data. The multivariate data matrix was subjected to similarity matrix using Morisita index of similarity. The similarity matrix was explored using data ordination technique. Ordination refers to projection of multivariate data sets in a two dimensional space to detect patterns upon visual inspection (Pielou, 1984). Principal coordinate analysis (PCOa) or also known as metric multidimensional scaling was implemented as data ordination (Gower, 1966). PCOa reduces the dimensionality of the data similar to principal component analysis but the advantage of PCoA is that it may be used with all types of variables (Legendre and Legendre, 1998). Cluster analysis was performed combining quantitative data into clusters in constructing a dendrogram. The resulting pattern generated from multivariate analysis detected variations in leaf morphometric traits. An inference on the sources of leaf morphometric variation was tested using two way analysis of variance (Anova). It was investigated if the significant variation could be attributed by the leaf characters, the species, and the interaction of leaf characters and species. Post hoc analysis was conducted when $P<0.05$ using Tukey's test. The PAST (Paleontological Statistical Software) software (Hammer *et al.*, 2009) was used in analyzing univariate and multivariate analysis.

Results

The univariate statistics of the morphometric traits of the plant groups are shown in Table 2 and Fig. 2. The results indicated that the leaf area and leaf length were the most variable among the measured leaf characters. The least variable traits were petiole width and leaf ratio. The rest of the leaf traits were relatively similar to the four plant groups.

Table 2. Univariate statistics of the nine morphometric traits of the four plant groups.

	Parameter	LL	LW	PW	WL	WR	VL	VR	LR	LA
H. halconensis, (n=30)	Min	81	30	1	15	15	3	3	1.62319	1800.01
	Max	156	75	4	37	38	6	7	3	7336.7
	Mean	115.367	53.7667	2.43333	26.6333	27.3	4.43333	4.5	2.18327	4249.33
	St. dev.	19.2703	11.4189	0.727932	5.70833	5.57797	0.773854	0.937715	0.292643	1520.62
H. buotii, (n=30)	Min	63	22	1	10	11	4	4	1.90909	953.338
	Max	119	47	3	27	23	9	9	3.47826	3332.02
	Mean	97	35.4667	1.83333	16.4667	16.7667	6.23333	6.5	2.77783	2328.99
	St. dev.	13.5341	6.27383	0.592093	4.40793	3.81181	1.0063	1.27982	0.385205	638.016
H. mindorensis (red), (n=30)	Min	63	24	2	12	12	4	4	2.17241	1218.01
	Max	174	58	4	27	31	9	8	3.83333	6573.37
	Mean	108.9	39.0667	3.03333	19.2333	19.8333	5.8	5.46667	2.77893	2979.37
	St. dev.	30.0899	8.31672	0.718395	4.02307	4.36351	1.15669	1.04166	0.387058	1490.31
H. indorensis (yellow), (n=30)	Min	56	32	3	15	15	4	4	1.55769	1194.67
	Max	157	67	5	33	34	8	7	2.74419	7012.7
	Mean	100.4	48.3	3.86667	24.1667	24.1333	5.7	5.73333	2.05773	3381.31
	St. dev.	27.7148	9.3593	0.62881	4.41848	5.07008	0.987857	0.868345	0.288655	1505.66

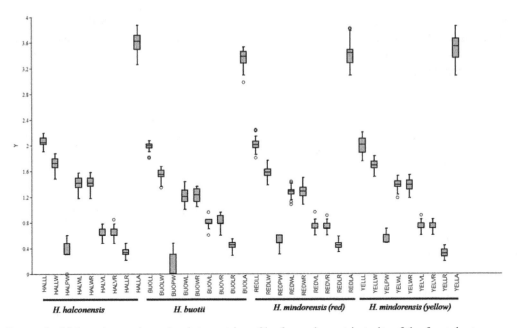

Fig. 2. Box and whisker plot on the univariate metrics of leaf morphometric traits of the four plant groups.

The results in PCOa accounted 4 effective coordinate axis with a total of 68.09% of the cumulative variance (Table 3). Axis 1 and axis 2 contributed 36.30% and 59.02% of variances respectively. The components in axis 1 and axis 2 were analyzed and used to project into a two dimensional plane (Table 4 and Fig. 3). The highest values generated in axis 1 and axis 2 was largely attributed to the leaf morphometric traits of *H. mindorensis* red bearing plant. The lowest values on the other hand were attributed to leaf traits of *H. mindorensis* yellow bearing plant. It can be viewed that leaf traits of the two *H. mindorensis* plants were highly variable. The leaf traits from *H. buotii* and *H. halconensis* were less variable.

Table 3. The eigenvalue of the principal coordinate axis and the respective accounted variance in the PCOa.

Axis	Eigenvalue	Percent variance	Cumulative variance
1	0.033984	36.301	36.301
2	0.021271	22.721	59.022
3	0.0047677	5.0928	64.1148
4	0.003722	3.9759	68.0907

The data ordination of PCOa clearly displayed the variation of the leaf morphometric traits between the two *H. mindorensis* plant groups (Fig. 3). The red bearing flower *H. mindorensis* occupied quadrant 1 and quadrant four of the orthogonal plane. The yellow bearing flower *H. mindorensis* largely occupied quadrant 2. The rest of the values were distributed closely in quadrants 3 and 4.

Another technique employed in the exploratory analysis of the morphometric data was cluster analysis. The dendrogram also revealed a similar pattern observed in PCOa. Again, the two *H. mindorensis* plant groups were located on the opposite ends of the dendrogram. It indicated that the two plant groups were highly variable. The *H. mindorensis* yellow bearing flower was more morphometrically similar *H. buotii* but their Euclidean distance was still far. This indicated variability of leaf morphometric traits. The *H. mindorensis* red bearing flower was very similar to *H. halconensis* in terms of the measured leaf traits. In general, the four plant groups revealed distinct leaf characteristics as indicated by the clusters in the dendrogram (Fig. 4).

Table 4. The principal coordinate scores of the two highest accounted variances in axis1 and axis 2 on the nine morphometric traits derived from the four plant groups. (Legend: The first three letters comprise the plant groups as HAL = *H. halconensis* ; BUO = *H. buotii*; RED = *H. mindorensis* (red); and YEL = *H. mindorensis* (yellow). The last two letters comprise the code characters of the leaf in Table 1).

Leaf traits	axis 1	axis 2	Leaf traits	axis 1	axis 2
REDLA	0.1130	0.0470	BUOLL	-0.0025	-0.0007
REDLL	0.0374	0.0153	HALWL	-0.0059	-0.0333
BUOPW	0.0336	0.0051	YELVR	-0.0068	0.0120
REDWR	0.0296	0.0097	HALWR	-0.0083	-0.0357
REDPW	0.0294	-0.0007	HALLR	-0.0085	0.0114
REDLW	0.0287	0.0102	HALLW	-0.0092	-0.0361
REDWL	0.0281	0.0107	HALLL	-0.0101	-0.0223
BUOWR	0.0167	-0.0050	YELVL	-0.0112	0.0137
HALPW	0.0166	-0.0258	YELLR	-0.0141	0.0086
BUOWL	0.0138	-0.0076	BUOLR	-0.0177	-0.0064
BUOLA	0.0101	-0.0027	HALLA	-0.0206	-0.0796
REDVR	0.0082	-0.0100	YELWL	-0.0244	0.0156
REDVL	0.0077	-0.0126	HALVR	-0.0248	-0.0034
BUOLW	0.0065	0.0001	YELPW	-0.0276	0.0078
REDLR	0.0020	0.0025	YELLW	-0.0279	0.0154
BUOVL	0.0001	-0.0040	YELWR	-0.0321	0.0154
HALVL	-0.0018	-0.0019	YELLL	-0.0385	0.0268
BUOVR	-0.0024	-0.0087	YELLA	-0.0869	0.0691

The emerging pattern generated in data exploration using multivariate analysis suggested variability and resemblances on the leaf morphometric traits. To detect if the variation was significant or not, two way Anova was conducted. The sources of variation were generated from the leaf characters, four plant groups and the interaction between leaf characters and four plant groups. The two way Anova detected a highly significant differences among the mentioned sources of variation. Post hoc analysis was conducted using Tukey's test in the four plant groups only. It was not conducted to the leaf characters as it may give irrelevant output (e.g. leaf area is obviously significantly different to petiole width). The Tukey's test revealed that *H. halconensis* and *H. mindorensis* red bearing flower were more similar compared to other comparison. The *H. mindorensis* red bearing flower and *H. mindorensis* yellow bearing flower was highly significantly different. Other combinations of plant group comparisons showed highly significant differences in their leaf morphometric traits. The summary on the two way Anova table was shown in Table 5 and 6.

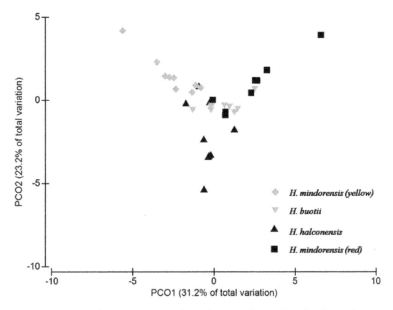

Fig. 3. The data ordination on the nine morphometric traits contributed by the four plant groups using PCOa.

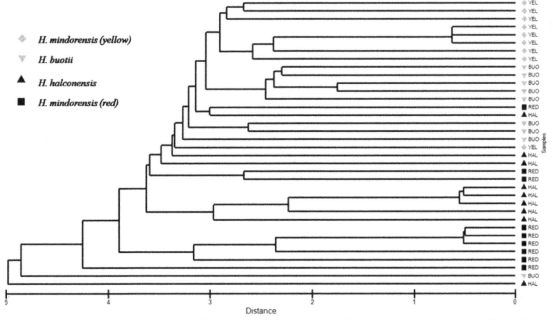

Fig. 4. Single linkage or nearest neighbor cluster dendrogram on the nine morphmetric traits from the leaves of the four plant groups.

Table 5. Two way Anova table on the sources of variation in leaf morphometric traits.

Source	Sum of squares	df	Mean squares	F ratio	P value
Leaf characters (A)	985.8	8	123.2	$F(8, 232) = 2530$	$P < 0.0001$
Species (B)	9.887	3	3.296	$F(3, 87) = 25.86$	$P < 0.0001$
Interaction: A x B	62.51	24	2.605	$F(24, 696) = 52.17$	$P < 0.0001$
Residual	34.75	696	0.04993		

Table 6. Pair wise comparison on the different plant groups using Tukey's test.

Multiple Comparisons Test	Mean diff.	95% CI of diff.	Summary
H. halconensis vs. *H. buotii*	0.1244	0.04389 to 0.2048	***
H. halconensis vs. *H. mindorensis* (red)	-0.0312	-0.1116 to 0.04931	ns
H. halconensis vs. *H. mindorensis* (yellow)	-0.1443	-0.2248 to -0.06386	****
H. buotii vs. *H. mindorensis* (red)	-0.1555	-0.2360 to -0.07506	****
H. buotii vs. *H. mindorensis* (yellow)	-0.2687	-0.3492 to -0.1882	****
H. mindorensis (red) vs. *H. mindorensis* (yellow)	-0.1132	-0.1936 to -0.03269	**

Discussion

The study was conducted to examine the variations of the leaf morphometric traits of the four plant groups consisting *Hoya* species. The univariate statistical analysis showed pronounced variability in leaf area and laminar length. On the other hand, less variability was observed in petiole width and leaf ratio. Leaf variation was more conspicuous to parts of the leaf with bigger morphometric values. The multivariate analysis generated a pattern that the *H. mindorensis* red bearing flower and *H. mindorensis* yellow bearing flower were highly variable. This was supported by results on cluster analysis and the comparison using Tukey's test as post hoc analysis to two way Anova. The detected variability of the two plant groups, although belonging to the same species could be attributed to environmental stress (Van Valen, 1962). The effects of environmental stress can eventually lead to developmental instability manifested in variability of leaf traits (Valentine and Soule, 1973). *Hoya* species were known also to exhibit phenotypic plasticity (Tungmunnithum *et al.*, 2011). Leaf variations of the two plant groups could be phenotypic variation attributed to plasticity. On the other hand, a closer investigation should be conducted to consider other leaf parameters or another vegetative part of the plant. The variation of leaf traits between the two plant groups was highly significant, as also indicated by difference in the colour of the flowers. The possibility that they were taxonomically different, as a subspecies probably is at large. Hence, it is recommended that a separate study should be conducted that would incorporate many characters in the analysis.

In general, the multivariate analysis was able to detect variations of the morphometric traits on the leaves of selected Philippine *Hoya* species. The detected variation was statistically tested to discriminate one plant group to another. This study confirms the distinct separation of *Hoya buotii* and *Hoya halconensis* that had been always mistaken to be one species (Aurigue, 2013). This study demonstrated the applicability of multivariate analysis as a tool in numerical taxonomy. The technique detected variations in leaf morphometric traits and can further be employed to demonstrate likelihood of relationship among the *Hoya* species.

Acknowledgement

The authors would like to thank the Philippine government agency DOST-ASTHRDP for the support in the conduct of the study.

References

Aurigue, F.B. 2013. A collection of Philippine Hoyas and their culture. Philippine Council for Agriculture, Aquatic and Natural Resources and Development DOST, Laguna, Philippines 195p.

Aurigue, F.B., Sahagun, J. R., and Suarez, W. M. 2013. *Hoya cutis-porcelana* (Apocynaceae): A New Species from Samar and Biliran Islands, Philippines. Journal of Nature Studies. **12**(1): 12-17

Dilcher, D.L.1974. Approaches to the identification of angiosperm leaf remains. Bot. Rev. **40**: 1-156.

Forster, P.I. and Little, D. J. 1996. Flora of Australia. **28**: 231-237. CSIRO, Canberra.

Fuller, D.Q. 2005. Systematics and leaf architecture of the Gunneraceae. The Botanical Review **71**(3): 295-353.

Gower, J.C. 1966. Some distance properties of latent root and vector methods used in multivariate analysis. Biometrika **53**: 325-338.

Hammer, O., Harper, D.A.T., and Ryan, P.D. 2009. Past version 1.91: Paleontological Statistical Software package for education and data analysis. Paleontologia Electronica **4** (1):9- 4.

Hickey, L.J. and Taylor, D.W. 1991. The leaf architecture of *Ticodendron* and application of foliar characters in discerning its relationships. Ann. Missouri. Bot. Gard., **78**: 105-130.

Kleijn, D. and Van Donkelaar, R. 2001. Notes on the taxonomy and ecology of the genus *Hoya* (Asclepiadaceae) in Central Sulawesi. Blumea **46**: 457–483.

Kloppenburg, R.D., Guevarra, M.L.D., Carandang, J.M. and Maranan, F.S. 2012. New Species of *Hoya* R. Br. (Apocynaceae) from the Philippines. Journal of Nature Studies **11**(1&2): 34-48.

Kloppenburg, R.D. and Siar S.V. 2008. Three new species of *Hoya* R.Br. (Apocynaceae) from the Philippines. Asia Life Sciences **17**(1):57-70.

Kunze, H. and Wanntorp, L. 2008. Corona and anther skirt in *Hoya* (Apocynaceae, Marsdenieae). Plant Syst. Evol. **271**: 9–17.

Legendre, P. and Legendre, L. 1998. Numerical Ecology. 2nd English edition. Elsevier, Amsterdam.

Lipscomb, D., Platnick, N. and Wheeler, Q.D. 2003. The Intellectual Content of Taxonomy: A Comment on DNA Taxonomy. Trends in Ecology and Evolution **18**(2): 65–66.

Leaf Architecture Working Group. 1999. Manual of Leaf Architecture – morphological description and categorization of dicotyledonous and net-veined monocotyledonous angiosperms. Washington, DC.

Maranan, F.S. and Diaz, M.G.Q. 2013. Molecular Diversity and DNA Barcode Identification of Selected Philippine Endemic Hoya Species (Apocynaceae). The Philippine Agricultural Scientist **96** (1): 86-92.

Omlor, R. 1996. Notes on Marsdenieae (Asclepiadaceae) - A new, unusual species of *Hoya* from northern Borneo. Novon **6**: 288–294.

Otte, D. and Endler, J.A. 1989. Speciation and its consequences. Sunderland, Massachusetts. Sinauer Associates, pp 28-59.

Pielou, E. C. 1984. The interpretation of ecological data: A primer on classification and ordination. Wiley, New York.

Rodda, M. and Juhoneweb, N. S. 2013. The taxonomy of *Hoya micrantha* and *Hoya revoluta* (Apocynaceae, Asclepiadoideae). Webbia: Journal of Plant Taxonomy and Geography **68**: (1) 7–16.

Tungmunnithum, D., Kidyoo, M. and Khunwasi, C. 2011. Morphological variations in *Hoya siamica* Craib (Asclepiadaceae) in Thailand. Tropical Natural History **11**(1): 29-37.

Valentine, D.W., and Soule, M. 1973. Effect of p,p'-DDT on developmental stability of pectoral fin rays in the Grunion, Leuresthes tenuis. Fisheries Bulletin **71**: 921-926.

Van Valen, L. 1962. A study of fluctuating asymmetry. Evolution **16**: 125-142.

Wanntorp, L., Kocyan, A. and Renner, S.S. 2006. Wax plants disentangled: A phylogeny of *Hoya* (Marsdenieae, Apocynaceae) inferred from nuclear and chloroplast DNA sequences. Molecular Phylogenetics and Evolution **39**: 722–733.

Wheeler, Q.D. 2004. Taxonomic triage and the poverty of phylogeny. Philosophical Transactions of the Royal Society of London, B **359**: 571–583.

CAMELLIA (THEACEAE) CLASSIFICATION WITH SUPPORT VECTOR MACHINES BASED ON FRACTAL PARAMETERS AND RED, GREEN, AND BLUE INTENSITY OF LEAVES

W. Jiang[1], Z.M. Tao[1], Z.G. Wu[1], N. Mantri[2], H.F. Lu[*] and Z.S. Liang

College of Life science, Zhejiang Sci-Tech University, Hangzhou 310018, China

Keywords: *Camellia*; Classification; Fractal analysis; RGB; SVM.

Abstract

Leaf traits are commonly used in plant taxonomic applications. The aim of this study was to test the utility of fractal leaf parameters analysis (FA) and leaf red, green, and blue (RGB) intensity values based on support vector machines as a method for accurately discriminating *Camellia* (68 species from five sections, 11 from sect. *Furfuracea*, 13 from sect. *Paracamellia*, 15 from sect. *Tuberculata*, 24 from sect. *Theopsis* and 5 from sect. *Camellia*). The results showed that the best classification accuracy was up to 96.88% using the RBF SVM classifier ($C = 16$, $g = 0.5$). The linear kernel overall accuracy was 90.63%, and the correct classification rates of 40.63% and 93.75% were achieved for the sigmoid SVM classifier ($C = 16$, $g = 0.5$) and the polynomial SVM classifier ($C = 16$, $g = 0.5$, $d = 2$), respectively. A hierarchical dendrogram based on leaf FA and RGB intensity values was mostly on agreement with the generally accepted classification of the *Camellia* species. SVM combined with FA and RGB may be used for rapidly and accurately classifying *Camellia* species and identifying unknown genotypes.

Introduction

Camellia L. is a commercially important genus of family Theaceae. It is cultivated globally, particularly in tropical and subtropical regions of East and Southeastern Asia (Ming, 2000; Gao *et al.*, 2005; Lu *et al.*, 2012). Some *Camellia* species are used to produce tea, others are cultivated as ornamental plants, and the seeds of some species are used for making edible oils (Chen *et al.*, 2005; Vijayan *et al.*, 2009; Jiang *et al.*, 2012). Currently, there are number of discrepancies in relation to classification of species from this important genus. There are three popular *Camellia* monographs developed by Sealy (1958), Chang (1998) and Ming (2000) that differ significantly in species, section and subgenus arrangement. All these taxonomic classifications are based on the morphology. Many studies have shown that classifications purely based on the traditional morphological characteristics are insufficient for closely related species because low divergence prevents having reasonable qualitative features to support the taxonomic systems (Bari *et al.*, 2003; Lu *et al.*, 2008a,b; Pandolfi *et al.*, 2009; Jiang *et al.*, 2010). As a result, there is no concordance in the method for classification of *Camellia* and further taxonomic research is necessary (Pi *et al.*, 2009).

Leaf characters have been successfully exploited to solve plant taxonomy problems (Plotze *et al.*, 2005, Lin *et al.*, 2008; Ye and Weng, 2011). Traditionally, leaf traits such as shape (Ming, 2000), morphology research (Barthlott *et al.*, 2009), and leaf anatomy (Pi *et al.*, 2009; Jiang *et al.*, 2010) have been used for classification. Recently, several researchers have used fractal parameters

---

[*] Corresponding author. Email: luhongfei0164@163.com
[1] Zhejiang Institute of Subtropical Crops, Zhejiang Academy of Agricultural Sciences   Wenzhou 325005, China.
[2] School of Applied Sciences, Health Innovations Research Institute, RMIT University, Melbourne 3000, Victoria, Australia.

for plant identification (Mancuso *et al.*, 2003; Azzarello *et al.*, 2009; Pandolfi *et al.*, 2009). Mancuso (1999) highlighted the importance of the leaf fractal geometry for fingerprinting plants. In addition, leaf colour information provides useful data for judging maturity of agricultural products (Gunasekaran *et al.*, 1985), detecting diseases (Howaith *et al.*, 1990), and fruit sorting (Harrell *et al.*, 1989). Thus, the leaves really provide plenty of characteristics that can be used as a source of data for plant taxonomy (Yang and Lin, 2005).

Supervised techniques are one of the most effective analysis tools in classification field currently (Lu *et al.*, 2012). These tools apply available information about a category membership of samples to developed model for classification of the genus. Support vector machines (SVM) is a supervised pattern recognition technology which has the algorithm developed in the machine learning community and is capable of learning in high-dimensional feature spaces (Cortes and Vapnik, 1995; Lu *et al.*, 2011). The standard SVM takes a set of input data and predicts, for each given input, which of two possible classes the input is a member of, which makes the SVM is a non-probabilistic binary linear classifier. Recently, SVM has been used in a variety of areas like information retrieval (Jain *et al.*, 1999), object recognition (Pontil and Verri, 1998), food bruise detection (Lu *et al.*, 2011), qualitative assessment of tea (Chen *et al.*, 2008), and fruit classification (Zheng *et al.*, 2010). Chen *et al.* (2007) demonstrated that SVM fixes the classification decision function based on structural risk minimum mistakes instead of the minimum mistake of the misclassification on the training set to avoid over-fitting problem. Compared to other pattern recognition tools such as artificial neural networks (ANNs), SVM is a powerful method with a higher training speed and can avoid overtraining (Jack and Nandi, 2002; Kumar *et al.*, 2011). In addition, Burges (1998) suggested that SVM could get the best solution of data set with better ability of generalization.

So far there is no knowledge about the utility of leaf image analysis and machine learning as a taxonomic toolkit for classification of genus *Camellia*. In this study, we combine the fractal leaf parameters and leaf red, green, and blue intensity values with SVM to analyze the taxonomical classification of *Camellia* plants. The main objective of this work was to (a) develop and evaluate the effectiveness of SVM for identifying 68 species in genus *Camellia*, and (b) confirming these relationships based on fractal parameters and red, green, and blue (RGB) intensity values of leaves. Our purpose is to provide a potential tool for accurate classification of *Camellia* species.

Material and Methods

Materials

All plant materials were collected from the International *Camellia* Garden in Jinhua, Zhejiang Province (29°07' N, 119°35' E, 40 m in altitude) in July 2011. All plants share the same environment in this garden which reduces the major effect of geographical distribution on leaf development. Healthy leaf samples following Chang's taxonomic treatment (1998), 11 species from sect. *Furfuracea*, 13 species from sect. *Paracamellia*, 15 species from sect. *Tuberculata*, 24 species from sect. *Theopsis*, and five species from sect. *Camellia*, for a total of 68 species were examined, and split into two groups: 36 for training phase of SVM model construction and the other 32 for the validation phase (Table 1). All samples were taken from the third mature leaves that was fully exposed to sunlight and horizontally arranged on the two-year-old branches of the plants. At least three plants per species were selected. Means of data were obtained using SAS version 9.0 (SAS Institute, Cary, NC, USA). Voucher specimens for all species were deposited in the Chemistry and Life Science College of Zhejiang Normal University (ZJNU) (see Appendix 1 for voucher details).

Image acquisition and fractal parameters

A Canon EOS 50D camera with a Canon EF-S 18-55 mm f/3.5-5.6 IS lens at 50 mm, was used to acquire leaf images. All image acquisition was carried out at least in five and the lighting for images was entirely from natural light on a sunny summer morning. Leaf fractal parameters were calculated using fractal image analysis software (HarFA, Harmonic and Fractal Image Analyzer 5.4) as previously described by Mancuso (2002), Pandolfi *et al.* (2009) and Zheng *et al.* (2011). Briefly, Figure 1 shows schematic diagram of HarFA output and five parameters in detail. The basic procedure was as follows: (1) each *Camellia* leaf image was split into the constituent color channels (red, green, blue); (2) each channel was set for a threshold color value between 0 and 255; (3) the fractal dimension (D) for red, green, and blue channel was calculated by box counting method; (4) then the D which is presented as a function of thresholding condition in fractal spectrum was plotted against the colour intensity to obtain the fractal spectra of the three channels; (5) determining the baseline ($D = 1$) that separates the fractal ($D > 1$) from the non-fractal ($D < 1$) zone of the spectrum. For this study, we selected $D = 1.2$ as the baseline. (6) Finally, the five fractal parameters (X_1, X_2, X, Y, and S) were determined by Origin Lab (version 8.0). Additionally, average RGB intensity values from *Camellia* images were assessed using the colour histogram tool of Image J (National Institutes of Health, Bethesda, MD).

Cluster analysis

As a method of grouping data based on attributes of given population into similar and dissimilar groups, we conducted clustering analysis to classify 68 species in genus *Camellia* based on 15 fractal parameters and average RGB intensity values of leaf and compared it to Chang's (1998) results. A hierarchical dendrogram was constructed using Unweighted Pair-Group Method with Arithmetic Mean analysis (UPGMA). The Gower General Similarity Coefficient was applied to address multi-dimensional scaling. The multivariate statistical package (Version 3.13n, Kovach Computing Services) was used to conduct the cluster analysis.

SVM analysis

Support vector machine (SVM) was first proposed for pattern recognition applications by Vapnik (1995) based on statistical learning theory. The classification mechanism of SVM can be described as simple as: SVM tries to create an appropriate boundary (hyperplane) that meets the requirements of classification, the distance between the boundary and the nearest data points (support vectors) are maximal while the classification precision is also guaranteed. Theoretically, SVM can realize the optimal classification of linearly separable data. In order to solve non-linear problem, SVM converts the data from a low dimension input space to a high dimension feature space through a transformation function (kernel function).

All SVM algorithms are implemented with LIBSVM (Version 3.0) under MATLAB software (The Mathworks, Inc., Natick, MA, USA, version 7.9 R2009b). The LIBSVM is a library for support vector machines (2001).

Results

The fractal dimension and RGB intensity values of species

As shown in the flow chart (Fig. 1), for each species, the five fractal parameters (X_1, X_2, X, Y, S) were derived from the fractal spectra of each (red, green, and blue) colour channels (15 variables). The fractal values obtained for different *Camellia* species belonging to sections *Furfuracea, Paracamellia, Tuberculata, Theopsis,* and *Camellia* are shown in Figs. 2-4). These RGB intensity values were shown in Table 2. Thus, 18 input variables were obtained for modeling.

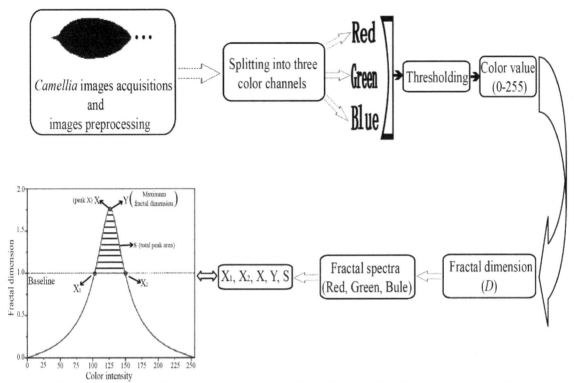

Fig. 1. Schematic diagram of the experimental protocol used to get fractal parameters and RGB intensity values from the image analysis of *Camellia* leaves.

Unsupervised cluster analysis

The relationship between the 68 *Camellia* species was examined by constructing a dissimilarity dendrogram using the 18 variables described above (Fig. 5). The species classified under sect. *Theopsis* by Chang (1998) clustered together (number 40 to 63) in the current study. Further, species number 12 to 24 and number 28 grouped together as an independent branch, which is also mostly congruous with Chang's treatment of sect. *Paracamellia*. Species number 1 to 11 belonging to sect. *Furfuracea* according to Chang's taxonomy also clustered together. However, two species, viz. *C. tuberculata* and *C. obovatifolia* from sect. *Tuberculata* also clustered with them. The other sect. *Tuberculata* species clustered together apart from *C. rhytidophylla* that clustered with sect. *Paracamellia*. Finally, species from sect. *Camellia* clustered together apart from *C. xiafongensis* that clustered with sect. *Theopsis*.

Support vector machine (SVM) classification accuracy

The training set and test set of SVM model is presented in Table 1. The class designation is important for training of SVM algorithms. The 68 species analyzed in the current study were divided into five categories, so the class designation followed the predefined Chang's (1998) taxonomy. Two SVM parameters namely regularization parameter (C) and kernel parameter (g), which are the keys to obtain good model performance, are optimized by cross validation. In current work, $\log_2 C$ and $\log_2 g$ were distributed from -5 to 5 with increments of 0.5. As seen in Fig. 6, the highest average accuracy of 83.33% was achieved when $C = 16$ and $g = 0.5$ for the training data set. The parameter of polynomial SVM were the combinations of another polynomial degree (d) with $d \in \{2,3,4,5,6,7,8,9\}$. The classification results of linear, radial basis function (RBF), and

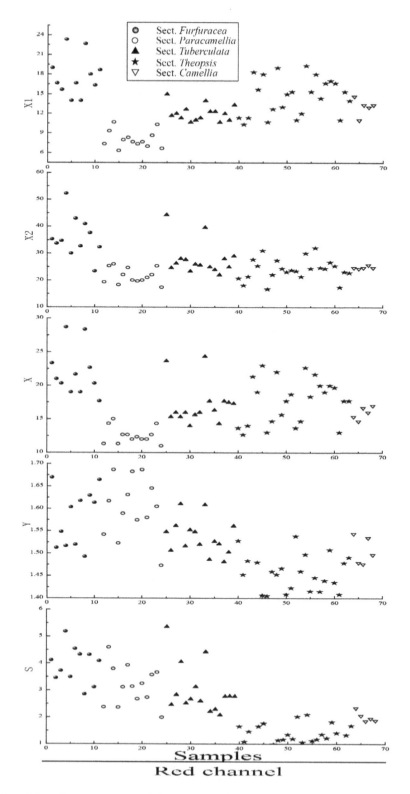

Fig. 2. Scatter plot of fractal parameters used for SVM models in this study. The five fractal parameters (X1, X2, X, Y, S) derived from the samples using **red** channel are shown. Numbers in the figure correspond to the species numbers in Table 1.

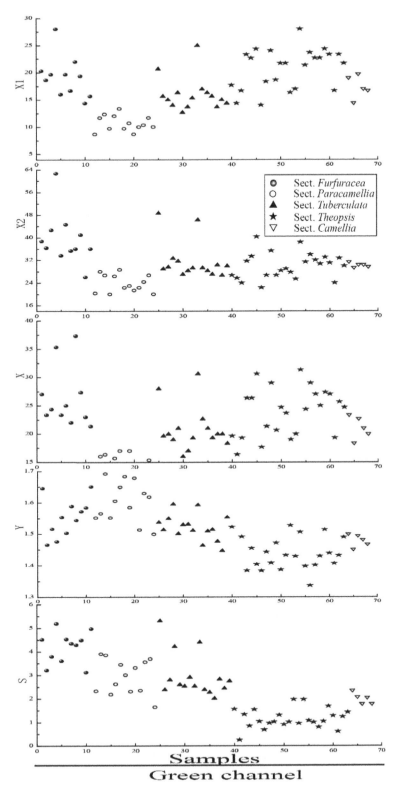

Fig. 3. Scatter plot of fractal parameters used for SVM models in this study. The five fractal parameters (X1, X2, X, Y, S) derived from the samples using **green** channel are shown. Numbers in the figure correspond to the species numbers in Table 1.

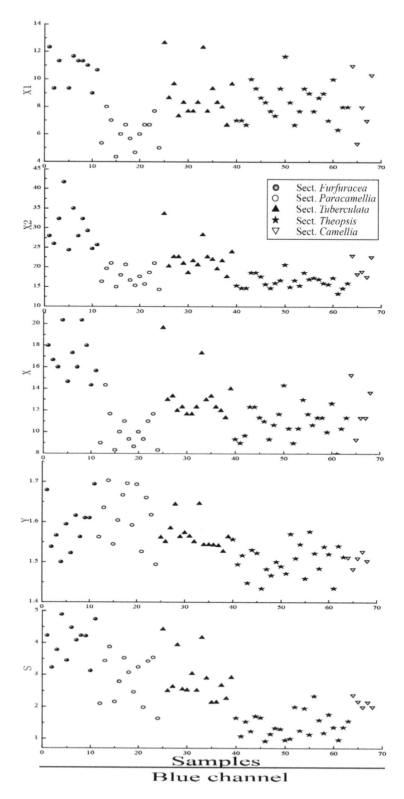

Fig. 4. Scatter plot of fractal parameters used for SVM models in this study. The five fractal parameters (X1, X2, X, Y, S) derived from the samples using **blue** channel are shown. Numbers in the figure correspond to the species numbers in Table 1.

Fig. 5. UPGMA dendrogram of genus *Camellia* based on fractal parameters and RGB intensity values. sect. *Furfuracea* (●), sect. *Paracamellia* (○), sect. *Tuberculata* (▲), sect. *Theopsis* (□), sect. *Camellia* (▽).

sigmoid SVM models, with optimal parameters of C and g are presented in Fig. 7. The RBF SVM classifier offers the best conformance to Chang's classification with 96.88% accuracy rate (sect. *Furfuracea*-100%, sect. *Paracamellia*-100%, sect. *Tuberculata*-85.71%, sect. *Theopsis*-100%, sect. *Camellia*-100%). The only misclassification was in sect. *Tuberculata*, it suggested species number 12 (*C. grijsii*) belongs to sect. *Furfuracea*. Table 3 reveals that the classification results obtained by RBF SVM classifier approach in the training set is 100%, which highlights the good

Table 1. Species assessed, as classified by Chang (1998). Notes: [1, 2, 3, 4, 5] represent the sample labels (categories) used in SVM model. Species without parenthesis are training set, which are followed by parenthesis are test set. The numbers in parenthesis is the number of species in test set.

	Samples			
Sect. Furfuracea[1]	1. *C. pubifurfuracea*	2. *C. latipetioata*	3. *C. crapnalliana*	4. *C. multibracteata*
	5. *C. furfuracea*	6. *C. oblata*	7. *C. gaudichaudii* (1)	8. *C. gigantocarpa* (2)
	9. *C. octopetala* (3)	10. *C. parafurfuracea* (4)	11. *C. connatistyla* (5)	
Sect. Paracamellia[2]	12. *C. grijsii*	13. *C. yuhsienensis*	14. *C. confusa*	15. *C. kissi*
	16. *C. brevistyla*	17. *C. hiemalis*	18. *C. maliflora*	19. *C. shensiensis* (6)
	20. *C. puniceiflora* (7)	21. *C. miyagii* (8)	22. *C. weiningensis* (9)	23. *C. odorata* (10)
	24. *C. phaeoclada* (11)			
Sect. Tuberculata[3]	25. *C. tuberculata* (12)	26. *C. lipingensis* (13)	27. *C. rhytidocarpa*	28. *C. rhytidophylla*
	29. *C. leyeensis*	30. *C. anlungensis*	31. *C. rubituberculata*	32. *C. atuberculata*
	33. *C. obovatifolia*	34. *C. rubimuricata*	35. *C. parvimuricata* (14)	36. *C. hupehensis* (15)
	37. *C. zengii* (16)	38. *C. pyxidiacea* (17)	39. *C. crassifolia* (18)	
Sect. Theopsis[4]	40. *C. macrosepala* (19)	41. *C. cuspidatevar. synapidate* (20)	42. *C. cuspidata*	43. *C. forerrestii*
	44. *C. lipoensis*	45. *C. buxifolia*	46. *C. minutiflora*	47. *C. acutissima*
	48. *C. dubia*	49. *C. handelii*	50. *C. costei*	51. *C. tsaii*
	52. *C. rosthorniana*	53. *C. euryoides*	54. *C. trichoclada* (21)	55. *C. parvilimba* (22)
	56. *C. parvilimba var. brevipes* (23)	57. *C. septempetala* (24)	58. *C. elongate* (25)	59. *C. campanisepala* (26)
	60. *C. parvi-ovata* (27)	61. *C. lancicalyx* (28)	62. *C. parvicaudata* (29)	63. *C. tsofui* (30)
Sect. Camellia[5]	64. *C. jinshajiangica*	65. *C. semoserrata var. albiflora*	66. *C. xiafongensis*	67. *C. chekiangoleosa* (31)
	68. *C. lienshanensis* (32)			

Table 2. The RGB intensity values derived from samples used for SVM models in this study.

	Sect. *Furfuracea*		Sect. *Paracamellia*		Sect. *Tuberculata*		Sect. *Theopsis*		Sect. *Camellia*	
RGB intensity	Range (min-max)	Mean ± SD	Range (min-max)	Mean ± SD	Range (min-max)	Mean ± SD	Range (min-max)	Mean ± SD	Range (min-max)	Mean ± SD
R	15.61-33.39	23.54±5.25	12.16-17.85	14.25±1.77	15.52-26.63	18.81±3.38	13.34-25.08	19.22±3.55	15.65-19.19	17.35±1.26
G	18.28-39.97	27.65±6.10	13.52-18.99	15.86±2.01	17.89-32.94	22.54±4.15	18.36-34.07	25.82±4.31	19.69-23.95	22.08±1.78
B	12.54-24.96	18.02±3.77	9.91-13.76	11.55±1.21	11.04-20.82	14.03±2.49	9.6-14.57	12.14±1.45	10.32-15.68	12.80±2.20

Table 3. The classification results in the training set of RBF SVM classifier.

Samples	Sample number	Classification results					Total accuracy
		Sect. Furfuracea	Sect. Paracamellia	Sect. Tuberculata	Sect. Theopsis	Sect. Camellia	
Sect. Furfuracea	6	6	0	0	0	0	
Sect. Paracamellia	7	0	7	0	0	0	
Sect. Tuberculata	8	0	0	8	0	0	100%
Sect. Theopsis	12	0	0	0	12	0	
Sect. Camellia	3	0	0	0	0	3	

Table 4. The predicted classification of the polynomial SVM under different degrees with the optimal parameters ($C=16$, $g=0.5$).

Subset	Samples Number	Polynomial degree							
		2	3	4	5	6	7	8	9
Sect. Furfuracea	5	100%	100%	100%	100%	100%	80%	80%	80%
Sect. Paracamellia	6	100%	83.33%	66.67%	66.67%	66.67%	66.67%	66.67%	66.67%
Sect. Tuberculata	7	71.43%	71.43%	57.14%	57.14%	57.14%	57.14%	57.14%	57.14%
Sect. Theopsis	12	100%	100%	100%	100%	91.67%	91.67%	91.67%	91.67%
Sect. Camellia	2	100%	100%	100%	100%	100%	100%	100%	100%
Total accuracy (%)		93.75%	90.63%	84.38%	84.38%	81.25%	78.13%	78.13%	78.13%

performance of the RBF SVM classifier. The linear SVM classifier for five sections shows correct classification rate of 90.63% (sect. *Furfuracea*-100%, sect. *Paracamellia*-100%, sect. *Tuberculata*-57.14%, sect. *Theopsis*-100%, sect. *Camellia*-100%), but the sigmoid kernel overall accuracy for the test data set is worse than any other classifiers with only 40.63% (sect. *Furfuracea*-0%, sect. *Paracamellia*-33.33%, sect. *Tuberculata*-0%, sect. *Theopsis*-91.67%, sect. *Camellia*-0%). For polynomial classifiers, in fact, it is a linear classifier when polynomial degree $d = 1$. The classification results of polynomial SVM classifier with different degrees from 2 to 9 are shown in Table 4. The polynomial SVM classifiers with $d = 2$ achieved the best overall classification accuracies (93.75%) of the five sections (sect. *Furfuracea*-100%, sect. *Paracamellia*-100%, sect. *Tuberculata*-71.43%, sect. *Theopsis*-100%, sect. *Camellia*-100%). In addition, the active effect on the classification accuracies was very less when d was greater than 2, with increasing polynomial degree, the classification accuracies take on a descending trend (Table 4).

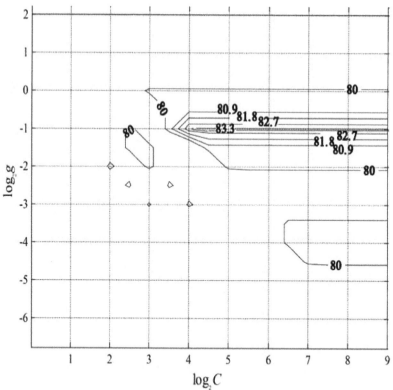

Fig. 6. Average classification accuracy in different kernel parameter (*C*) and regularization parameter (*γ*) by cross-validation.

Table 5. Summary of the supervised techniques, materials, factors, and accuracies for Chang (1998)'s *Camellia* classification.

Classification techniques	Materials	Factors	Accuracy	Reference	Demerit
Cluster analysis	63 species and 2 varieties in 4 sections	Fourier transform infrared data of leaves	84.7%	Lu et al. 2008a	Expensive
Cluster analysis	21 species from 4 sections	Fourier transform infrared data combined with leaf anatomy	85.7%	Lu et al. 2008b	Wasting time, money and low-efficiency
Particle swarm optimization-aided fuzzy cloud classifier	24 species from 3 sections	23 quantitative features cover the characters of flower, fruit and leaf	98.0%	Lu et al. 2009	Laborious and time consuming
Pattern recognition techniques	93 species from 5 sections	31 variables from Leaf morphological and venation characters	LVQ1-ANN for 60% LVQ2-ANN for 91.11% DAN2 for 91.11% SVM for 97.78%	Lu et al. 2012	Heavy workload
Cluster analysis and principal coordinate analysis	19 species from 2 sections	28 variables from floral morphology characters	84.2%	Jiang et al. 2012	Poor repeatability
Back-propagation neural networks	47 species from 3 sections	7 leaf anatomy attributes	86.36%	Jiang et al. 2013	Time consuming

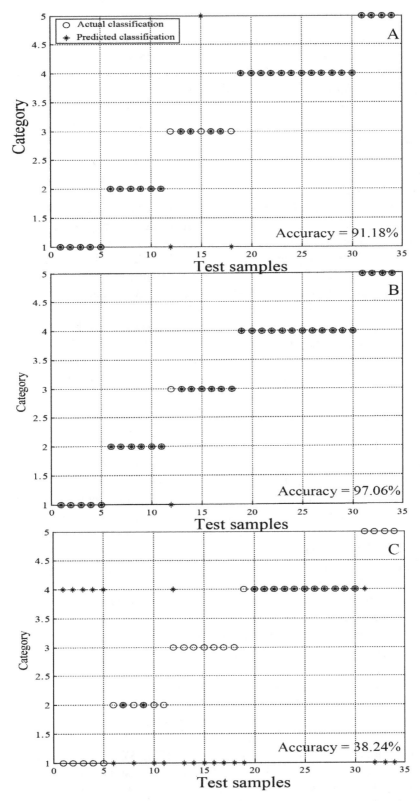

Fig. 7. The classification results of linear (A), RBF (B) and sigmoid (C) SVMs with the optimal parameters.

Discussion

Plant numerical taxonomy applies numerical methods or supervised techniques like SVM in the classification of taxonomic units. It converts the information content of taxa to numerical quantitative and its aim is in its objectivity. Thus, developing a taxonomic toolkit is becoming an indispensable aid in modern systematics. Traditionally, leaf characters have been used as a basis for plant taxonomy and they have been successfully used to solve plant classification problems (Linnaeus, 1753). Contemporary classification especially for genus *Camellia*, have involved use of advanced technology tools. Some examples are, classification within genus level based on simulated annealing aided cloud classifier (Pi *et al.*, 2011); use of genetic information with molecular biotechnology tools; fourier transform infrared spectroscopy (FTIR) combined with shape and anatomy analysis of *Camellia* leaves (Lu *et al.*, 2008b; Shen *et al.*, 2008), which suggested that the chemical method also had important taxonomic significance. However, as shown in Table 5, some of these methods are laborious and expensive, and do not always guarantee satisfactory results. Moreover, a defect common to all the approaches (Table 5) is that they get quantitative features of plant is based on damaging leaves. However, fractal analysis and RGB intensity values combined with support vector machine (SVM) used in our study are not only non-destructive, but are simple, and easily performed. The fractal spectrum was introduced as a botanical identification key by Mugnai *et al.* (2008). Actually, leaf colour is a very special characteristic but often ignored by taxonomists. *Camellia* species are both trees and shrubs, and plant height and leaf feature may interfere with plant photosynthesis. The chlorophyll content in turn is correlated to the leaf colour (Du *et al.*, 2009). Moreover, the long-term evolution of *Camellia* species have made them a stable system, therefore they can be classified based on leaf traits like chlorophyll content.

Chang (1998) and Ming (2000) are two comprehensive floras prominently used by *Camellia* researchers. People often turn to flora to identify a new species; however, traditional information retrieval processes is frequently cumbersome. Further, some basic characteristics can only be manually identified which needs experience and is often subjective. These limitations can be overcome by developing an automated method of plant identification which is rapid and efficient. We have developed an automated method using leaf fractal parameters in SVM model to classify 68 *Camellia* species. The taxonomic results are very encouraging allowing us to achieve accuracy of up to 96.88% using the RBF fractal values. As a modern pattern recognition tool, the SVM is advantageous over other methods like back-propagation artificial neural network (BP-ANN). The common problem with neural networks is the networks structure; BP-ANN may suffer from the over-fitting problem because its approaches are based on the empirical risk minimization principles. Comparatively, the over-fitting can be easily controlled in SVM by choosing a suitable margin to get the best resolution of entire data set (Burges, 1998). In addition, SVM does not need a great quantity of training sets for developing model.

Our results were mostly congruent with Chang's (1998) classification of *Camellia* species with some differences. However, it should be noted that other researchers have also reported deviations from Chang's classification. For example, when our results are compared to *Camellia* classification by Vijayan *et al.* (2009), the general agreement in classification of the 68 *Camellia* species indicates the usefulness of fractal parameters and RGB intensity in detecting phylogenetic relationships. For the plants from sect. *Furfuracea* and sect. *Theopsis*, all collected species from two sections were joined and intermixed respectively (Fig. 5), which is in agreement with the classification by Vijayan *et al.* (2009). In addition, our results support the grouping of *C. yuhsienensis* (No. 13, from sect. *Paracamellia*) and *C. rhytidophylla* (No. 28, from sect. *rhytidophylla*) together as reported by Vijayan *et al.* (2009). This is however different from Chang's (1998) treatment of these two species. Further, as shown in Fig. 5, species from sect.

Paracamellia grouped together, whilst Vijayan *et al.* (2009) taxonomic treatment advocates these species as three clades. In analyzing results from the SVM classifiers, we found that the species number 12 (*C. grijsii*) from sect. *Paracamellia* was incorrectly classified as a species from sect. *Furfuracea* by all SVM classifiers [linear, RBF, sigmoid, and polynomial ($d = 2$) classifiers]. The deviation from this classification needs further investigation to see if this misclassification is due to the underlying algorithm's fitting of the data, or *C. grijsii* really has a close relationship with sect. *Furfuracea*.

In addition, high quality seeds are the key to develop the modern agriculture, it is necessary to select good seed varieties for improvement of crops yield. An elite variety with greater benefits should replace the variety with inferior quality seeds. Bacchetta *et al.* (2011) identified Sardinian species of *Astragalus* section *Melanocercis* by seed image analysis. Developing countries are still using traditional manual seed separation method. In this context, the application of SVM based on fractal leaf parameters analysis (FA) and leaf red, green, and blue (RGB) intensity values used in the present study is not only proposed as a complementary method for botanical identification, but also proposed as a modern method of good seed selection. The SVM-FA-RGB system is very simple to establish and requires only a personal computer and an optical scanner. Therefore it could potentially replace old methods that are complicated, labour-intensive and expensive.

Conclusion

We have developed a system for automatic binary classification of 68 *Camellia* species into five sections based on SVM and discussed the important features of this classification. The hierachical dendrogram based on fractal parameters and RGB intensity values confirms the morphological classification of the five sections proposed by Chang's (1998) research. The linear, polynomial ($d = 2$), RBF SVM classifier with $C = 16$, $g = 0.5$ work well in the classification of the genus *Camellia*. Especially RBF SVM classifier showed encouraging results that obtaining a correct classification rate of 96.88%. The above results indicate that fractal parameters and RGB intensity values analysis using SVM, particularly RBF kernel, can be effectively used to distinguish the *Camellia* at genus level, or even at higher taxa level. In addition, the SVM-FA-RGB system could be used to select high quality seeds in agriculture breeding programs.

Acknowledgements

The study was partially supported by grants from the Science and Technology research Plan of Jinhua city, China (No. 2009-2-020). The authors thank Mr. Bin Wang, Ms. Jingjing Lou and Ms. Zhihui Zhu (Zhejiang Normal University) for collecting living species and assistance with the experiments.

References

Azzarello, E., Mugnai, S., Pandolfi, C., Masi, E., Marone, E. and Mancuso, S. 2009. Comparing image (fractal analysis) and electrochemical (impedance spectroscopy and electrolyte leakage) techniques for the assessment of the freezing tolerance in olive. Tress-Stucture Function **23**: 159–167.

Bacchetta, G., Fenu, G., Grillo, O., Mattana, E. and Venora, G. 2011. Identification of Sardinian species of *Astragalus* section *Melanocercis* (Fabaceae) by seed image analysis. Ann. Bot. Fenn. **48**: 449–454.

Barthlott, W., Wiersch, S., Colic, Z. and Koch, K. 2009. Classification of trichome types within species of the water fern Salvinia, and ontogeny of the eggbeater trichomes. Botany **87**: 830–836.

Burges, C.J.C. 1998. A tutorial on support vector machines for pattern recognition. Data Min. Knowl. Disc. **2**: 121–167.

Chang, H.T. 1998. Theaceae. In: Flora of China Editorial Committee (Ed.), Vol. **49**. Flora of China. Science Press, Beijing.

Chen, L., Wang, S. and Nelson, M. 2005. Genetic diversities within Camellia species confirmed by random amplified polymorphic DNA (RAPD) markers. J. Am. Soc. Hortic. Sci. **40**: 993–1147.

Chen, Q.S., Guo, Z.M. and Zhao, J.W. 2008. Identification of green tea's (*Camellia sinensis* L.) quality level according to measurement of main catechins and caffeine contents by HPLC and support vector classification pattern recognition. J. Pharmaceut. Biomed. **48**: 1321–1325.

Chen, Q.S., Zhao, J.W., Fang, C.H. and Wang, D.M. 2007. Feasibility study on identification of green, black and Oolong teas using near-infrared reflectance spectroscopy based on support vector machine (SVM). Spectrochim. Acta. A **66**: 568–574.

Cortes, C. and Vapnik, V. 1995. Support-vector networks. Mach. Learn. **20**: 273–297.

Du, P., Ling, Y.H., Sang, X.C., Zhao, F.M. and Xie, R. 2009. Gene mapping related to yellow green leaf in a mutant line in rice (*Oryza sativa* L.). Genes Genom. **31**: 165–171.

Gao, J.Y., Parks, C.R. and Du, Y.Q. 2005. Collected species of the genus *Camellia*, an illustrated outline. Zhejiang Science and Technology Press, Hangzhou.

Gunasekaran, S., Paulsen, M.R. and Shove, G.C. 1985. Optical methods for nondestructive quality evaluation of agricultural and biological materials. J. Agri. Eng. Res. **32**: 209–241.

Harrell, R.C., Slaughter, D.C. and Adsit, P.D. 1989. A fruit-tracking system for robotic harvesting. Mach. Vision Appl. **2**: 69–80.

Howaith, M.S., Searcy, S.W. and Birth, G.S. 1990. Reflectance characteristics of fresh market carrots. T. ASABE **33**: 961–964.

Jack, L.B. and Nandi, A.K. 2002. Fault detection using support vector machines and artificial neural networks: augmented by genetic algorithms. Mech. Syst. Signal Pr. **16**: 373–390.

Jain, A.K., Murty, M.N. and Flynn, P.J. 1999. Data Clustering: A Review. ACM Comput. Surv. **31**: 264–323.

Jiang, B., Peng, Q.F., Shen, Z.G., Moller, M., Pi, E.X. and Lu, H.F. 2010. Taxonomic treatments of *Camellia* (Theaceae) species with secretory structures based on integrated leaf characters. Plant Syst. Evol. **290**: 1–20.

Jiang, W., Nitin, M., Jiang, B., Zheng, Y.P., Hong, S.S. and Lu, H.F. 2012. Floral morphology resolves the taxonomy of *Camellia* L. (Theaceae) Sect. *Oleifera* and Sect. *Paracamellia*. Bangladesh J. Plant Taxon. **19**: 155–165.

Kumar, K.S., Jayabarathi, T. and Naveen, S. 2011. Fault identification and location in distribution systems using support vector machines. Eur. J. Sci. Res. **51**: 53–60.

Lin, X.Y., Peng, Q.F., Lu, H.F., Du, Y.Q. and Tang, B.Y. 2008. Leaf anatomy of *Camellia* Sect. Oleifera and Sect. *Paracamellia* (Theaceae) with reference to their taxonomic significance. J. Syst. Evol. **46**: 183–193.

Linnaeus, C. 1753. Species Plantarum. Stockholm: Salvias.

Lu, H.F., Jiang, B., Shen, Z.G., Shen, J.B., Peng, Q.F. and Cheng, C.G. 2008a. Comparative leaf anatomy, FTIR discrimination and biogeographical analysis of *Camellia* section *Tuberculata* (Theaceae) with a discussion of its taxonomic treatments. Plant Syst. Evol. **274**: 223–235.

Lu, H.F., Shen, J.B., Lin, X.Y. and Fu, J.L. 2008b. Relevance of Fourier transform infrared spectroscopy and leaf anatomy for species classification in *Camellia* (Theaceae). Taxon **57**: 1274–1288.

Lu, H.F., Pi, E.X., Peng, Q.F., Wang, L.L. and Zhang, C.J. 2009. A particle swarm optimization-aided fuzzy cloud classifier applied for plant numerical taxonomy based on attribute similarity. Expert Syst. Appl. **36**: 9388–9397.

Lu, H.F., Zheng, H., Hu, Y., Lou, H.Q. and Kong, X.C. 2011. Bruise detection on red bayberry (*Myrica rubra* Sieb. & Zucc.) using fractal analysis and support vector machine. J. Food Eng. **104**: 149–153.

Lu, H., Jiang, W., Ghiassi, M., Lee, S. and Nitin, M. 2012. Classification of *Camellia* (Theaceae) species using leaf architecture variations and pattern recognition techniques. PLoS ONE **7**(1): e29704.

Mancuso, S. 1999. Fractal geometry-based image analysis of grapevine leaves using the box counting algorithm. Vitis **38**: 97–100.

Mancuso, S. 2002. Discrimination of grapevine (*Vitis vinifera* L.) leaf shape by fractal spectrum. Vitis **41**: 137–142.

Mancuso, S., Nicese, F.P. and Azzarello, E. 2003. The fractal spectrum of the leaves as a tool for measuring frost hardiness in plants. J. Hortic. Sci. Biotech. **78**: 610–616.

Ming, T.L. 2000. Monograph of the Genus *Camellia*. Yunnan Science and Technology Press, Kunming.

Mugnai, S., Pandolfi, C., Azzarello, E., Masi, E. and Mancuso, S. 2008. *Camellia japonica* L. genotypes identified by an artificial neural network based on phyllometric and fractal parameters. Plant Syst. Evol. **270**: 95–108.

Pandolfi, C., Messina, G., Mugnai, S., Azzarello, E. and Masi, E. 2009. Discrimination and identification of morphotypes of *Banksia integrifolia* (Proteaceae) by an Artificial Neural Network (ANN), based on morphological and fractal parameters of leaves and flowers. Taxon **58**: 925–933.

Pi, E.X., Lu, H.F., Jiang, B., Huang, J., Peng, Q.F. and Lin, X.Y. 2011. Precise plant classification within genus level based on simulated annealing aided cloud classifier. Expert Syst. Appl. **38**: 3009–3014.

Pi, E.X., Peng, Q.F., Lu, H.F., Shen, J.B., Du, Y.Q., Huang, F.L. and Hu, H. 2009. Leaf morphology and anatomy of *Camellia* section *Camellia* (Theaceae). Bot. J. Linn. Soc. **159**: 456–476.

Plotze, R.D.O., Falvo, M., Pádua, J.G., Bernacci, C. and Vieira, M. 2005. Leaf shape analysis using the mutiscale Minkowski fractal dimension, a new morphometric method: A study with *Passiflora* (Passifloraceae). Can. J. Bot. **83**: 287–301.

Pontil, M. and Verri, A. 1998. Support vector machines for 3-D object recognition. IEEE T Pattern Anal. **20**: 637–646.

Sealy, J.R. 1958. A Revision of the Genus *Camellia*. The Royal Horticultural Society, London.

Shen, J.B., Lu, H.F., Peng, Q.F., Zheng, J.F. and Tian, Y.M. 2008. FTIR spectra of *Camellia sect.* Oleifera, sect. *Paracamellia*, and sect. *Camellia* (Theaceae) with reference to their taxonomic significance. J. Syst. Evol. **46**: 194–204

Vapnik, V. 1995. The Nature of Statistical Learning Theory. NY, New York, USA: Springer-Verlag.

Vijayan, K., Zhang, W.J. and Tsou, C.H. 2009. Molecular taxonomy of *Camellia* (Theaceae) inferred from nrITS sequences. Am. J. Bot. **96**: 1348–1360.

Yang, Z.R. and Lin, Q. 2005. Comparative morphology of the leaf epidermis in *Schisandra* (Schisandraceae). Bot. J. Linn. Soc. **148**: 39–56.

Ye, P. and Weng, G.R. 2011. Classification and recognition of plant leaf based on neural networks. Key Eng. Mater. **464**: 38–42.

Zheng. H., Lu, H.F., Zheng, Y.P., Lou, H.Q. and Chen, C.Q. 2010. Automatic sorting of Chinese jujube (*Zizyphus jujuba* Mill. cv. 'hongxing') using chlorophyll fluorescence and support vector machine. J. Food Eng. **101**: 402–408.

POLLEN CHARACTERS AS TAXONOMIC EVIDENCE IN SOME SPECIES OF DIPSACACEAE FROM IRAN

EBADI-NAHARI MOSTAFA[1], NIKZAT-SIAHKOLAEE SEDIGHEH[2] AND EFTEKHARIAN ROSA[2]

Department of Biology, Faculty of Science, Azarbaijan Shahid Madani University, Tabriz, Iran.

Keywords: Dipsacaceae; Palynological characters; SEM; UPGMA.

Abstract

Pollen morphology of nine species representing four genera: *Cephalaria* Schrad, *Dipsacus* L., *Pterocephalus* Vaill. and *Scabiosa* L. of the family Dipsacaceae in Iran has been investigated by means of scanning electron microscopy (SEM). The results showed that pollen grains were triporate and tricolpate. The pollen type of *Scabiosa rotata* Bieb. (tri- and tetraporate) is the first report in the world. The sizes of pollen grains fall into the classification group magna (pollen grain diameter 50–100 μm). Pollen shapes vary from preoblate to prolate and their polar views were triangulate and lobate. The exine ornamentation varies from gemmate in *S. rotata* to spinulate in the rest studied species. Species of *Scabiosa* have been dispersed in UPGMA tree that this confirmed the previous studies about taxonomic problems and species complexity in this genus. These results show the transfer of the some *Scabisoa* species to *Lomelosia* Raf. based on palynological characters. Pollen morphology of the family is helpful at the generic and specific level.

Introduction

The family Dipsacaceae consists of around 10-13 genera and more or less 300 species (Ehrendorfer, 1965; Verlaque, 1977; Mabberley, 2008) of annual to perennial herbs and shrubs that occur primarily in the Mediterranean Basin, with about 20% distributed in Asia and Africa. In Iran, Dipsacaecae is represented by 54 species belonging to five genera distributed in different regions (Jamzad, 1993). The family has long been regarded as belonging to the Dipsacaceae, whereas according to APG III it is included within the larger family Caprifoliaceae (Reveal and Chase, 2011). Delimitation of taxa within the family has always been subject to argument; accordingly, circumscription of genera and tribes has repeatedly changed over of the overall morphological similarity among the taxa in the family. The family was divided into two tribes by De Candolle (1830), viz. Morineae (including a single genus, *Morina* L.) and Scabioseae (including *Cephalaria*, *Dipsacus*, *Knautia*, *Pterocephalus*, and *Scabiosa*). Verlaque (1984) divided this family into three tribes with nine genera. Caputo and Cozzolino (1994) divided Dipsacaceae into two major clades (based on morphological and palynological characters), one includes *Dipsacus* and *Cephalaria*, the other contain the remaining genera.

The significance of pollen morphology in Plant Systematics has been stressed by various researchers. Stuessy (2009) state that data from pollen grains are known to be useful at all levels of the taxonomic hierarchy (generic, subgeneric, inter-specific and intraspecific levels), and can often be helpful in suggesting a relationship. Some studies (for example, Feng *et al*., 2000; Khalik, 2010; Perveen, 2011) showed that pollen morphological characteristics play a major role in solving taxonomic problems. Palynological characteristics have been able to reposit several disputed genera and interpret problems related to the origin and evolution of many taxa (Nair, 1980) and to derive a classification of angiosperms (Cronquist, 1981).

---

1 Corresponding author: ebadi2023@yahoo.com
2 Faculty of Biological Sciences, Shahid Beheshti University, Tehran, Iran.

The works of Mayer and Ehrendorfer (2000) on the pollen morphology of *Pterocephalus* and Feng *et al.*, (2000) on the palynology of the genus *Dipsacus* show that the study of pollen grains provides useful data for the taxonomy of different genera.

There is a modicum of information on the pollen morphology of the family Dipsacaceae, most especially in Iran. This study reports the pollen morphology of some species in the family Dipsacaceae from Iran in order to establish their availability for future taxonomic works.

Materials and Methods

Pollen grains of 9 species, representing 4 genera of Dipsacaceae distributed in Iran were studied by means of scanning electron microscope (SEM). The studied plant samples were collected from natural populations in different regions during spring and summer in 2014-2015. The voucher specimens were deposited in Azarbaijan Shahid Madani University Herbarium (ASMUH). The list of voucher specimens and details of localities are given in Table 1.

Table 1. List of species used in the study along with localities and vouchers.

Genus	Species	Locality	Voucher No.
Cephalaria Schrad	*C. kotschyi* Boiss.	Mazndaran, Chalus	ASMUH95001
	C. procera Fisch.	Ardebil, Khalkhal	ASMUH95002
Pterocephalus Vaill.	*P. plumosus* (L.) Coulter	Mazndaran, Chalus	ASMUH95003
	P. canus Coulter	Tehran, Ab-ali	ASMUH95004
Dipsacus L.	*D. strigosus* Willd	Mazndaran, Chalus	ASMUH95005
Scabiosa L.	*S. caucasica* M. B.	Ardebil, Khalkhal	ASMUH95006
	S. amoena Jacq.	Gilan, Masuleh	ASMUH95007
	S. koelzii Rech. f.	Khorasan, Bojnord	ASMUH95008
	S. rotata Bieb.	Khorasan, Mashhad	ASMUH95009

Pollen grains were separated from anthers by using binocular microscope. For each taxon, three specimens were used, and from each specimen at least five anthers were examined. Pollen grains for scanning electron microscopy were mounted on standard aluminum stubs using double-sided adhesive tape and then photographed using Phenomprox scanning electron microscope at 10 KV voltages.

Palynological characters such as equatorial diameter (E), polar axis length (P), P/E, exine ornamentation etc. were measured (at least 30 pollen grains) by using Image Tools software with high accuracy and confidence degree. For grouping of the studied taxa, data were standardized (mean = 0, variance = 1) and used for the multivariate analyses using unweighted pair-group method with arithmetical mean (UPGMA) based on Euclidean Distances and principal component analysis (PCA) by means of PAST Package (Hammer *et al.*, 2001). The terminology used is in accordance with Erdtmann (1952) and Punt *et al.* (2007).

Results

The pollen grains of the studied species revealed some variations. All palynological structures and measurements for the examined species concerning pollen type from polar view, polar (P) and equatorial (E) measurements, P/E ratio, pollen shape, polar view and exine ornamentation were shown in Table 2.

Table 2. Measurements of pollen-morphological character in the studied taxa.

	Pollen type	Polar axis (P)	Equatorial axis (E)	(P/E)	Pollen shape	Polar view	Spine length	Exin ornamentation
C. kotschyi	Triporate	57.52 ± 1.36	86.90 ± 0.65	0.66	Oblate	Triangulate	1.60 ± 0.11	Spinulate-Spinuloid
C. procera	Triporate	55.73 ± 1.71	70.05 ± 0.56	0.79	Suboblate	Triangulate	1.72 ± 0.15	Spinulate-Spinuloid
P. plumosus	Tricolpate	78.51 ± 0.27	54.93 ± 0.81	1.42	Prolate	Lobate	1.37 ± 0.09	Spinulate-Spinuloid
P. canus	Tricolpate	75.52 ± 0.17	60.36 ± 1.76	1.25	Subprolate	Lobate	2.40 ± 0.08	Spinulate-Spinuloid
D. strigosus	Triporate	90.43 ± 1.81	74.59 ± 0.28	1.21	Subprolate	Triangulate	1.14 ± 0.09	Spinulate
S. caucasica	Triporate	50.97 ± 1.67	92.47 ± 1.01	0.55	Oblate	Triangulate	1.83 ± 0.14	Spinulate-Spinuloid
S. amoena	Tricolpate	92.1 ± 0.14	61.70 ± 0.19	1.49	Prolate	Triangulate	1.30 ± 0.11	Spinulate-Spinuloid
S. koelzii	Tricolpate	91.53 ± 0.26	65.57 ± 0.2	1.39	Prolate	Lobate	1.55 ± 0.15	Spinulate-Spinuloid
S. rotata	Tri and tetraporate	28.97 ± 0.59	75.05 ± 1.47	0.38	Preoblate	Triangulate	-	Gemmate-Spinuloid

Generally, there are two major types of pollen grain apertures, varying from triporate to tricolporate among studied species (Fig 1). The pollen grains of *S. rotata* are triporate and tetraporate (Fig 1. I, J).

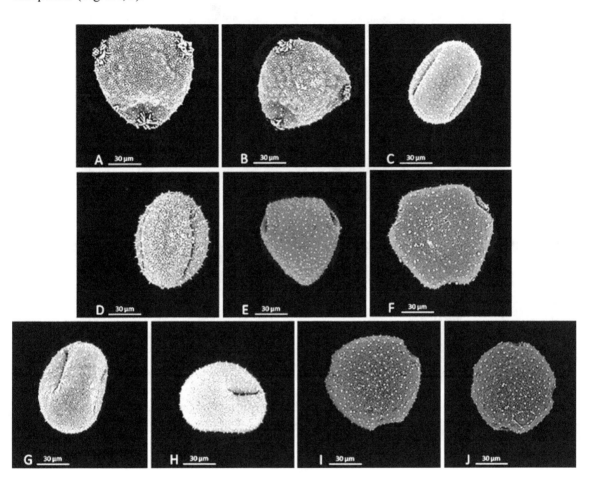

Fig. 1. Scanning electron microscope photographs of pollen grains. A: *C. kotschyi*, B: *C. procera*, C: *P. plumosus*, D: *P. canus*, E: *D. strigosus*, F: *S. caucasica*, G: *S. amoena*, H: *S. koelzii*, I: *S. rotata* (triporate), J: *S. rotata* (tetraporate)

Polar axis (P) length of pollen grains showed large variation, ranging from the smallest size for *S. rotata* (28.97 μm) to the largest size for *S. amoena* (92.1 μm). Equatorial axis (E) length of pollen grains ranged from the smallest size in *P. plumosus* (54.93 μm) to the largest size in *S. caucasica* (92.47μm). The shape classes are based on the ratio between the length of polar axis (P) and equatorial diameter (E). The P/E ratio ranged from 0.38 μm to 1.49 μm, therefore the pollen shape is subprolate to prolate in *Seabiosa amoena*, *S. koelzii*, *Pterocephalus plumosus*, *P. canus* and *D. strigosus* but preoblate to oblate in the rest studied species. However, the pollen shape in polar view varies from lobate in *P. plumosus*, *P. canus* and *S. koelzii* to triangulate in the rest of the species (Fig. 1).

The exine sculpturing showed a complex structure. The SEM showed that the outer surface of the tectum is a solid layer which is covered by numerous similar small conical spinuloid (Fig. 2). The exine surface with gemmate (wart-like pegs) was found in *S. rotata* but spinulate (more than 3 μm long) in the rest studied species.

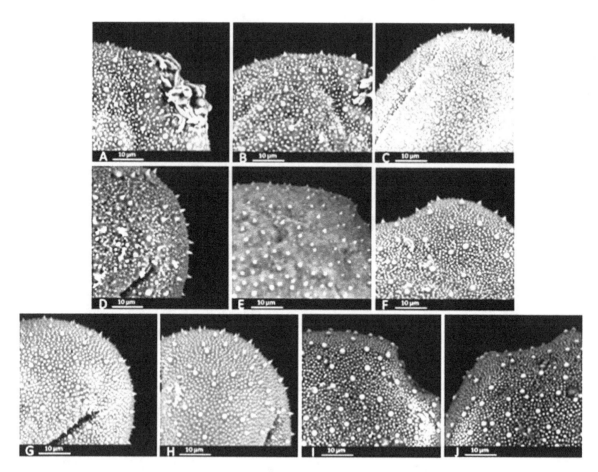

Fig. 2. Scanning electron microscope photographs of pollen surface ornamentation. A: *C. kotschyi*, B: *C. procera*, C: *P. plumosus*, D: *P. canus*, E: *D. strigosus*, F: *S. caucasica*, G: *S. amoena*, H: *S. koelzii*, I: *S. rotata* (triporate), J: *S. rotata* (tetraporate)

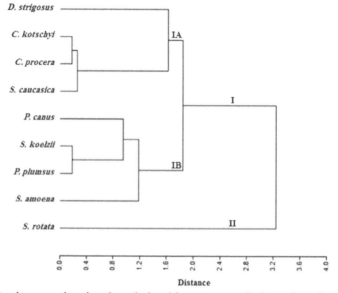

Fig. 3. UPGMA dendrogram showing the relationship among studied taxa based on pollen characters.

The studied taxa were separated from each other in a UPGMA tree based on palynological characters. Cluster analysis showed that the studied species placed in two majorclusters (Fig. 3). One contained *S. rotata* (cluster II) and remainders were clustered in other major branch (cluster I) of two sub-clusters assigned as IA and IB. Sub-cluster IA included *D. strigosus*, *C. kotschyi*, *C. procera* and *S. caucasica*. *Pterocephalus canus*, *P. plumsus*, *S. koelzii* and *S. amoena* grouped together within the sub-cluster IB (Fig. 3).

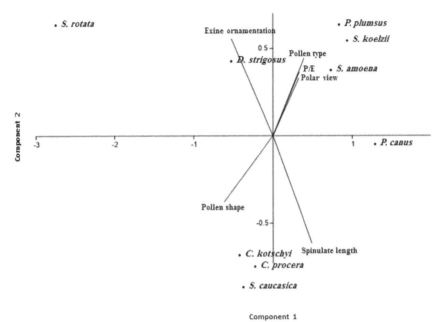

Fig. 4. Principal Component Analysis (PCA) among studied taxa based on pollen characters.

Principal component analysis (PCA) showed that the pollen type, pollen view, P/E and exine ornamentation have main role in grouping of species belonging to the clade IA. While the pollen shape and spine length are importance characters in grouping of species belonging to the clade IB (Fig. 4). According to the exine ornamentation, *S. rotata* was separated from the remaining species.

Discussion

The present study shows a palynological polymorphism within the family Dipsacaceae. Generally, the interspecific differences within a genus are often trivial, but there are remarkable differences among the various genera (Khalik, 2010). Various palynological investigations on different species of Dipsacaceae confirmed importance of pollen traits for distinguishing taxa. For instance, Khalik (2010) studied pollen characteristics in nine species belonging to four genera of Dipsacaceae in Egypt and showed that pollen character can be used to delimit the species.

Statistical analysis showed that most of the qualitative characters were useful in classification of studied species and had taxonomic value. Aperture types, polar view and pollen shape can be used as diagnostic evidence in palynological studies.

In this study, two types of pollen apertures (porate and colpate) were found in this family. In addition, we distinguished the pollen with tetraporate aperture in *S. rotata* for the first time. Erdtmann (1952) and Clarke and Jones (1981) studied the pollen morphology of the Dipsacaceae, distinguished two types of pollen apertures: porate and colpate.

Considering the exine ornamentation, it is obvious that this palynological character cannot be used to delimit studied species, because all of them were spinulate except *S. rotata* that was gemmate. Interestingly, spine length varied between studied species and it could be suitable as a diagnostic character. Based on spine length, *P. canus* can be distinguished from the other taxa. In our study, *S. rotata* has specific palynological character such as gemmate of exine ornamentation and tri-tetraporate pollen that helps to distinguish it.

The findings of the present study on the size of pollen are in agreement with previous study (Ebadi-Nahari and Nikzat-Siahkolaee, 2016). Pollen grains have been classified into groups according to their sizes by Erdtman (1952) as perminuta (diameter less than 10 μm), minuta (diameter 10-25 μm), media (diameter 25-50 μm), magna (diameter 50- 100 μm), permagna (diameter 100-200 μm) and giganta (diameter greater than 200μm). Based on this classification, the pollen grains of the species studied belong to group magna (diameter 50-100μm). Khalik (2010) reported nine species belonging to four genera of the Dipsacaceae in the group size magna (diameter 50-100 μm). The pollen grain size reported for this family supports the fact that the flowers in the genera are more insect-and-bird pollinated than by wind.

Ehrendorfer (1965) first studied phylogeny of Dipsacaceae at generic and infrageneric levels. The morphology and anatomy of flowers and the phylogeny, palynology, and karyology of Dipsacaceae were studied by Verlaque (1977, 1986). The work of Verlaque demonstrated that the evolution within the Dipsacaceae followed complex paths and that several genera were polyphyletic. There has been much discussion about infrageneric taxonomy of the genus *Scabiosa* (Bobrov, 1957; Jasiewicz, 1976), and in spite of the fact that the genus has been the subject of many taxonomic studies.

In this survey, *Scabiosa* s. l Rechinger and Lack (1991) is considered. On the basis of Greuter and Raus (1985) studies, Iranian species of *Scabiosa* divided into two genera: *Lomelosia* (= *Scabiosa* sec. *Astrocephalus* and sec. *Olivoerinae*) and *Scabiosa* s.s (*Scabiosa* s. l sec. *Scabiosa*) that this classification was not accepted in Flora Iranica (Rechinger and Lack, 1991) and Flora of Iran (Jamzad, 1993). Existence of eight pits on the epicalyx tube is a distinguishing character of *Lomeloisa* genus that separated this genus from other related genera (de Castro and Caputo, 1999). Regarding to this classification, *S. rotata* and *S. caucasica* grouped in *Lomeloisa* and *S. koelzii* and *S. amoena* maintained in *Scabiosa* s. s. Considering this point of view, studied species of *Scabisa* (*S. koelzii* and *S. amonea*) were closely related to *Pterocephalus* that supported Verlaque (1986) regarding *Scabiosa* s.s and *Ptercephalus* have closely related than *Lomelosia*, *Sixalix* and *Pycnocomon*.

There is high degree of homoplasy at generic level in Dipsacaceae (de Castro and Caputo, 1999). As shown in this study and previous study (Ebadi-Nahari and Nikzat-Siahkolaee, 2016), all studied *Scabiosa* species with 8 groove on the epicalyx (*S. columbaria*, *S. koelzii*, *S. amoena*) have colpate pollen apertures and all *Scabiosa* species with 8 pits on the epicalyx (*S. micrantha*, *S. persica*, *S. calocephala*, *S. olivieri*, *S. flavida*, *S. rotata*, *S. caucasica*) have porate pollen apertures. These results show the transfer of the some *Scabisoa* species to *Lomelosia* based on palynological characters. However, there is parallelism in the genus *Lomelosia* with related genera (*Sixalix*, *Scabiosa*) (de Castro and Caputo, 1999) that confusing relationships within and between specimens.

References

Bobrov, E.G. 1957. Dipsacaceae. In: Shishkin, R.K. (Ed.). Flora of the USSR, Vol. XXIV. Akademii Nauk SSSR, Moscow.

Caputo, P. and Cozzolino, S. 1994. A cladistics analysis of Dipsacaceae (Dipsacales). Plant Sys. Evol. **189**: 41–61.

Clarke, G.C.S. and Jones, M.R. 1981. The Northwest European Pollen Flora, 21. Dipsacaceae. Rev. Palaeobot Palynol. **33**(1-2): 1–25.

Cronquist, A. 1981. An Integrated System of Classification of Angiosperms. Columbia University Press, New York.

De Candolle A.P. 1830. Prodromus Systematis Naturalis Regni Vegetabilis 4. Paris, France: Treuttel and Wurtz (in Latin).

De Castro, O. and Caputo, P. 1999. A phylogenetic analysis of genus *Lomelosia* Rafin. (Dipsacaceae) and allied taxa. Delpinoa. **41**: 29–45.

Ebadi-Nahari, M. and Nikzat-Siahkolaee, S. 2016. Palynological study of some Iranian species of *Scabiosa* L. (Caprifoliaceae). Bangladesh J Plant Taxon. **23**(2): 215–222.

Ehrendorfer, F. 1965. Evolution and karyotype differentiation in a family of flowering plants: Dipsacaceae. Genetics Today. **2**: 399–407.

Erdtmann, G. 1952. Pollen Morphology and Plant Taxonomy. Chronica Botanica Co., Massachusettes; Copenhagen.

Feng, X.F., Ai, T.M. and Xu, H.N. 2000. A study on pollen morphology of *Dipsacus*. Zhongguo Zhong Yao Za Zhi. **25**(7): 394–401.

Greuter, W. and Burdet, H.M. 1985. *Dipsacaceae.* In: Greuter, W. and Raus, T., (Eds): Med Checklist notulae, 11. Willdenowia **15**: 71-77.

Hammer, Ø, Harper, D.A.T. and Ryan, P.D. 2001. PAST: Paleontological Statistics Software Package for Education and Data Analysis. Palaeontologia Electronica 4: 9. http://palaeo-electronica.org/2001_1/past/issue1_01.htm.

Jamzad, Z. 1993. Scabiosa L. *In:* Assadi, M., Khatamsaz, M. and Maassoumi, AA, (Eds). Flora of Iran, Vol. **8**. Islamic Republic of Iran, Ministry of Jahad-e Sazandegi, Research Institute of Forests and Rangelands, Tehran, pp. 63–106.

Jasiewicz, A. 1976. Scabiosa. Flora Europaea, Vol. 4 (ed. by T.G. Tutin, V.H. Heywood, N.A. Burges, D.H. Valentine, S.M. Walters and D.A. Webb), Cambridge University Press, Cambridge, pp. 68–74.

Khalik, K.A. 2010. A palynological study of the family Dipsacaceae in Egypt and its taxonomic significance, J. Bot. Taxonomy & Geobotany **121**(3-4): 97–111.

Mabberley, D.J. 2008. The Plant Book, a Portable Dictionary of Higher Plants. Cambridge University Press, Cambridge.

Mayer, V. and Ehrendorfer, F. 2000. Fruit differentiation, palynology, and systematics in *Pterocephalus* Adanson and *Pterocephalodes,* gen. nov. (Dipsacaceae). Bot. J. Linn. Soc. **132**: 47–78.

Nair, P.K.K. 1980. Pollen Morphology of Angiosperms. Vikas Publications, New Delhi.

Perveen, A. and Qaiser, M. 2011. Pollen flora of Pakistan - Dipsacaceae LXVIII., Pak. J. Botany **42**(6): 2825-2827.

Punt, W., Hoen, P.P., Blackmore, S., Nilsson, S. and Le Thomas, A. 2007. Glossary of pollen and spore terminology. Rev. Palaeobot Palynol. **143**: 1–81.

Rechinger, K.H. and Lack, H.W. 1991. Dipsacaceae. In: Rechinger, K.H., (Ed.), Flora Iranica, Vol. 168, pp. 1–67.

Reveal, J.L. and Chase, M.W. 2011. APG III: Bibliographical information and synonymy of Magnoliidae. Phytotaxa **19**: 71-134.

Stuessy, T.F. 2009. Plant Taxonomy: the systematic evaluation of comparative data. Columbia University Press. 568 pp.

Verlaque, R. 1977. Rapports entre les Valerianaceae, les Morinaceaeet les Dipsacaceae. Bull. Soc. Bot. **124**: 475–482 (in French).

Verlaque, R. 1984. A biosystematic and phylogenetic study of the Dipsacaceae. In: Grant, R. (Ed.), Plant Biosystematics. Toronto. pp. 307–320.

Verlaque, R. 1986. Etude biosystématiqueetphylogénétique des Dipsacaceae. Rev Cytol. Biol. Veg Le Botaniste. **9**: 5–72 (in French).

CHECKLIST OF MOSSES (BRYOPHYTA) OF GANGETIC PLAINS, INDIA

KRISHNA KUMAR RAWAT[1], AFROZ ALAM[2] AND PRAVEEN KUMAR VERMA[3]

CSIR-National Botanical Research Institute, Lucknow, India

Keywords: Bryophyta; Gangetic plains; Uttar Pradesh; Bihar; West Bengal

Abstract

An updated account of 79 taxa of mosses of Gangetic plains, representing 40 genera and 19 families, is provided. The family Pottiaceae with 17 taxa belonging to 9 genera appears most dominant and diversified family in the area while at generic level, the genus *Fissidens* (Fissidentaceae) with 19 species shows maximum diversity, followed by *Hyophila* and *Physcomitrium* each with five species.

Introduction

In our earlier publications, boundaries of 'Central Indian bryo-geographical zone' and 'Panjab and Rajasthan plains bryo-geographic zone' were proposed along with a checklist of mosses in these areas (Alam *et al.*, 2015; Rawat *et al.*, 2015). In present paper the boundaries of 'Gangetic plains bryo-geographic zone' is redefined for ease in distributional analysis, and provided updated checklist of mosses recorded from here.

Materials and Methods

The present compilation is based on the all available literature on mosses of Gangetic plains till date and gets its foundation from extra-ordinary work of Gangulee (1969-72; 1972-76; 1976-78), who has provided the most elaborative, informative and reliable data of moss diversity of India. In enumeration, the taxa reported earlier without specific epithet, have been excluded. The summary of various families and genera is followed by alphabetical list of taxa. The accepted names are cited in bold. Format of the enumeration of taxa follows Alam *et al.* (2015). The classification scheme broadly follows Goffinet *et al.* (2008).The extant of 'Gangetic plains bryo-geographic zone' described here is broadly based on Erenstein *et al.* (2007), with some modifications.

Results and Discussion

Earlier, entire Uttar Pradesh, Delhi, Bihar (including Jharkhand) and almost entire West Bengal (except Darjeeling) have been broadly treated as part of Gangetic Plains bryo-geographic zone. However, some parts of this region, due to somewhat different geo-physical properties and climatic conditions, shows more affinity to neighbouring zones, hence need re-assessment and refinement of boundaries of this unique bryo-geographical zone. The extant of 'Gangetic Plains bryo-geographic zone' described in present work is broadly followed Erenstein *et al.* (2007), however, some modifications are made on account of the unique geo-climatic conditions meet in the region. The southern region of Uttar Pradesh [particularly the Bundelkhand region (Jalaun,

---

[1] Corresponding author. Email: drkkrawat@rediffmail.com
[2] Department of Bioscience and Biotechnology, Banasthali University, Rajasthan, India.
[3] Forest Research Institute, Dehra Dun, India.

Jhansi, Lalitpur, Hamirpur, Mahoba, Banda, Chitrakoot districts) and Sonbhadra district] is a plateau and hence shows contrasting difference from the alluvial plains in geo-physical properties, hence treated in Central Indian bryo-geographic zone. Similarly, Purulia district of west Bengal, consisting of easternmost segments of Chhota nagpur plateau is now treated under Central Indian bryo-geographical zone (Alam *et al.,* 2015). Entire Uttarakhand state, on the other hand is being treated in Western Himalayan zone, however, most of the part of Haridwar and Udham Singh Nagar districts of Uttarakhand is either plain or terai (foot hills), hence need to be incorporated in Gangetic Plain region. Hilly areas of Darjeeling districts has been considered as part of Eastern Himalayan zone, however, Jalpaiguri and Koch Bihar district of West Bengal are proposed to be included in Brahmputra plains which itself can be treated as a separate bryo-geographic zone, subject to further studies. Earlier reports of mosses from Gangetic Plains (from Delhi) have now been transferred to Rajasthan and Panjab plains zone (Rawat *et al.,* 2015).

Therefore, in present work we propose Haridwar and Udham Singh district of Uttarakhand, Shahdara zone or East Delhi, Uttar Pradesh (except southern plateau region of Uttar Pradesh including Bundelkhand region and Sonbhadra district), Bihar and West Bengal (excluding Purulia, Kooch Bihar, Jalpaiguri districts and plains of Darjeeling district) as parts of 'Gangetic plains bryo-geographic' zone (Fig. 1).

Fig. 1. Map of India showing the proposed extant of 'Gangetic plains bryo-geographic zone'.

The present document provides updated accounts of 105 taxa of mosses reported so far from this zone, out of which 79 are still valid. Distributional details with relevant references are also given, which may be useful for future workers and will encourage bryo-floristic studies in neglected areas.

Taxonomic Enumeration

Archidium birmannicum Mitt. *ex* Dixon, J. Indian Bot. 2: 175. 1921. UTTAR PRADESH: Allahabad (Lal, 2007; Singh, 2013), Lakhimpur-Kheri, Pilibhit, Shahjahanpur (Sahu and Asthana, 2015); WEST BENGAL: Hoogly, Ramnagar (Gangulee, 1969-72; Lal, 2007).

Aulacopilum luzonense E.B. Bartram, Philipp. J. Sci. 68: 169. 1939. WEST BENGAL: Kolkata, Shantiniketan, Midnapore (Lal, 2007).

Barbula arcuata Griff., Calcutta J. Nat. Hist. 2: 491. 1842. WEST BENGAL: Kolkata, BIHAR (Lal, 2007).
Barbula consanguinea (Thwaites & Mitt.) A. Jaegr. → ***Hydrogonium consanguineum*** (Thwaites & Mitt.) Hilp.
Barbula constricta Mitt. → ***Didymodon constrictus*** (Mitt.) Saito
Barbula gangetica C. Muell. → ***Hydrogonium arcuatum*** (Griff.) Wijk. & Margad

Barbula indica (Hook.) Spreng. Nomencl. Bot. 2: 72. 1824. *Tortula indica* Hook., Musci Exot. 2, 135. 1819. BIHAR (Lal, 2007); UTTAR PRADESH: Allahabad, Pratapgarh, Raebareli, Saharanpur, Varanasi (Aziz and Vohra, 2008) Lakhimpur-Kheri, Pilibhit (Sahu and Asthana, 2015); WEST BENGAL (Lal, 2007).
Barbula javanica Dozy & Molk. → ***Hydrogonium javanicum*** (Doz. & Molk.) Hilp.

Bartramidula roylei (Hook. f.) Bruch & Schimp., Bryol. Eur. 4: 55. 1846. Gangetic plains (Lal, 2005).

Brachymenium indicum (Dozy & Molk.) Bosch & Sande Lac. Bryol. Jav. 1: 141. 1860. *Bryum indicum* Dozy ex Molk., Musci Fr. Ined. Archip. Indici 1: 22.1845. WEST BENGAL: Namkhana, Sundarban (Gangulee, 1974-78; Lal, 2007).
Bryum apiculatum Schwaegr. → ***Bryum mildeanum*** Jur.

Bryum coronatum Schwaegr., Sp. Musc. Frond, suppl. 1(2): 103.1816. WEST BENGAL: Burdwan, Howrah, Hoogli, Kolkata, Midnapore (Gangulee, 1974-78; Lal, 2007).
Bryum indicum Doz. ex Molk., → ***Brachymenium indicum***(Dozy & Molk.) Bosch & Sande Lac.

Bryum kliggraeffii Schimp., Höh. Crypt. Preuss. 81.1858. UTTAR PRADESH: Allahabad (Lal, 2007).

Bryum mildeanum Jur., Verh. Zool.-Bot. Ges. Wien 12: 967.1862. *Bryum apiculatum* Schwaegr., Sp. Musc. Frond., suppl. 1, 2: 102.1816. Gangetic plains (Lal, 2005 as *B. apiculatum*).
Bryum plumosum Dozy & Molk. → ***Gemmabryum apiculatum***(Schwägr.) J.R. Spence & H.P. Ramsay
Calymperes calcuttense E.B. Bartram & Gangulee → ***Syrrhopodon burmensis*** (Hamp.) Reese & Tan
Calymperes sundarbanense Gangulee → ***Heliconema peguense*** (Besch.) L.T. Ellis & A. Eddy

Calymperes tenerum var. ***tenuicola*** Gangulee, Mosses E. India 1: 600.1972, WEST BENGAL: Kolkata (Gangulee 1969-72; Lal, 2007).

Campylodontium flavescens (Hook.) Bosch. & Sande Lac., Bryol. Jav. 2: 128.1865. Gangetic plains (Lal, 2005).

Ceratodon purpureus (Hedw.) Brid., Bryol. Univ. 1: 480.1826. UTTAR PRADESH: Pilibhit (Sahu and Asthana, 2015); Raebareli (Sinha et al., 1990; Lal, 2007; Singh, 2013).

Ceratodon stenocarpus Bruch & Schimp., Bryol. Eur. 2: 146.1849. UTTAR PRADESH: Raebareli (Kumar and Kazmi, 2004, 2006; Singh et al., 2005; Singh, 2013).
Conomitrium bengalense Hamp.→ **Fissidens xiphioides Fleisch.**

Diaphanodon blandus (Harv.) Renauld & Cardot, Bull. Soc. Roy. Bot. Belgique 38(1): 23.1900. WEST BENGAL: Kolkata (Chopra, 1975, Lal, 2007).

Diaphanodon procumbens (Müll.) Renauld & Cardot, Bull. Soc. Roy. Bot. Belgique 38(1): 24.1900. WEST BENGAL: Kolkata (Chopra, 1975; Lal, 2007).

Dicranella macrospora Gangulee, Nova Hedwigia 8: 145.1964. UTTAR PRADESH: Lakhimpur-Kheri (Sahu and Asthana, 2015).

Didymodon constrictus (Mitt.) Saito, J. Hattori Bot. Lab. 39: 514.1975. *Barbula constricta* Mitt., J. Proc. Linn. Soc. Bot., suppl. 1: 33. 1859. UTTAR PRADESH: Faizabad (Singh and Kumar 2003 as *B. constricta*).

Entodontopsis tavoyense (Hook. ex Harv.) W.R. Buck & R.R. Ireland, Nova Hedwigia 41: 105. 1985. UTTAR PRADESH: Pilibhit (Sahu and Asthana, 2015).

Entosthodon nutans Mitt., J. Proc. Linn. Soc. Bot., suppl. 1: 55. 1859. UTTAR PRADESH: without locality (Gangulee, 1974-78; Lal, 2007); WEST BENGAL: Champadanga, Howrah, Hoogli, Kalyani, Konnagar, Nadia, Ranaghat (Gangulee, 1974-78; Lal, 2007).

Entosthodon wichurae M. Fleisch., Musci Fl. Buitenz., 2:481.1904. UTTAR PRADESH: Pilibhit (Sahu and Asthana, 2015).

Erpodium mangifereae Müll. Hal., Linnaea 37: 178.1872. UTTAR PRADESH: Allahabad, Saharanpur (Gangulee, 1974-78; Lal, 2007; Singh, 2013), Lakhimpur-Kheri (Sahu and Asthana, 2015); WEST BENGAL: Birbhum, Hoogli, Kalyani, Kolkata, Midnapore, Nadia (Gangulee, 1974-78).
Fissidens bengalensis Par. →*Fissidens zollingeri* Mont.

Fissidens bilaspurense Gangulee, Bull. Bot. Soc. Beng. 11: 66. 1957. WEST BENGAL: Midnapore (Gangulee 1969-72; Lal, 2007).

Fissidens bryoides Hedw., Sp. Musc. Frond. 153.1801. UTTAR PRADESH: Pilibhit (Sahu and Asthana, 2015); WEST BENGAL: Kalyani, Kolkata (Gangulee 1969-72; Lal, 2007).

Fissidens ceylonensis Dozy & Molk., Ann. Sci. Nat., Bot., ser. 3, 2: 304. 1844. *Fissidens perpusillus* Dozy & Molk, J. Proc. Linn. Soc., Bot., suppl. 2: 141. 1859. *Fissidens bicolor* Thwaites & Mitt., J. Linn. Soc. Bot. 13: 322. 1873. *Fissidens pennatulus* Thwaites & Mitt., J. Linn. Soc. Bot. 13: 325. 1873. *Fissidens ceylonensis* var. *jhargramii* Gangulee, Bull. Bot. Soc. Bengal 11: 72. 1957. WEST Bengal: Jhargram, Tarapheni (Gangulee, 1969-72; Lal, 2005).

Fissidens crenulatus Mitt., Musc. Ind. Or. 140. 1859 var. **crenulatus**. UTTAR PRADESH: Pilibhit, Sahajahanpur (Sahu and Asthana, 2015).

Fissidens crenulatus Mitt. var. *titalyanus* (Müll. Hal.) Gangulee, Mosses E. India 3: 506. 1972. *Fissidens titalyanus* Müll. Hal., Linnaea 37: 165.1872. WEST BENGAL: 24-Pargana, Dooars, Jhargram (Gangulee, 1969-72; Lal, 2007).

Fissidens crispulus var. *crispulus* Brid., Musc. Rec. suppl. 4: 187. 1819. *Fissidens sylvaticus* var. *zippelianus* (Dozy & Molk.) Gangulee, Mosses E. India 1: 537. 1972. WEST BENGAL: Jhargram (Gangulee, 1969-72; Lal, 2007; both reported as *F. sylvaticus* var. *zippelianus*).

Fissidens curvatoinvolutus Dixon, Notes Roy. Bot. Gard. Edinburgh 19: 279. 1938. UTTAR PRADESH: Raebareli, Saharanpur (Sinha et al., 1990; Singh et al., 2005; Lal, 2007; Singh, 2013).

Fissidens diversifolius Mitt., J. Proc. Linn. Soc., Bot., suppl. 2: 140. 1859. BIHAR: North Bihar, WEST BENGAL: Thakuranpahari (Gangulee 1969-72; Lal, 2005).

Fissidens flaccidus Mitt., Trans. Linn. Soc. London 23: 56. 1860. *Fissidens splachnobryoides* Broth. in Schum. & Lauterb., Fl. Deutsch. Schutzgeb. Suedsee: 81. 1900. UTTAR PRADESH: Lakhimpur-Kheri, Pilibhit (Sahu and Asthana, 2015); WEST BENGAL: 24-Parganas, Bolpur, Kolkata (Gangulee 1969-72; Lal, 2007 as *Fissidens splachobryoides*).

Fissidens involutus Wilson ex Mitt., J. Proc. Linn. Soc., Bot., suppl. 2: 138.1859. UTTAR PRADESH: Pilibhit, Saharanpur (Sahu and Asthana, 2015).

Fissidens kurzii Müll. Hal., Linnaea 37: 163. 1872. North Bengal plains (Gangulee, 1969-72; Lal, 2005).

Fissidens orishae Gangulee, Nova Hedwigia 8: 140. 1964. WEST BENGAL: Belpahari, Jhargram, Midnapore (Gangulee, 1969-72; Lal, 2007).

Fissidens ranchiensis Gangulee, Bull. Bot. Soc. Beng. 11: 68. 1957. WEST BENGAL: Belpahari, Jhalgram (Gangulee, 1969-72).

Fissidens semperfalcatus Dixon, J. Siam Soc., Nat. Hist. suppl. 10: 2.1935.WEST BENGAL: Jhargram (Gangulee, 1969-72; Lal, 2005).
Fissidens splachnobryoides Broth.→*Fissidens flaccidus* Mitt.

Fissidens subpalmatus C. Muell., Linnaea 37: 164.1872. BIHAR: Purnea. WEST BENGAL: 24-Parganas, Bolpur, Kharagpur (Gangulee 1969-72; Lal, 2007).

Fissidens sylvaticus Griff. var. *calcuttense* Gangulee, Mosses E. India, 2: 538. 1971. WEST BENGAL: Jhargram, Kolkata (Gangulee, 1969-72; Lal, 2007).

Fissidens sylvaticus var. *teraicola* (Müll. Hal.) Gangulee, Mosses E. India, 2: 539.1971. *Fissidens teraicola* Müll. Hal., Linnaea 37: 164.1872. WEST BENGAL: Kalyani (Gangulee, 1969-72; Lal, 2007 as *Fissidens teraicola*).
Fissidens sylvaticus var. *zippelianus* (Dozy & Molk.) Gangulee, →*Fissidens crispulus* Brid. var. *crispulus*
Fissidens teraicola Müll. Hal. →*Fissidens sylvaticus* var. *teraicola* (C. Muell.) Gangulee
Fissidens titlyanus Müll. Hal. → *Fissidens crenulatus* var. *titlyanus*

Fissidens virens Thwaites & Mitt., J. Linn. Soc., Bot. 13: 324.1873. WEST BENGAL: Jhargram (Gangulee, 1969-72).
Fissidens xiphioides Fleisch. →*Fissidens zollingeri* Mont.

Fissidens zollingeri Mont. Ann. Sci. Nat. ser. 3,4: 114.1845. *F. xiphioides* Fleisch., Hedwigia 38: 125. 1899. *Fissidens bengalensis* Par., Index Bryol. 461. 1896. *Conomitrium bengalense* Hamp.,

Linnaea 39: 364.1896. UTTAR PRADESH: Lakhimpur-Kheri, Pilibhit, Shahjahanpur (Sahu and Asthana, 2015); WEST BENGAL: Bengal plains (Gangulee 1969-72; Lal, 2007 as *Fissidens xiphioides*).

Funaria hygrometrica Hedw., Sp. Musc. Frond. 172.1801. UTTAR PRADESH: Raebareli (Singh et al., 2005; Kumar and Kazmi, 2006).

Garckea flexuosa (Griff.) Margad. & Nork., J. Bryol. 7:440. 1973. WEST BENGAL: Midnapore (Lal, 2007).

Garckea phascoides(Hook.) C. Muell. Bot. Zeit. 3:865. 1845. WEST BENGAL: Midnapore (Gangulee, 1969-72).

Gemmabryum apiculatum (Schwägr.) J.R. Spence & H.P. Ramsay, Phytologia 87: 65. 2005. *Bryum plumosum* Dozy & Molk., Ann. Sci. Nat., Bot., ser. 3(2): 301. 1844. WEST BENGAL: Kolkata, Howrah (Gangulee, 1974-78; Lal, 2007 as *Bryum plumosum*).

Glossadelphus zollingeri (Müll. Hal.) M. Fleisch., Musci Buitenzorg 4: 1355. 1923. WEST BENGAL: Howrah (Lal, 2007).

Gymnostomiella vernicosa (Hook. ex Harv.) M. Fleisch., Musci Buitenzorg 1: 310. 1904. UTTAR PRADESH: Allahabad, WEST BENGAL: 24-Pargana, Howrah, Kolkata (Gangulee, 1974-78; Lal, 2007; Singh 2013).

Gymnostomum calcareum Nees & Hornsch., Bryol. Germ. 1: 153. 1823. UTTAR PRADESH: Pilibhit (Sahu and Asthana, 2015).

Heliconema peguense (Besch.) L.T. Ellis & A. Eddy, J. Bryol. 15: 730. 1989. *Calymperes sundarbanense* Gangulee, Mosses E. India 3: 611. 1972. WEST BENGAL: Sundarban (Gangulee 1969-72 as *C. sunderbanense*).

Hydrogonium arcuatum(Griff.) Wijk. & Marg., Taxon 7: 289.1958. *Barbula gangetica* C. Muell. Linnaea, 37: 177.1872. *Hydrogonium gangeticum* (C. Muell.) Chen, Hedwigia 80: 237.1941. UTTAR PRADESH: Upper Gangetic Plains (Gangulee1969-72; Lal 2005 as *H. gangeticum*), Lakhimpur-Kheri, (Sahu and Asthana, 2015). WEST BENGAL: Lower Bengal, Birbhum (as *H. gangeticum*) (Gangulee1969-72).

Hydrogonium consanguineum (Thwaites *et* Mitt.) Hilp., Beih. Bot. Centralbl. 50(2): 626.1933 - *Barbula consanguinea* (Thwaites & Mitt.) A. Jaegr., Ber. Thätigk. St. Gallischen Naturwiss. Ges. 1877-78: 490.1880. BIHAR: Darbhanga (Aziz and Vohra, 2008); UTTAR PRADESH: Upper Gangetic plains (Ganguleee 1969-72), Raebareli (Kumar and Kazmi, 2004, 2006 as *B. consanguinea*), Varanasi (Aziz and Vohra, 2008); WEST BENGAL: Howrah, Kolkata, Midnapore, Titlaya (Aziz and Vohra, 2008).

Hydrogonium gangeticum (C. Muell.) Chen → ***Hydrogonium arcuatum*** (Griff.) Wijk. & Marg.

Hydrogonium javanicum (Doz. & Molk.) Hilp., Beih. Bot. Centralbl. 50(2): 632.1933. *Barbula javanica* Doz. & Molk. in Ann. Sci. Nat. Bot. ser. 3,2: 300.1884. UTTAR PRADESH: Allahabad, Kanpur (Jajmau), Raebareli, WEST BENGAL: North Bengal plains (Gangulee, 1969-72; Sinha et al, 1990; Singh et al., 2005; Kumar et al., 2007; Lal, 2007; Singh, 2013, all reported as *Barbula javanica*, however, Aziz & Vohra, 2008 treated *B. javanica* as synonym of *Hydrogonium javanicum*).

Hyophila involuta (Hook.) A. Jaegr., Ber. S. Gall. Naturew. Ges. 1871-72: 356.1873. BIHAR: North Bihar; UTTAR PRADESH: Upper Gangetic Plains (Gangulee 1969-72), Allahabad (Lal,

2007; Aziz and Vohra, 2008; Singh 2013); Lucknow (Aziz and Vohra, 2008; Nath et al., 2010; Singh 2013); WEST BENGAL: Lower Bengal (Gangulee 1969-72).

Hyophila nymaniana (M. Fleisch.) M. Menzel, Willdenowia 22: 198.1992. UTTAR PRADESH: Allahabad, Lakhimpur-Kheri, Pilibhit (Sahu and Asthana, 2015).

Hyophila rosea Williams, Bull. NewYork Bot. Gard. 8: 341.1941. UTAR PRADESH: Allahabad (Lal, 2007; Singh 2013).

Hyophila spathulata (Harv.) A. Jaegr., Ber. Thätigk. St. Gallischen Naturwiss. Ges. 1871-72: 353.1873. UTTAR PRADESH: Allahabad (Lal, 2007; Singh 2013), Lakhimpur-Kheri, Shahjahanpur (Sahu and Asthana, 2015).

Hyophila walkeri Broth., Rec. Bot. Surv. India 1: 317.1899. UTTAR PRADESH: Faizabad (Singh and Kumar, 2003).

Octoblepharum albidum Hedw., Sp. Musc. Frond. 50.1801. WEST BENGAL: Howrah (Gangulee, 1969-72; Lal, 2007).

Philonotis falcata (Hook.) Mitt., J. Proc. Linn. Soc., Bot., suppl. 1: 62.1859. WEST BENGAL: Midnapore (Lal, 2007).

Philonotis hastata (Duby) Wijk. & Margad., Taxon 8: 74.1959.WEST BENGAL: Kolkata, Howrah (Gangulee, 1974-78; Lal, 2007).

Philonotis mollis (Dozy & Molk.) Mitt., J. Proc. Linn. Soc., Bot., suppl. 1: 60, 1859. UTTAR PRADESH: Lakhimpur-Kheri, Pilibhit, Shahjahanpur (Sahu and Asthana, 2015).

Physcomitrium coorgense Broth., Rec. Bot. Surv. India 1(12): 319. 1899. UTTAR PRADESH: Allahabad (Lal, 2007; Singh, 2013).
Physcomitrium cyathicarpum Mitt. →*Physcomitrium immersum* Sull.

Physcomitrium eurystomum Sendtn., Denkschr. Bayer. Bot. Ges. Regensburg 3: 142. 1841. UTTAR PRADESH: Allahabad (Lal, 2007; Singh 2013), Lakhimpur, Pilibhit (Sahu and Asthana, 2015); WEST BENGAL: Burdwan, Hoogli (Gangulee, 1974-78).

Physcomitrium immersum Sull., Manual 648, 1848. *Physcomitrium cyathicarpum* Mitt., J. Proc. Linn. Soc., Bot., suppl. 1: 54. 1859. BIHAR: Patna; UTTAR PRADESH: Allahabad; WEST BENGAL: Barasat, Hoogli, Kolkata, Nadia (Gangulee, 1974-78; Lal, 2007; Singh 2013; all as *Physcomitrium cyathicarpum*).

Physcomitrium indicum (Dix.) Gangulee, Bull. Bot. Soc. Bengal 23: 131. 1969. *Physcomitrellopsis indica* Dix. in Gupta. J. Indian Bot. Soc. 13: 122. 1933. UTTAR PRADESH: Unnao (Shuklaganj), Pratapgarh (Kalakankar) (Lal, 2007, Sinha et al., 1990; Kumar and Kazmi 2004, 2006; Singh et al., 2005), Varanasi (Gupta, 1933 as *Physcomitrellopsis indica*); WEST BENGAL: Hoogli, Nadia, Maldah, Murshidabad (Gangulee, 1974-78; Lal, 2007).

Physcomitrium japonicum (Hedw.) Mitt., Trans. Linn. Soc. London, Bot. 3: 164. 1891. UTTAR PRADESH: Gorakhpur (Gangulee, 1974-78; Lal, 2007), Pratapgarh (Kalakankar), Raebareli, Unnao (Shuklaganj) (Lal, 2007, Sinha et al., 1990; Kumar et al., 2007; Singh et al., 2005; Kumar and Kazmi, 2006; Singh, 2013).

Pinnatella alopecuroides (Mitt.) M. Fleisch. var. *culcutensis* (M. Fleisch.) Gangulee. Mosses E India 5: 1440. 1976. *Urocladium calcutense* Müll. Hal., J. Bot. 50: 152. 1912. *Pinnatella calcutensis* M. Fleisch., Hedwigia 45: 84. 1906. Gangetic plains (Lal, 2005 as *Pinnatella*

calcutensis M. Fleisch.) WEST BENGAL: Kolkata "Culcutta" (type of *Urocladium calcutense*, however, locality doubted by Gangulee 1974-78).
Pinnatella calcutensis M. Fleisch. →*Pinnatella alopecuroides* (Mitt.) M. Fleisch. var. *culcutensis* (M. Fleisch.) Gangulee

Pohlia flexuosa Hook., Icon. Pl. Rar. 1: 19. 1836. WEST BENGAL: Kolkata (Gangulee, 1974-78; Lal, 2007).

Semibarbula orientalis (F. Weber.) Wijk. & Margad, Taxon 8: 75. 1959. UTTAR PRADESH: Lucknow (Nath et al., 2010); Bengal plains (Gangulee, 1969-72).

Splachnobryum bengalense Gangulee, Mosses E India 4:865. 1974. WEST BENGAL: Konnagar (Gangulee, 1974-78; Lal, 2007).
Splachnobryum indicum Hampe & Müll. Hal.→*Splachnobryum obtusum* (Brid.) Müll. Hal.

Splachnobryum obtusum (Brid.) Müll. Hal., Verh. K.K. Zool.-Bot. Ges. Wien 19: 504. 1869. *Splachnobryum indicum* Hampe & Müll. Hal., Linnaea 37: 174. 1972. UTTAR PRADESH: Allahabad (Gangulee, 1974-78; Lal, 2007; Singh 2013; all as *Splachnobryum indicum*), Lakhimpur-Kheri, Pilibhit (Sahu and Asthana, 2015); WEST BENGAL: Hoogli, Howrah, Kolkata (Gangulee, 1974-78; Lal, 2007; Singh 2013; all as *Splachnobryum indicum*).

Stereophyllum tavoyense (Hook. *ex* Harv.) A. Jaegr., Ber. Thätigk. St. Gallischen Naturwiss. Ges. 1877-78: 279. 1880. BIHAR: tropical plains (Gangulee, 1978-80; Lal, 2007)

Stereophyllum wightii (Mitt.) A. Jaegr, Ber. Thätigk. St. Gallischen Naturwiss. Ges. 1877-78: 279. 1880. WEST BENGAL: Bengal plains, 24-pargana, Maldah (Gangulee, 1978-80; Lal, 2007)

Syrrhopodon burmensis (Hamp.) Reese & Tan, Taxon 35(4): 693. 1986. *Calymperes calcuttense* Bartr. & Gangulee, J. Bombay Nat. Hist. Soc. 60: 632. 1963. WEST ENGAL: Sonarpur near Kolkata (Gangulee 1969-72; type of *Calymperes calcuttense*).

Taxithelium nepalense (Schwägr.) Broth., Monsunia 1: 51. 1899. WEST BENGAL: Kolkata, Howrah, Hoogli, Nadia (Gangulee, 1978-80; Lal, 2007).

Trachyphyllum inflexum (Harvey) Gepp. in Hiren, Cat. Weln. Afr. Pl. 2, 21: 299. 1901. UTTAR PRADESH: Pilibhit (Sahu and Asthana, 2015).

Trachypodopsis serrulata (P. Beauv.) Fleisch., Hedwigia 45: 67. 1906. WEST BENGAL: Kolkata (Chopra, 1975; Lal, 2007).

Trematodon capillifolius Müll. Hal. ex G. Roth., Aussereur. Laubm. 296: 28. 1911. UTTAR PRADESH: Unnao (Shuklaganj), Kanpur (Jajmau), Raebareli (Dalmau) (Sinha et al., 1990; Lal, 2007).
*Urocladium calcutense*Müll. Hal. → *Pinnatella alopecuroides* (Hook.) Fleisch. var. *culcutensis* (M. Fleisch.) Gangulee

Vesicularia montagnei (Schimp.) Broth., Nat. Pflanzenfam. 1(3): 1094. 1908. WEST BENGAL: Howrah, Kolkata (Gangulee, 1978-80; Lal, 2007).

Weissia controversa Hedw., Spec. Musc. Frond. 67, 1801. UTTAR PRADESH: Shahjahanpur (Sahu and Asthana, 2015).

Summary of Family wise representation of mosses of Gangetic plains:

Archidiaceae: *Archidium* (1)

Bartramiaceae: *Bartramidula* (1), *Philonotis* (3)

Bruchiaceae: *Trematodon* (1)

Bryaceae: *Brachymenium* (1), *Bryum* (3), *Gemmabryum* (1)

Calymperaceae: *Calymperes* (1), *Heliconema* (1), *Octoblepharum* (1), *Syrrhopodon* (1)

Dicranaceae: *Dicranella* (1)

Ditrichaceae: *Ceratodon* (2), *Garckea* (2)

Erpodiaceae: *Aulacopilum* (1), *Erpodium* (1)

Fabroniaceae: *Campylodontium* (1)

Fissidentaceae: *Fissidens* (19)

Funariaceae: *Entosthodon* (2), *Funaria* (1), *Physcomitrium* (5)

Hypnaceae: *Glossadelphus* (1), *Vesicularia* (1)

Meteoriaceae: *Diaphanodon* (2), *Trachypodopsis*(1),

Mniaceae: *Pohlia* (1)

Neckeraceae: *Pinnatella* (1)

Plaisiadelphaceae: *Taxithelium* (1)

Pottiaceae: *Barbula* (2), *Didymodon* (1), *Gymnostomiella* (1), *Gymnostomum* (1) *Hydrogonium* (3), *Hyophila* (5), *Semibarbula* (1), *Splachnobryum* (2), *Weissia* (1)

Pterigynandraceae: *Trachyphyllum*(1)

Stereophyllaceae: *Entodontopsis* (1) *Stereophyllum* (2)

Doubtful records:

Barbula tenuirostris Brid., Bryol. Univ. 1: 826.1827. Lal (2005) listed it in Gangetic plains without locality, however, Aziz & Vohra (2008) did not reported its occurrence in Indian region.

Ganguleea angulosa (Broth. & Dix.) Zander, Phytologia 65: 427.1989. UTTAR PRADESH: Lucknow (Bansal et al., 2015; reported without specimen number and herbarium name, which makes the record dubious)

Excluded record:

Archidium birmanicum var. *pariharii* Lal, in Nath & Asthana, Current Trends in Bryology, 133, 2008. UTTAR PRADESH: Allahabad, (Lal, 1995, 2007; Singh 2013) nom. inval. (ICN art.39.1, no Latin diagnosis; art. 40.1, no Type; art. 40.7, no Herbarium specified)

Acknowledgements

The authors are thankful to Drs. A.K. Asthana and Vinay Sahu, Bryology laboratory, CSIR-National Botanical Research Institute, Lucknow for their kind help during the study. KKR wishes to acknowledge the financial support from GAP-3356 by Ministry of Water Resources, Govt. of India. One of the authors (AA) is also grateful to Professor Aditya Shastri, Vice Chancellor, Banasthali Vidyapith and Professor Vinay Sharma, Dean, Faculty of Science and Technology, Banasthali University, Rajasthan, India, for their kind support for this research work.

References

Alam, A., Rawat, K.K., Verma, P.K., Sharma, V. and Sengupta D. 2015. Moss flora of Central India. Plant Science Today **2**(4): 159-171. DOI: http://dx.doi.org/10.14719/pst.2015.2.4.126

Aziz, N. and Vohra, J.N. 2008. Pottiaceae (Musci) of India. Bishen Singh Mahendra Pal Singh, Dehradun, India.

Bansal, P., Srivastava, A. and Nath, V. 2015. Occurrrence of *Ganguleea angulosa* (Broth. & Dix.) Zand. in India. Geophytology **45**(2): 273-276

Chopra, R.S. 1975. Taxonomy of India mosses. New Delhi, pp. 1-631,

Erenstein, O., Hellin, J. and Chandra, P. 2007. Livelihood, poverty and targeting in the Indo-Gangetic Plains: a spatial mapping approach. CIMMYT and Rice-Wheat Consortium for the Indo-Gangetic Plains (RWC), New Delhi, India.

Gangulee, H.C. 1969-72. Mosses of Eastern India and adjacent regions. vol. **I**, Culcutta, India.

Gangulee, H.C. 1974-78. Mosses of Eastern India and adjacent regions. vol. **II**, Culcutta, India.

Gangulee, H.C. 1978-80. Mosses of Eastern India and adjacent region. vol. **III**. Calcutta, India.

Goffinet, B, Buck, W.R. and Shaw, A.J. 2008. Morphology and classification of Bryophyta. *In:* Bryophyte Biology, Goffinet, B and Shaw, A.J. (Eds.) 2nd edition, Cambridge University Press. pp. 55-138.

Gupta, K.M. 1933. On the structure of a new species of Indian mosses *Physcomitrellopsis indica* Dixon, sp. nov. from Benaras. Journal of Indian Botanical Society **12**: 122-128.

Kumar, A. and Kazmi, S. 2004. Bryophytes from Unchahar, Raebareli, U.P. Geophytology **34**: 121-123.

Kumar, A. and Kazmi, S. 2006. Leaf area indices of mosses from Unchahar, Raebareli, Uttar Pradesh. Geophytology **36**: 23-26.

Kumar, A., Shukla, M.and Kumar, D. 2007. Effect of polluted water on chlorophyll concentration of Bryophytes growing in Raebareli. *In*: Nath, V. and Asthana, A.K. (Eds.), Current Trends in Bryology. Bishen Singh Mahendra Pal Singh, Dehradun, pp. 189-205.

Lal, J. 1995. *Archidium birmensis* Dix. var. *pariharii* J. Lal var. nov. from Gangetic plain (Musci: Archidiaceae). National Conference on Bryology and Symposium on Recent Advances in Bryology, N.B.R.I., Lucknow, Abstract, pp. 87.

Lal, J. 2005. A checklist of Indian Mosses. Bishen Singh Mahendra Pal Singh, Dehra Dun, India.

Lal, J. 2007. Mosses of Gangetic plains – A neglected Biogeographic zone of India. *In*: Nath, V. and Asthana, A.K. (Eds.), Current Trends in Bryology, Bishen Singh Mahendra Pal Singh, Dehradun, India. pp. 131-147.

Nath, V., Sinha, S., Sahu, V., Govind, G., Srivastava, M. and Asthana, A.K. 2010. A study on metal accumulation in two mosses of Lucknow (U.P.). Indian Journal of Applied and Pure Biology **25**: 25-29.

Pande, S.K. 1958. Some aspects of Indian Hepaticology. Journal of the Indian Botanical Society **37**(1): 1-27.

Rawat, K.K., Alam, A. and Verma, P.K. 2015. Moss flora of Rajasthan and Punjab plains. Plant Science Today **2**(4): 154-158.

Sahu, V. and Asthana, A.K. 2015. Bryophyte diversity in Terai regions of Uttar Pradesh, India with some new additions to the state. Tropical Plant Research **2**(3): 180-191.

Singh, M., Nath, V. and Kumar, A. 2005. The ecological studies on bryophytes, growing on the bank of polluted river Sai (Raebareli), India. Proceedings of National Academy of Sciences, India **75**(B): 41-50

Singh, S.K. 2013. A checklist of liverworts, hornworts and mosses of Uttar Pradesh, India. Geophytology **42**(2): 163-166.

Singh, S.K. and Kumar, S. 2003. A note on bryophytes of Ram Nagri (Ayodhya), Faizabad, Uttar Pradesh, India. Phytotaxonomy **3**: 108-111.

Sinha, A.K., Pandey, D.C., Kumar, A. and Sinha, A. 1990. Moss flora of the banks of river Ganga between Shuklaganj (Unnao) and Kalakankar (Pratapgarh). Geophytology **20**(1): 37-40.

TAXONOMIC REVISION OF SAUDI ARABIAN *TETRAENA* MAXIM. AND *ZYGOPHYLLUM* L. (ZYGOPHYLLACEAE) WITH ONE NEW VARIETY AND FOUR NEW COMBINATIONS

DHAFER AHMED ALZAHRANI[1] AND ENAS JAMEEL ALBOKHARI[2]

Department of Biological Sciences, Faculty of Science, King Abdulaziz University, Jeddah, Saudi Arabia

Keywords: Taxonomic revision; *Tetraena*; *Zygophyllum*; New variety; New combination; Saudi Arabia.

Abstract

The genera *Tetraena* Maxim. and *Zygophyllum* L. (Zygophyllaceae) present different morphological characters, *viz.* growth habit, leaf features, flower traits and fruit shape, and have a high diversity of species in Africa, Australia and Asia. Six species of *Tetraena* [*T. alba* (L.f.) Beier & Thulin, *T. coccinea* (L) Beier & Thulin, *T. decumbens* (Delile) Beier & Thulin, *T. hamiensis* (Schwein f.) Beier & Thulin, *T. propinqua* (Decne.) Ghaz. & Osborne and *T. simplex* (L.) Beier & Thulin], and one species of *Zygophyllum* (*Z. fabago* L.) have been identified in Saudi Arabia, most of which grow in sandy soils and saline habitats as shrubs and herbs. One new endemic variety (*T. alba* var. *arabica* Alzahrani & Albokhari) along with four new combinations [*T. alba* var. *amblyocarpa* (Baker) Alzahrani & Albokhari, *T. hamiensis* var. *qatarensis* (Hadidi *ex* Beier & Thulin) Alzahrani & Albokhari, *T. hamiensis* var. *mandavillei* (Hadidi *ex* Beier & Thulin) Alzahrnai & Albokhari, and *T. propinqua* subsp. *migahidii* (Hadidi *ex* Beier & Thulin) Alzahrani & Albokhari] are proposed. Descriptions, illustrations, distribution maps and a key for identification of the taxa are presented. Conservation status has been proposed for the new variety and combinations.

Introduction

The genus *Zygophyllum* L. distributed in Saudi Arabia received much attention based on morphological and anatomical characters (Hadidi, 1978; Migahid, 1978, 1996; Hosny, 1988; Mandaville, 1990; Chaudhary, 2001; Soliman *et al.*, 2010; Waly *et al.*, 2011). However, according to the most recent taxonomic proposal of *Tetraena* Maxim. and *Zygophyllum* presented by Beier *et al.* (2003), the most Saudi Arabian taxa of *Zygophyllum* were transferred to *Tetraena*. *Zygophyllum* and *Tetraena* have similar morphological characters, *viz.* growth habit, leaf features, flower traits and fruit shape. Beier *et al.* (2003) showed that *Zygophyllum* and *Tetraena* could easily be distinguished from each other by the characters of fruit dehiscence and staminal appendages. *Zygophyllum propinquum* was not included in Beier *et al.* (2003) and it was not mentioned as a synonym under any species in their study. However, Ghazanfar and Osborne (2015) transferred this species to *Tetraena propinqua*.

Taxonomically, Zygophyllaceae R. Br. was placed in different orders by several authors. Engler (1964) placed the family in the order Geraniales, while Cronquist (1968) and Hutchinson (1969) positioned Zygophyllaceae in the order Sapindales. Dahlgren (1980) deposited it in the order Geraniales, following Engler (1964), but Takhtajan (1980) positioned the family within Rutales. Currently, APG III (2009) placed Zygophyllaceae with Krameriaceae within a new order

---

[1]Corresponding author. Email: dalzahrani@kau.edu.sa; dhaferalzahrani@hotmail.com
[2]Department of Biological Sciences, Faculty of Applied Sciences, Umm Al-Qura University, Makkah, Saudi Arabia.

Zygophyllales. These two families are placed with strong support as sister to a clade containing more than two orders. Sheahan and Chase (2000) analysed both *rbc*L and *trn*L-F sequences from 36 taxa of Zygophyllaceae and their results were supported by previous classification of Zygophyllaceae into the five subfamilies (Zygophylloideae, Larreoideae, Seetzenioideae, Tribuloideae and Morkillioideae). Moreover, Sheahan and Chase (2000) have indicated that *Tetraena* is nested within the large and variable *Zygophyllum*.

The genus *Zygophyllum* was first described by Linnaeus (1753) and has been accepted by several authors who worked broadly on systematics of this genus (Zumbruch, 1931; Van Huyssteen, 1937; Oltmann, 1971; Hadidi, 1978; Sheahan and Chase, 1996, 2000; Van Zyl, 2000). Linnaeus (*l.c.*) classified six species within *Zygophyllum*, namely *Z. fabago* L., *Z. morgsana* L., *Z. sessilifolium* L., *Z. fulvum* L., *Z. coccineum* L. and *Z. spinosum* L. Based on growth habit, androecium characters and dehiscence of the capsule Van Huyssteen (1937) split *Zygophyllum* into two subgenera, *viz. Zygophyllotypus* Huysst. (= subgenus *Zygophyllum*) and *Agrophyllum*, and further subdivided *Zygophyllotypus* into eight sections and *Agrophyllum* into five sections. In this classification, *Zygophyllum coccineum*, *Z. album* and *Z. aegyptium* were placed in section *Mediterranea* Engl., and *Z. simplex* and *Z. decumbens* in section *Bipartita* Huysst. *Z. dumosum* was placed in section *Alata* Huysst (Van Huyssteen, 1937).

Hadidi (1977) recognized eight species of *Zygophyllum* in Arabia and all of them belong to section *Mediterranea*, including two new species, i.e. *Zygophyllum mandavillei* Hadidi, and *Z. migahidii* Hadidi. *Z. migahidii* is closely related to *Z. propinquum*, but they differ in flower and fruit characters. In *Z. migahidii*, the flowers and fruit are solitary at each node, while they are grouped in clusters in *Z. propinquum*, conversely, *Z. mandavillei* can be easily recognized from other species of its section by its glabrous, large long-stalked flowers and sausage-shaped capsule. Later, Hadidi (1978) described *Z. qatarense* from Qatar as a new species. Hosny (1988) reported 13 species and three varieties of *Zygophyllum* in Arabia and among them 10 species and one variety were distributed in Saudi Arabia. She classified them into two subgenera *Zygophyllum* and *Agrophyllum* (Necker) Endl. *ex* Van Huyssteen following Engler (1931) and Van Huyssteen (1937). According to Van Huyssteen (1937), two of the Saudi species belong to section *Bipartita*, seven species belong to section *Mediterranea*, including *Z. boulosii* A. Hosny as a new species, and two species belong to section *Hamiensia* Engl.

Sheahan and Chase (1996) studied the phylogenetic relationships of Zygophyllaceae based on morphology, anatomy and the *rbc*L sequence and found *Fagonia* as sister to the rest of the subfamily, while *Zygophyllum fabago* (type species of *Zygophyllum*) is a sister to genus *Augea*, and *Z. simplex* is a sister to genus *Tetraena* and concluded that *Z. simplex* might not belong to *Zygophyllum*. Later, Sheahan and Chase (2000) investigated the phylogenetic relationships of 36 taxa of Zygophyllaceae including 15 species of *Zygophyllum* from Africa, Australia and southwest Asia using nucleotide sequences of the plastid gene *rbc*L and non-coding *trn*L-F. Their results agreed with a high support to the previous results that stated Zygophyllaceae needs to be divided into five subfamilies and the subfamily Zygophylloideae was further classified into five clades with a high support of bootstrap. Also, they concluded that *Zygophyllum* is polyphyletic. *Zygophyllum fabago* was nested with another Asian species, *Z. xanthoxylum*, whereas *Z. simplex* was placed in a strong clade with genus *Tetraena* and other species of *Zygophyllum*: *Z. cylindrifolium*, *Z. decumbens*, *Z. album* and *Z. coccineum* (last three are distributed in Saudi Arabia). Moreover, molecular studies have indicated that *Tetraena* is nested within the large and paraphyletic *Zygophyllum* (Sheahan and Chase, 2000).

Van Zyl (2000) made a revision for 54 species of south African *Zygophyllum* and classified these species into two subgenera *Zygophyllum* and *Agrophyllum*, based on morphological characters, more particularly capsule dehiscence, seed attachment and presence of spiral threads in the seed mucilage. The result agreed with classification provided by Endlicher (1841) and Van Huyssteen (1937). Takhtajan (1987) separated the genus *Tetraena* from subfamily Zygophylloideae and erected Tetraenoideae based on morphology of pistil, fruit, pollen grains and chromosomes. Species within the genus *Tetraena* can be distinguished by growth habit, plant colour, leaf structure, flower colour, and fruit type and shape (Van Huyssteen, 1937; Hosny, 1988; Van Zyl, 2000; Chaudhary, 2001; Beier *et al.*, 2003).

Based on *trn*L plastid DNA sequences and morphological characters Beier *et al.* (2003) showed that Zygophylloideae is monophyletic, whereas the genus *Zygophyllum* is paraphyletic, since it was placed with the genera of *Augea* Thunb., *Tetraena* and *Fagonia* L. In addition, they proposed a new classification for *Tetraena* and *Zygophyllum* which is supported by combination of morphological and molecular data, transferring 35 species from *Zygophyllum* to *Tetraena* as new combinations. These species are known from Africa and Asia. *Zygophyllum* is characterized by a loculicidal capsule and undivided staminal appendages, while *Tetraena* is distinguished by a schizocarp and sometimes bipartite staminal appendages. Subsequently, many authors agreed with this transfer and used the combinations proposed by Beier *et al.* (2003) as valid in their works (Norton *et al.*, 2009; Louhaichi *et al.*, 2011; Mosti *et al.*, 2012; Sakkir *et al.*, 2012; Azevedo, 2014; Symanczik *et al.*, 2014).

Recently, Alzahrani and Albokhari (2017a) studied phylogenetic relationships of 44 specimens representing seven taxa of Saudi Arabian *Tetraena* Maxim. and *Zygophyllum* L., based on individual and combined chloroplast DNA data of *rbc*L and *trn*L-F. Molecular phylogenetic of the cpDNA analysis of this study, divided Saudi Arabian *Tetraena* plants into six groups: *T. hamiensis* (Schweinf.) Beier & Thulin, *T. propinqua* (Decne.) Ghazanfar & Osborne, *T. alba* (L. f.) Beier & Thulin, *T. coccinea* (L.) Beier & Thulin, *T. simplex* (L. f.) Beier & Thulin, and *T. decumbens* (Delile) Beier & Thulin and one species of *Zygophyllum* (*Zygophyllum fabago* L.).

In Saudi Arabia, Zygophyllaceae is represented by eight genera including *Balanites* Del., *Fagonia* L., *Nitraria* L., *Peganum* L., *Seetzenia* R. Br., *Tetraena* Maxim., *Tribulus* L. and *Zygophyllum* L. (Collenette, 1985, 1998, 1999; Mandaville, 1990; Migahid, 1996; Chaudhary, 2001). The genus *Tetraena* is widespread in Saudi Arabia, while the genus *Zygophylum* is only represented by *Z. fabago* in northern parts of Saudi Arabia. In Saudi Arabia, a few taxonomic studies on the genera *Tetraena* and *Zygophyllum* using morphological and anatomical characters have been carried out (Soliman *et al.*, 2010; Al-Arjany, 2011; Waly *et al.*, 2011) and new combination have been made (Alzahrani, 2017; Alzahrani and Albokhari, 2017b,c). The objective of the present study is to revise the Saudi Arabian genera *Tetraena* and *Zygophyllum* belonging to the family Zygophyllaceae with detailed taxonomic notes.

Materials and Methods

The present revisionary study of *Tetraena* Maxim. and *Zygophyllum* L. in Saudi Arabia is based on extensive field survey, literature and analysis of more than 348 specimens, including types and images of types from different herbaria, *viz.* BM, CAI, CAIM, E, K, KAUH, KSU and RIY. Field surveys were carried out in c. 31 localities, between 2013 and 2014, and 72 collected samples were deposited in KAUH (King Abdulaziz University Herbarium, Jeddah, Saudi Arabia). The collected specimens were critically studied and examined, and identifications were confirmed using standard literature (Hosny, 1988; Chaudhary, 2001; Beier *et al.*, 2003). In each case, several duplicate voucher specimens were made and these were complemented with fresh material preserved in 70% ethanol and stored for further research.

A total of 74 morphological traits were found effective to determinate species and new combinations from field collections and herbarium specimens. These characters, including both vegetative and reproductive features, were examined and scored using a Novex dissecting microscope and X10 hand lens. Differences between species, subspecies and varieties were supported by distribution maps, diagnostic traits placed in the identification key, and drawings, and evaluation of synonyms. The conservation status for the new variety and combinations was assessed following the guidelines of IUCN (IUCN, 2014).

Results and Discussion

Zygophyllum is represented by only one species in Saudi Arabia, *Z. fabago*. It clearly differs from *Tetraena* species by several morphological characters such as size, shape and colour of leaves, flowers and fruits. The present study reveals that *Z. fabago* is characterized by 2-foliolate, flat leaves, up to 4 cm long, and 2.5 cm wide, creamy flowers, and loculicidal, oblong-cylindrical capsule, up to 3 cm long. This is congruent with Beier's results (Beier *et al.*, 2003).

Based on the morphological traits and morphometric analysis (Alzahrani, 2017; Alzahrani and Albokhari, 2017b, c), this study reported six species of *Tetraena* (e.g. *T. alba*, *T. coccinea*, *T. decumbens*, *T. hamiensis*, *T. propinqua* and *T. simplex*) distributed in Saudi Arabia.

Key to the Saudi Arabian species of *Zygophyllum* and *Tetraena*

1.	Fruit a loculicidal capsule; leaves 2-foliolate, flat, obovate-elliptic.	*Zygophyllum fabago*
-	Fruit a schizocarp; leaves simple, 1- or 2-foliolate, mostly cylindrical and fleshy, seldom flat.	2
2.	Staminal appendages bipartite; stems and leaves glabrous.	3
-	Staminal appendages undivided; stems and leaves mostly pubescent.	4
3.	Fruits obovoid, 5-lobed; flowers yellow; leaves simple, sessile, cylindrical, fleshy.	*Tetraena simplex*
-	Fruits obconical, 5-ridged; flowers white creamy; leaves 2-foliolate, petiolate, obovate, flat.	*T. decumbens*
4.	Leaves 1-foliolate; fruits oblong-obovate, 5-angled at the upper end to cylindrical.	*T. hamiensis*
-	Leaves 2-foliolate; fruits oblong-ovate, obconial or cylindrical, with or without ridges or lobes at the upper end.	5
5.	Flowers arranged in clusters. Fruits obconical, 5-ridged at the upper end.	*T. alba*
-	Flowers 1-3 at each node.	6
6.	Fruits cylindrical, without lobes or angled at the upper end (sausage shaped).	*T. coccinea*
-	Fruits ovate-oblong to obconical, 5-angled at the upper end.	*T. propinqua*

Tetraena alba (L. f.) Beier & Thulin, Pl. Syst. Evol. 240: 35 (2003). (**Fig. 1**).
Diagnosis: *T. alba* differs from other species of the genus by its pubescent stem, 2-foliolate, cylindrical with acute apex, fleshy leaves, mostly arranged flowers in clusters, undivided staminal appendages and fruit shape. It occurs in western Saudi Arabia. Three morphological variations are recognized.

Small shrub, perennial, green or greenish grey, 50–60 cm tall, 40 cm wide. Stem pubescent, with unicellular simple trichomes. Leaves 2-foliolate, 7–12×3.0–5.5 mm, fleshy, cylindrical or elliptic, apex acute; petiole 10–18 mm; stipules triangular, herbaceous, 1.0×1.5 mm, pubescent. Flowers arranged in clusters, sometimes solitary, bisexual, white, 4.0–5.5×3–5 mm, pedicel 1–2 mm long. Sepals 5, rounded-obtuse at the apex, herbaceous, yellowish green, obovate, 3–4×2–3 mm, pubescent, aestivation imbricate. Petals 5, white, spathulate, 3.5–6.0×1–2 mm, aestivation valvate. Stamens 10, 3–4 mm long, staminal appendages undivided, 2.0–2.5×1.0 mm; anthers 2-lobed, yellow, dorsifixed, dehiscent longitudinally; disc smooth. Ovary 5-locular, pubescent; style single, c. 1 mm long. Fruit a schizocarp, obconical, oblong- obconical or star shaped, acute, with keeled lobes 7–12×(2–6)8–13 mm, pubescent, pericarp extended as wings, peduncle 2–6 mm long, pubescent.

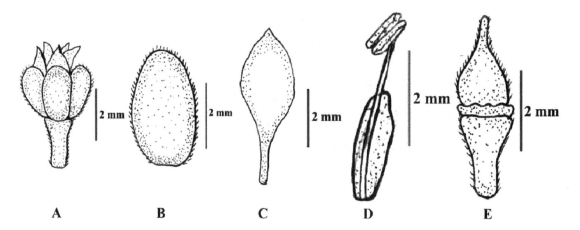

Fig. 1. *Tetraena alba*: A. Flower; B. Sepal; C. Petal; D. Stamen; E. Ovary (Alzahrani and Albokhari, 2017b).

Tetraena alba consists of three varieties, viz. *T. alba* (L. f.) Beier & Thulin var. *alba*, *T. alba* (L.f.) Beier & Thulin var. *arabica* Alzahrani & Albokhari, **var. nov.** and *T. alba* (L. f.) Beier & Thulin var. *amblyocarpa* (Baker) Alzahrani & Albokhari, **comb. nov**. Taxonomy of these varieties are summarized below:

Tetraena alba (L. f.) Beier & Thulin var. **alba**. *Zygophyllum album* L. f., Dec. Pl. Hort. Upsal.: 11 t. 6 (1762); *Z. proliferum* Forssk., Fl. Egypt. Arab.: 12 (1775). **(Figs 2A, D & G)**.
Diagnosis: This variety can be distinguished by the petiole of the leaflets up to 15 mm long, flowers 4.0–4.5×3–4.5 mm; schizocarps obconical, star-shaped, with thick broad lobes 8–10 mm long, 7–10 mm wide of upper end, 3–6 mm wide of lower end, pedicel up to 3 mm long.
Type: Linnaeus HL544-2 [LINN, lectotype designated by El-Hadidi in Webbia 33: 51(1978)].
Vernacular names: Rotreyt, Qarmal, Harm.
Phenology: February to June.
Distribution: Saudi Arabia: Along the Red Sea coast (Fig. 3). Worldwide: Egypt, Jordan, Tunisia, Palestine, Somalia, South Africa and Greece.
Habitat: Found in the salt marshy habitats, coastal and inland saline sandy soils, dunes and sheets, and in saline depressions.
Specimens examined: **SAUDI ARABIA:** Shuaiba (20°52'23"N 39°22'6"E), February 2013, *Alzahrani et Albokhari D&E110* (KAUH); Umluj (24°59'05"N 37°17'09"E), March 2013, *Alzahrani et Albokhari D&E132, D&E134, D&E139* (KAUH); Umluj (25°03'34.87"N 37°15'50.86"E), May 2014, *Alzahrani D148, D153* (KAUH); Coast 12 km north of Muweli (27°41'6.02"N 35°29'20.33"E), September 1983, *Collenette 4521* (RIY, K); near Umm Sidrah 75 km north of Jeddah, January 1980, *Collenette 1518* (K). **EGYPT**: Sallum east, April 1932, *Shabetai 1780* (CAIM); North of Helwan, February 1944, *Davis 6302B* (E); Helwan, March 1891,

Scott Elliot 3554 (E). **JORDAN**: Aqaba, October 1989, *Leonard 7468* (E). **TUNISIA**: Monastir, August 1968, *Davis 48050* (E); Southeast Tunisia, west of Oudref, February 1966, *Archibald 884* (E). **GREECE**: July 1950, *Davis 18109* (E); EP. Ierapetro, October 1966, *Greuter 7811* (E).

Tetraena alba (L. f.) Beier & Thulin var. **arabica** Alzahrani & Albokhari, **var. nov.**

(Figs 2B, E & H).

Diagnosis: This variety can be distinguished by its petiole of the leaflets, up to 18 mm long, flowers 5.5×5.0 mm, schizocarps oblong-obconical, star shaped, with narrow lobes, 11–13 mm long, upper end 8–10 mm wide, lower end 2–3 mm wide, pedicel up to 6 mm long.

Type: Saudi Arabia, Umluj (24°58'19"N 37°17'03"E), March 2013, Alzahrani *et* Albokhari 138 (*Holotype*: KAUH; *Isotype*: KSU).

Small shrub, perennial, green or greenish grey, 50–60 cm tall, 40 cm wide. Stem pubescent, with unicellular simple trichomes. Leaves 2-foliolate, 7–12×3.0–5.5 mm, fleshy, cylindrical or elliptic, apex acute; petiole up to 18 mm long; stipules triangular, herbaceous, 1.0×1.5 mm, pubescent. Flowers white, arranged in clusters, sometimes solitary, bisexual, 5.5×5.0 mm, pedicel 1–2 mm long. Sepals 5, rounded-obtuse at the apex, herbaceous, yellowish green, obovate, 3–4×2–3 mm, pubescent, aestivation imbricate. Petals white, 5, spathulate, 3.5–6.0×1–2 mm, aestivation valvate. Stamens 10, 3–4 mm long, staminal appendages undivided, 2.0–2.5×1 mm; anthers 2-lobed, yellow, dorsifixed, dehiscent longitudinally; disc smooth. Ovary 5-locular, pubescent; style single, 1 mm long. Schizocarp oblong-obconical, star shaped, with narrow lobes, 11-13×(2–3)8–10 mm, pubescent, pericarp extended as wings, peduncle pubescent, up to 6 mm long.

Vernacular names: Rotreyt, Qarmal, Harm.

Phenology: February to June.

Distribution: Endemic to Saudi Arabia and apparently restricted to its western cost, mainly in Umluj (Fig. 3).

Habitat: Found in coastal and inland saline sandy soils, and salt marshy areas.

Etymology: The varietal epithet is derived from Arabia, the area of its distribution.

Conservation status: Based on its known distribution (area of occupancy estimated to be less than 10 km^2) and abundance (number of mature individuals less than 50), the IUCN Red List Category (IUCN, 2014) "Critically Endangered" is here attributed to this variety.

Tetraena alba (L. f.) Beier & Thulin var. **amblyocarpa** (Baker) Alzahrani & Albokhari, **comb. nov.** *Zygophyllum amblyocarpum* Baker, Hooker's Icon. Pl. 24: t. 2358 (1895); *Z. amblyocarpum* Baker, Kew Bull. 1894: 339 (1894), *nom. nud.*; *Z. album* L. f. var. *amblyocarpum* (Baker) El-Hadidi in Webbia 33: 52 (1978); *Z. album* L. f. var. *amblyocarpum* (Baker) El-Hadidi in Bot. Not. 131: 441 (1978).

(Figs 2C, F & I).

Diagnosis: This variety is recognized by petiole of leaflets up to 10 mm long, flowers 4×4 mm, schizocarps obconical, acute with keeled lobes 9–13 mm long, 8–12 mm wide at upper end, 2–3 mm wide at lower end and pedicel up to 6 mm long.

Type: Hadramout, Al Mukalla, Shary Burrock Valley, December 1893; Lunt 51 (*Holotype*: K!; *Isotype*: BM).

Phenology: February to June.

Vernacular names: Rotreyt, Qarmal, Harm.

Distribution: Saudi Arabia: Shuaibah (Fig. 3). Worldwide: South Arabia (Yemen), tropical East and North Africa (Egypt).

Habitat: Salt marshy areas.

Conservation status: Least Concern (lc), locally common on the west coast of Saudi Arabia, cost of Yemen, Egypt and Somalia.

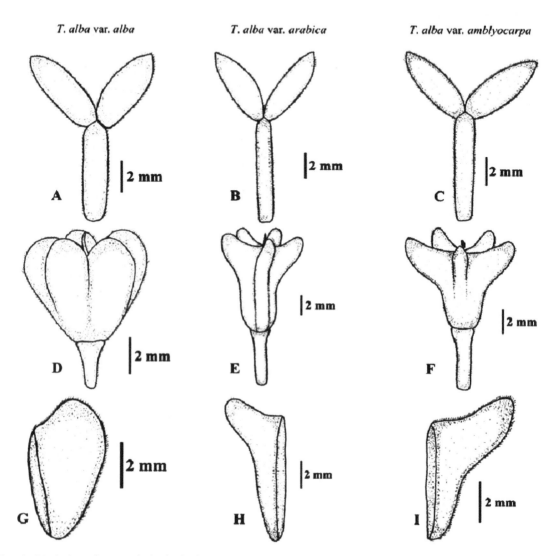

Fig. 2. Variations in morphological characters of *Tetraena alba* varieties: A. Leaf of *T. alba* var. *alba*; B. Leaf of *T. alba* var. *arabica*; C. Leaf of *T. alba* var. *amblyocarpa*; D. Fruit of *T. alba* var. *alba*; E. Fruit of *T. alba* var. *arabica*; F. Fruit of *T. alba* var. *amblyocarpa*; G. Schizocarp lobe of *T. alba* var. *alba*; H. Schizocarp lobe of *T. alba* var. *arabica*; I. Schizocarp lobe of *T. alba* var. *amblyocarpa* (Alzahrani and Albokhari, 2017b).

Specimens examined: **SAUDI ARABIA**: Shuaiba (20°51'10"N 39°23'47"E), February 2013, *Alzahrani et Albokhari D&E107* (KAUH). **YEMEN**: Hadramout, Al Mukalla, Shary Burrock valley, December 1893, *Lunt 51* (K!, holotype); Hadramout, 81 km from Qusayir along road to Sayhut, October 1992, *Thulin et al. 8247* (K). **EGYPT**: Jamailia, February 1948, *Shabetai 7730* (CAIM); Red Sea region, May 2005, *Abdel-Ghani et Abdel-Fattah s.n.* (CAIM); Safaga, May 2005, *Abdel-Ghani et Abdel-Fattah s.n.* (CAIM).

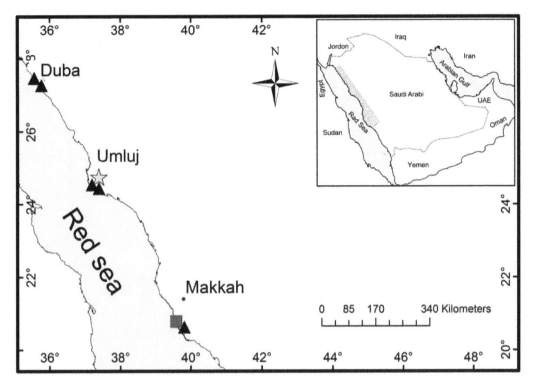

Fig. 3. Distribution map of *Tetraena alba* varieties in Saudi Arabia: ▲ *T. alba* var. *alba*, ☆ *T. alba* var. *arabica*, ■ *T. alba* var. *amblyocarpa* (Alzahrani and Albokhari, 2017b).

Tetraena coccinea (L) Beier & Thulin, Pl. Syst. Evol. 240: 35 (2003). *Zygophyllum coccineum* L., Sp. Pl. 1: 386 (1753); *Z. desertorum* Forssk., Fl. Aegypt.-Arab.: 87 (1775); *Z. berenicense* Scweinf., Fl. Egypt III: 65 (1887); *Z. coccineum* L. var. *berenicense* (Schweinf.) Muschl., Man. Fl. Egypt 1: 578 (1912). **(Fig. 4)**.

Diagnosis: *T. coccinea* with a shrubby habit can be recognized by its cylindrical fruits and persistent triangular stipules.

Type: Inter Kahiram & Sués, August 1762, Forsskål *s.n.* (*Holotype*: C; *Isotype*: BM! LD).

Small shrubs, perennial, green, up to 75 cm tall and 100 cm wide. Stem pubescent, with unicellular simple trichomes. Leaves 2-foliolate, cylindrical, up to 14.0×4.5 mm, fleshy, petiole up to 20 mm long; stipules triangular, herbaceous, 1.5×1.0 mm, pubescent. Flowers bisexual, white, 4–7×4–5 mm, pedicel up to 10 mm long. Sepals 5, rounded-obtuse at the apex, herbaceous, yellowish green, obovate, 4–6×2–3 mm, pubescent, aestivation imbricate. Petals 5, white, spathulate, 5–7×2.0–2.5 mm, aestivation valvate. Stamens 10, 3.0–4.5 mm long, staminal appendages undivided, 2–3×1.0–1.5 mm; anthers 2-lobed, yellow, dorsifixed, dehiscent longitudinally; disc smooth. Ovary 5-locular, pubescent; style single, 1 mm long. Fruit a schizocarp, cylindrical, 9–12×3–6 mm, glabrous, peduncle up to 11 mm long.

Vernacular names: Harm, Rotreyt, Batbat.

Phenology: February to June.

Distribution: Saudi Arabia: North-west to south-west Saudi Arabia (Fig. 5). Worldwide: Kuwait, Yemen, East and North Africa, and Palestine.

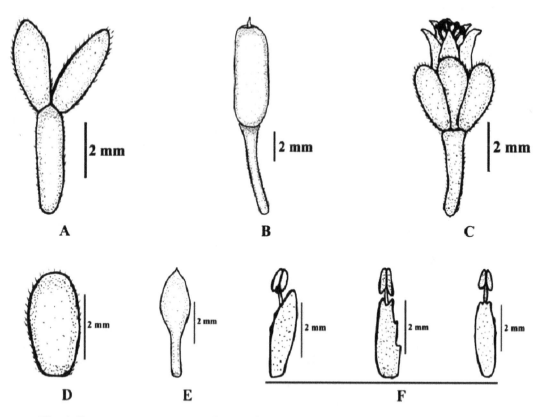

Fig. 4. *Tetraena coccinea*: A. Leaf; B. Fruit; C. Flower; D. Sepal; E. Petal; F. Stamens.

Habitat: Found in salt marshy areas.

Specimens examined: **SAUDI ARABIA**: South of Jeddah (22°31'11"N 39°10'41"E), February 2013, *Alzahrani et Albokhari D&E101, D&E102, D&E104, D&E105* (KAUH); Shuaibah (20°52'23"N 39°22'16"E), February 2013, *Alzahrani et Albokhari D&E108, D&E11, D&E113* (KAUH); North of Jeddah (21°50'23"N 39°07'05"E), February 2013, *Alzahrani et Albokhari D&E114, D&E115, D&E116, D&E117, D&E118* (KAUH); South of Alleith (19°56'15"N 40°31'17"E), February 2013, *Alzahrani D&E119, D&E120* (KAUH); near Rabigh (23°31'35"N 38°40'27"E), March 2013, *Alzahrani et Albokhari D&E123, D&E124, D&E125, D&E126* (KAUH); between Rabigh and Yanbu (23°53'08"N 38°27'08"E), March 2013, *Alzahrani et Albokhari D&E127, D&E128* (KAUH); Yanbu (24°07'23"N 38°02'02"E), March 2013, *Alzahrani et Albokhari D&E129, D&E130, D&E131, D&E135, D&E136, D&E*140, (KAUH); Farasan Island, February 1986, *Collenette 10357* (RIY); Tabouk, September 1983, *Chaudhary H8251* (RIY); Duqm Sabkha, February 1999, *Someya et Wutaid H19036* (RIY); Oasis near Duba, February 1999, *Someya et Wutaid H19035* (RIY); Rabigh, May 1998, *Someya et WutaidH19037* (RIY); Dumsaq, June 1988, *Chaudhary H14159* (RIY); North Hijaz, May 1978, *Collenette 747* (K); Red Sea near Jeddah, October 1983, *Collenette 5491* (K, E); Yanbu al Bahr, 1972, *Collenette 72-203* (K); Jabal Ohod north of Madinah, February 1945, *Khattab 323* (CAI). **EGYPT**: The desert road between Cairo-Faiyum at 40 km, August 1972, *Abbas et Abdel-Hay 838* (CAIM); Wadi Hamad, April 1944, *Davis 7165* (E); Lower Wadi Digla, October 1944, *Davis 7805* (E). **YEMEN**: January 1979, *Wood 2676* (K, E); Meidi, March 1944, *Khattab 681* (CAIM).

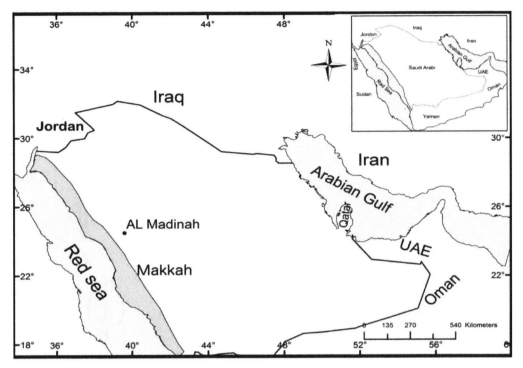

Fig. 5. Distribution map of *Tetraena coccinea* in Saudi Arabia.

Tetraena decumbens (Delile) Beier & Thulin, Pl. Syst. Evol. 240: 35 (2003). *Zygophyllum decumbens* Delile, Descr. Egypte, Hist. Nat. :221, t. 27, fig. 3 (1813); *Z.decumbens* Delile var. *megacarpum* Hosny, Bot. Not. 130: 467-468 (1977). (**Fig. 6**).

Diagnosis: *T. decumbens* can be distinguished from other *Tetraena* species by its shrubby habit, 2-foliolate flat leaves, white creamy flowers, bipartite staminal appendages and glabrous stem. It is sympatric to *T. alba* and *T. coccinea* in western Saudi Arabia, but these species have pubescent stem, undivided staminal appendages and different fruit features.

Type: Valée dans l'Egaroment; Delile 6967 (*Holotype*: MPU; *Isotype*: Fl).

Small shrub, perennial, green, 50 cm tall, 100 cm wide. Stem glabrous. Leaves 2-foliolate, obovate, flat, up to 21×12 mm, apex rounded, fleshy, petiole 12–15 mm long; stipules triangular, 1.0×1.5 mm, glabrous. Flowers bisexual, white-creamy, 5–6×4–5 mm, pedicel up to 7 mm long. Sepals 5, rounded-obtuse at the apex, herbaceous, yellowish green, obovate, glabrous, 3×2 mm, aestivation imbricate. Petals 5, white-creamy, spathulate, 4.0×1.5 mm, aestivation valvate. Stamens 10, 4.0–4.5 mm long, staminal appendages bipartite, 1.5×0.5 mm; anthers 2-lobed, yellow, dorsifixed, dehiscent longitudinally; disc smooth. Ovary 5-locular, glabrous; style single, c. 2 mm long. Fruit a schizocarp, obconical, 5-ridged, 3–6×1.5–5.0 mm, glabrous, peduncle up to 15 mm long.

Vernacular names: Harm, Rotreyt, Qarmal, Batbat.

Phenology: February to June.

Distribution: Saudi Arabia: Western to north-western region (Fig. 7). Worldwide: Oman, Yemen, Egypt, Sinai, Sudan, Somalia, Eritrea, and South Africa.

Habitat: Growing in sandy and gravels habitat.

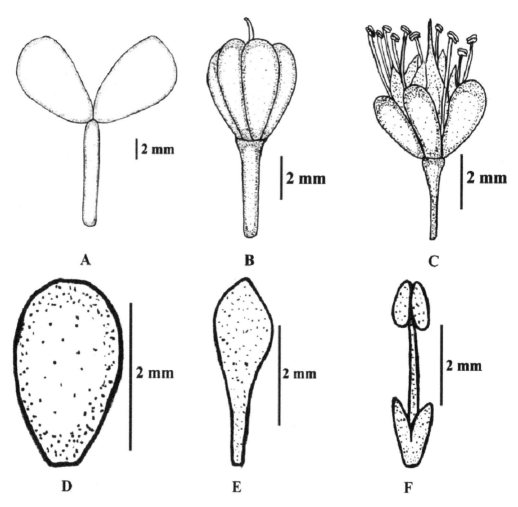

Fig. 6. *Tetraena decumbens*: A. Leaf; B. Fruit; C. Flower; D. Sepal; E. Petal; F. Stamen.

Specimens examined: **SAUDI ARABIA**: 30 km south of Umluj (24°45'06"N 37°19'56"E), March 2013, *Alzahrani et Albokhari D&E142* (KAUH); Al Wajh (26°20'36.6"N 36°23'13.5"E), May 2014, *Alzahrani D147, D152* (KAUH); Wadi Al Bayda, February 1998, *Someya et Wutaid H19029* (RIY); Jabal Hassan W coast, July 1998, *Someya et Wutaid H19028* (RIY); North of Muweli between Duba and Ash Sharma, September 1983, *Collenette 8846* (RIY); Jaziat Qummaah, February 1998, *Someya et Wutaid H19041* (RIY); Duba north of Hedjaz, January 1944, *Khattab K33* (CAI); 12 km north of Muweli Ash Sharma road, September 1983, *Collenette 4519* (K). **OMAN**: Dhofar, 25 km south of Thumrait on Salalah road, September 1984, *Miller 7654* (K); Ayun road 5 km east of turnoff to pools, September 1985, *Miller 7654* (K); Jabal Qamar 5 km northwest of Janook, October 1979, *Miller 2621* (K). **YEMEN**: Shabwah, 2 km northeast of Mahfis Wadi Bottom, October 1992, *Thulin et al. 7981* (K); Shabwa Wadi 5 km south of Ataq, January 1988, *Rowaished et al. 2792* (K); Hadramout, Central Plateau 19 km south of Sayun along the road to Al Mukalla, June 1987, *Boulos et al.17055* (K). **EGYPT**: Jabal Araqa, March 1944, *Davis 9799* (RIY); Wadi Quseib north Galala, March 1964, *Boulos s.n.* (K); in the desert of Elsaff south of Helwan, April 1959, *Boulos s.n.* (K).

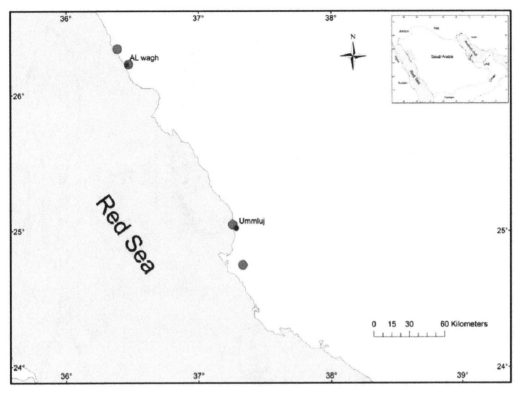

Fig. 7. Distribution map of *Tetraena decumbens* in Saudi Arabia.

Tetraena hamiensis (Schweinf.) Beier & Thulin, Pl. Syst. Evol. 240: 35 (2003). (**Fig. 8**).

Diagnosis: *T. hamiensis* can mostly be recognized by the presence of 1-foliolate leaves, but when they are 2-foliolate, usually are fleshy, terete or globular, pubescent stem, solitary white flowers and undivided staminal appendages.

Small shrubs, perennial, green, reddish or yellowish green, up to 80 cm tall, 90 cm wide. Stem pubescent, with unicellular simple trichomes. Leaves mostly 1-foliolate, sometimes 2-foliolate in upper branches, terete, globular, cylindrical or clavate, 4–9×3–6 mm, fleshy, pubescent or glabrous, petiole equal or longer than leaflets, up to 9 mm long; stipules triangular, herbaceous, 1.0×1.5 mm, pubescent. Flowers bisexual, solitary at each node, white, 4–6×3–5 mm, pedicel 3–5 mm long. Sepals 5, rounded-obtuse at the apex, herbaceous, yellowish green, obovate, 3–5×2–3 mm, pubescent, aestivation imbricate. Petals 5, white, spathulate, 4–6×1.5–2.0 mm, aestivation valvate. Stamens 10, 2–5 mm long, staminal appendages undivided, 2–3 mm long, 1.0–1.5 mm wide; anthers 2-lobed, yellow, dorsifixed, dehiscent longitudinally; disc smooth. Ovary 5-locular, pubescent; style single, 0.5–1.5 mm long. Fruit a schizocarp, oblong-obovate, oblong-obconical, 5-angled or cylindrical, 8–20×(2–3.5)2–5 mm, pubescent or glabrous, peduncle 5-10 mm long, pubescent or glabrous.

T. hamiensis comprises three varieties, namely, *T. hamiensis* (Scweinf.) Beier & Thulin var. *hamiensis*, *T. hamiensis* (Schweinf.) Beier & Thulin var. *qatarensis* (Hadidi *ex* Beier & Thulin) Alzahrani & Albokhari, **comb. nov.** and *T. hamiensis* (Schweinf.) Beier & Thulin var. *mandavillei* (Hadidi *ex* Beier & Thulin) Alzahrnai & Albokhari, **comb. nov.**

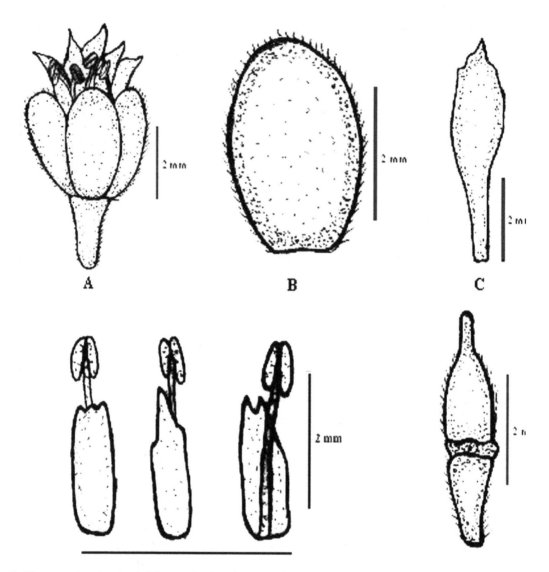

Fig. 8. *Tetraena hamiensis*: A. Flower; B. Sepal; C. Petal; D. Stamens; E. Ovary (Alzahrani and Albokhari, 2017c).

Tetraena hamiensis (Scweinf.) Beier & Thulin var. **hamiensis**. *Zygophyllum hamiense* Schweinf., Bull. Herb. Boissier VII. App. II : 277 (1899); *T. hamiensis* (Schweinf.) Beier & Thulin, Pl. Syst. Evol. 240: 36 (2003). **(Figs 9A & D)**.

Diagnosis: *T. hamiensis* var. *hamiensis* is distinguished by its green, clavate, 6–9×3–5 mm leaflets, petiole up to 9 mm, pedicel up to 5 mm long, schizocarp oblong-obconical, 5-angled, clearly lobed, 10–13×3–4 mm, pubescent, peduncle up to 10 mm long.

Type: El Hami, east Schehr. Schweinfurth 182 (*Isotype*: W).

Vernacular name: Harm.

Phenology: February to June and September to November.

Distribution: Saudi Arabia: Eastern and south-central region of Saudi Arabia (Fig. 10). Worldwide: United Arab Emirates, Oman, Kuwait, Yemen, Iran and Somalia.

Habitat: In sandy and saline soils.

Conservation status: *T. hamiensis* var. *hamiensis* appears to be distributed in some localities in the Eastern and south-central region of Saudi Arabia. At the international level, this variety is evaluated as Least Concern (LC) since it also grows in United Arab Emirates, Oman, Kuwait, Yemen, Iran and Somalia (IUCN, 2014).

Specimens examined: **SAUDI ARABIA**: Al Ahsa, Qatar road (25°16'30"N 49°41'09"E), May 2013, *Alzahrani D18* (KAUH); Al Ahsa, Qatar road (24°49'54"N 50°40'25"E), 25 km before Salwa, May 2013, *Alzahrani D19* (KAUH); Al Ahsa, Qatar road, 10 km before Alaudaidah (24°27'32"N 51°02'52"E), May 2013, *Alzahrani D24* (KAUH); Al Ahsa, Dammam road (25°37'33"N 49°31'11"E), May 2013, *Alzahrani D28* (KAUH); Alqateef, Alsharqia, July 1997, *Atar 5723* (KSU); Aflag, Layla, August 1998, *Atar 5834* (KSU); Dhahran, December 1953, *Baker XI* (K). **UNITED ARAB EMIRATES**: West side of jabal Hafit, January 1983, *Brown 439* (CAI). **OMAN**: Nizwa Agricult Inst. Firg., November 1981, *Maconochie 2948* (K); Bahala, March 1976, *Radcliffe-Smith 3790* (K); Dhufar, 50 km west of Mudhai, September 1985, *Miller 7621* (K). **YEMEN**: Hadramout, Sayun outside the town, weeds in field and road sides, June 1987, *Boulos et al. 17042* (CAI); Wadi Hajr, 100 km west of Mukalla, Howtah 11 km north of Meifa Haga, February 1989, *Miller et al. 8153* (K, E). **IRAN:** Southeast Iran, Zahedan province, 24 miles of Rask road to Chah Bahar, March 1971, *Grey-Wilson et Hewer 262* (K).

Tetraena hamiensis (Schweinf.) Beier & Thulin var. **qatarensis** (Hadidi *ex* Beier & Thulin) Alzahrani & Albokhari, **comb. nov.** *Zygophyllum qatarense* Hadidi, Webbia 32 (2): 394 (1978); *Z. hamiense* var. *qatarense* (Hadidi) Thomas & Chaudhary, Flora of the Kingdom of Saudi Arabia 2: 502 (2001); *T. qatarensis* (Hadidi) Beier & Thulin, Pl. Syst. Evol. 240: 36 (2003).

(Figs 9B & E).

Diagnosis: *T. hamiensis* var. *qatarensis* is distinguished by its reddish or olive green, globular, 4–6×4–6 mm leaflets, petiole up to 8 mm, pedicel up to 3 mm long, schizocarp oblong-obovate, 5-angled, 8–10×2–3 mm, pubescent, partly lobed, peduncle up to 7 mm long.

Type: Qatar, Um Slal Ali, c. 25 km north of Doha, March 1977, *Boulos 10953* (*Holotype*: K!; *Isotype*: CAI & Fl).

Vernacular name: Harm.

Phenology: February to June and September to November.

Distribution: Saudi Arabia: Eastern region and north-central part of Saudi Arabia (Fig. 10). Worldwide: Qatar, Kuwait, Bahrain, United Arab Emirates, Oman, Socotra, Samha Isl., Abd-al-Kuri Isl. (Yemen) and Iraq.

Habitat: Found in the saline sand, including beaches, coastal areas, and rocky habitat.

Conservation status: *T. hamiensis* var. *qatarensis* appears to be distributed in some localities in the eastern and north-central region of Saudi Arabia. At the international level, this variety is evaluated as Least Concern (LC) since it also grows in Qatar, Kuwait, Bahrain, United Arab Emirates, Oman, Socotra, Samha Isl., Abd-al-Kuri Isl. (Yemen) and Iraq (IUCN, 2014).

Specimens examined: **SAUDI ARABIA**: Al Ahsa, Qatar road (25°16' 29" N 49°41'07"E), May 2013, *Alzahrani D16* (KAUH); Al Ahsa, Qatar road (24°48'40"N 50°44'26"E), May 2013, *Alzahrani D20, D21, D22* (KAUH); Buraidah, March 1997, *Alfarhan et Thomas 766* (KSU); Al-Ahsa, March 1996, *Thomas 766* (KSU); Aljubail, Alsharqia, July 1997, *Atar H5729* (KSU); Alsafaneiyah, Dammam, February 1981, *Migahid et Alsheikh s.n.* (KSU, *H19992*); Umm Assahik, Alsharqia, July 1997, *Atar H5748* (KSU); Rocky coastal area near Batha check point, Salwa region, March 1990, *Chaudhary et al. H13357* (RIY); Dareen Island, May 1987, *Chaudhary H12190* (RIY); Abqaiq-Hofuf road 87 km from Dhahran, April 1982, *Podzorski 811* (RIY);

Nairyah, October 1983, *Jeha H8711* (RIY); 18 km north of Dammam, February 1982, *Naylor 5* (E). **QATAR**: Um slal Ali, c. 25 km north of Doha, March 1977, *Boulos 10953* (K!, holotype); Dukham Camp, 12 m waste ground, December 1970, *Wilcox 38* (K); Sheikh Khalifa Ibn Ali Al Thani Garden, April 1977, *Boulos 11179* (K). **KUWAIT**: Al-Khiran, March 1983, *Rawi et al. 1550* (CAI); Roadsides between Al-Ahmadi and mina Abdullah, March 1995, *Mathew 2531* (K). **BAHRAIN**: Near base of central hills of Bahrain main island, April 1984, *Rezk 103* (K); Jerdab, 1985, *Naguib 404* (K); Al-Areen Wild Life Park and Reserve, April 1985, *Boulos et Hasan 15687* (K). **UNITED ARAB EMIRATES**: Abu Dhabi, March 1981, *Western BW 20* (K). **OMAN**: Wahiba sands, January 1986, *Cope 36* (K); Nr Zukayt 10 km south-west of Izki, September 1979, *Miller et Whitcombe 2017* (K); Kuria Muria Island, Al Hallaniyah Island, February 1993, *McLeish 1587* (E). **IRAQ**: 25 km south-east of Zubair, March 1957, *Ghiust et al. 16871* (K); 25 km south-east of Zubair, March 1957, Ghiust, *Rawi et Rechinger 16872* (K); between Zubair and Safwan, March 1966, *Alizzi 34353* (K).

Tetraena hamiensis (Schweinf.) Beier & Thulin var. **mandavillei** (Hadidi *ex* Beier &Thulin) Alzahrnai & Albokhari, **comb. nov**. *Zygophyllum mandavillei* Hadidi, Publ. Cairo Univ. Herb. 7-8: 327 (1977); *Z. hamiense* var. *mandavillei* (Hadidi) Thomas & Chaudhary, Flora of the Kingdom of Saudi Arabia 2: 502 (2001); *T. mandavillei* (Hadidi) Beier & Thulin, Pl. Syst. Evol. 240: 36 (2003). **(Figs 9C & F)**.

Diagnosis: *T. hamiensis* var. *mandavillei* is distinguished by its yellowish green, cylindrical, 7–9×3–5 mm glabrous leaflets, petiole equal to the leaflet, up to 9 mm long, pedicel up to 4 mm long, schizocarp cylindrical, oblong, 16–20×3–4 mm, glabrous, peduncle up to 5 mm long.

Type: Saudi Arabia, ArRub' al-Khali, Camp Shaybah 9, June 1970, Mandaville, 2892 (*Holotype*: BM!; *Isotype*: CAI).

Vernacular name: Harm.

Phenology: February to June and September to November.

Distribution: Saudi Arabia: Central, eastern, north and eastern ArRub' al-Khali, Doshak Island, and south-west region of Saudi Arabia (Fig. 10). Worldwide: Oman, United Arab Emirates and Yemen (Aden Desert).

Habitat: Red sands, gravels or saline areas.

Conservation status: On the current evidence *T. hamiensis* var. *mandavillei* appears to be distributed in Eastern and Southwest region, north-west, northern and eastern of ArRub al Khali of Saudi Arabia. This species is evaluated as Least Concern (LC) since it also grows in Oman, United Arab Emirates and Yemen (IUCN, 2014).

Specimens examined: **SAUDI ARABIA**: Khurais, Al Ahsa road (25°13'55"N 48°36'16"E), May 2013, *Alzahrani D13* (KAUH); Al Ahsa, Qatar road (25°16'18"N 49°34'59"E), May 2013, *Alzahrani D15* (KAUH); Al Ahsa, Qatar road (25°16'29"N 49°41'07"E), May 2013, *Alzahrani D17* (KAUH); Al Ahsa, Qatar road 25 km before Alaudaidah (24°32'55"N 50°54'16"E), May 2013, *Alzahrani D23* (KAUH); Al Ahsa, Qatar road, 10 km before Alaudaidah (24°27'32"N 51°02'52"E), May 2013, *Alzahrani D25* (KAUH); Al Ahsa, Qatar road, Alaudaidah (24°26'07"N 51°07'01"E), May 2013, *Alzahrani D26* (KAUH); Shedgum, next to the cement factory, Al Ahsa-Dammam road (25°40'07"N 49°30'31"E), May 2013, *Alzahrani D30* (KAUH); Wadi Baysh, near Sabiya, June 1999, *Alfarhan et al. H19742* (KSU); Near Shabita (22°13'N 54°17'E), February 1990, *Chaudhary et al. H13312* (RIY); Doshak Island, June 1988, *Chaudhary H15762* (RIY); Layla lakes, sol Layla, March 1987, *Collenette 6046* (RIY, K); 10 km north-west of campus S-3, north-eastern ArRub' al-Khali, February 1979, *Mandaville 7085* (E); ArRub' Al-Khali, Camp Shaybah 9, June 1970, *Mandaville 2892* (BM!, holotype). **OMAN**: Near Wadi Tawsinat, north

Dhofar, May 1982, *Gallagher 6464/26* (E). **UNITED ARAB EMIRATES**: Sweehan, February 1996, *Boer 103* (RIY).

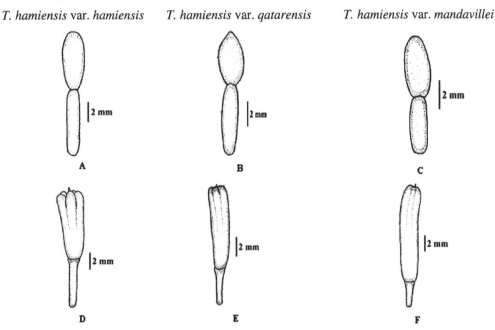

Fig. 9. Variation in morphological characters of *Tetraena hamiensis* varieties: A. Leaf of *T. hamiensis* var. *hamiensis*; B. Leaf of *T. hamiensis* var. *qatarensis*; C. Leaf of *T. hamiensis* var. *mandavillei*; D. Fruit of *T. hamiensis* var. *hamiensis*; E. Fruit of *T. hamiensis* var. *qatarensis*; F. Fruit of *T. hamiensis* var. *mandavillei* (Alzahrani and Albokhari, 2017c).

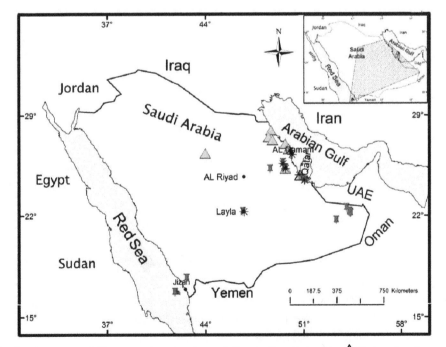

Fig. 10. Distribution map of *Tetraena hamiensis* varieties in Saudi Arabia. △ *T. hamiensis* var. *hamiensis*, ✳ *T. hamiensis* var. *qatarensis*, 📌 *T. hamiensis* var. *mandavillei* (Alzahrani and Albokhari, 2017c).

Tetraena propinqua (Decne.) Ghaz. & Osborne, Kew Bull. 70: 38 (2015). **(Fig. 11)**.

Small shrubs, perennial, green, 50 cm tall, 80–100 cm wide. Stem pubescent, with unicellular simple trichomes. Leaves 2-foliolate, cylindrical, up to 12×4 mm, apex rounded or acute, fleshy, pubescent, petiole up to 14 mm long; stipules triangular, herbaceous, 1.0×1.5 mm, pubescent. Flowers bisexual, white or white-creamy, 1–3 at each node, 4–7×3.5–5.0 mm, pedicel 7–14 mm long. Sepals 5, rounded-obtuse at the apex, herbaceous, yellowish green, obovate, 3–5×2–3 mm, pubescent, aestivation imbricate. Petals 5, white, spathulate, 2.5–6.0×1–3 mm, aestivation valvate. Stamens 10, 3–5 mm long, staminal appendages undivided, 2.0–3.5 mm long, 1 mm wide; anthers 2-lobed, yellow, dorsifixed, dehiscent longitudinally; disc smooth. Ovary 5-locular, pubescent; style single, 1–2 mm long. Fruit a schizocarp, ovate to oblong or obconical, 5-angled, 7–13 × 2.0–6.5 mm, pubescent, peduncle up to 14 mm long, pubescent.

Key to the subspecies of *Tetraena propinqua*

1. Leaflets apex acute; pedicel up to 7 mm long; schizocarp ovate oblong; peduncle up to 7 mm long. — subsp. *propinqua*
- Leaflet apex rounded; pedicel up to 14 mm long; schizocarp obconical; peduncle up to 14 mm long. — subsp. *migahidii*

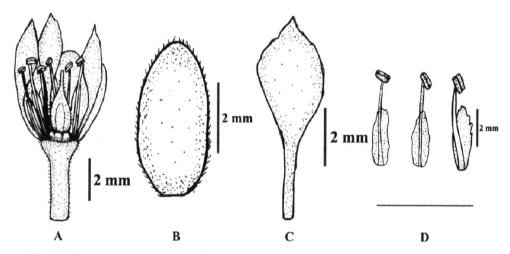

Fig. 11. *Tetraena propinqua*: A. Flower; B. Sepal; C. Petal; E. Stamens (Alzahrani, 2017).

Tetraena propinqua (Decne.) Ghaz. & Osborne subsp. **propinqua**. *Zygophyllum propinquum* Decne., Ann. Sci. Nat., Bot. Sér. 2, 3: 283 (1835). **(Figs 12A, C & E)**.

Diagnosis: *T. propinqua* subsp. *propinqua* can be distinguished by its acute leaflet apex, white flowers, up to 7 mm long pedicel, fruits ovate to oblong, 5- angled, 9-13 mm long, 4.0-6.5 mm wide at upper end, 2-4 mm wide at lower end, peduncle up to 7 mm long.

Type: Sinai, Gallam, Tor, June 1832, Bové 172 & 173 (*Isotype*: K!).

Vernacular names: Harm, Rotreyt.

Phenology: February to June.

Distribution: Saudi Arabia: Western to north-western Saudi Arabia (Fig. 13). Worldwide: Egypt, Sinai, Palestine, Iraq, Iran, Afghanistan, Pakistan and India.

Habitat: In sandy and gravel desert.

Specimens examined: **SAUDI ARABIA**: Shuaibah (20°52'23"N 39°22'16"E), February 2013, *Alzahrani et Albokhari D&E109* (KAUH); Umluj (24°59'05"N 37°17'09"E), March 2013, *Alzahrani et Albokhari D&E133, D&E137, D&E141* (KAUH); Dhallam, May 1998, *Thomas 5866* (KSU); 9 km south of Khaybar, October 1989, *Collenette 7287* (K). **IRAQ**: On the road near Karbala-Liwa, July 1962, *Al-Ani et Mohamed 12* (K); Karbala Musseiyib, May 1947, *Gillett 9968* (K). **EGYPT**: Sinai, Tor, April 1836, *Bové 274,275* (*Isotype*: K!); Gallam, Tor, June 1832, *Bové 172, 173* (*Isotype*: K!).

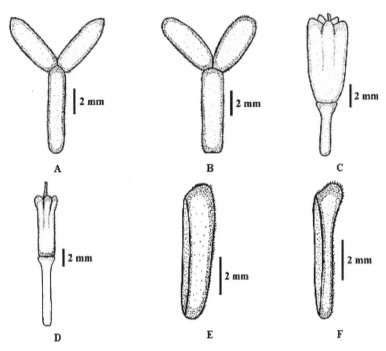

Fig. 12. Variation in morphological characters of *Tetraena propinqua* subspecies: A. Leaf of *T. propinqua* subsp. *propinqua*; B. Leaf of *T. propinqua* subsp. *migahidii*; C. Fruit of *T. propinqua* subsp. *propinqua*; D. Fruit of *T. propinqua* subsp. *migahidii*; E. Schizocarp lobe of *T. propinqua* subsp. *propinqua*; F. Schizocarp lobe of *T. propinqua* subsp. *migahidii* (Alzahrani, 2017).

Tetraena propinqua (Decne.) Ghaz. & Osborne subsp. **migahidii** (Hadidi *ex* Beier & Thulin) Alzahrani & Albokhari, **comb. nov.** *Zygophyllum migahidii* Hadidi, Publ. Cairo Univ. Herb. 7 & 8: 328 (1977); *Z. propinquum* subsp. *migahidii* (Hadidi) Thomas & Chaudhary, Flora of the Kingdom of Saudi Arabia 2: 501 (2001); *T. migahidii* (Hadidi) Beier & Thulin, Pl. Syst. Evol. 240: 36 (2003).   (**Figs 12B, D & F**).

Diagnosis: *T. propinqua* subsp. *migahidii* is recognized by its rounded apex of the leaflet, white-creamy flowers, pedicel up to 14 mm long, schizocarps obconical, 5-angled, 9-12 mm long, 3-5 mm wide at upper end, 2-5 mm wide at lower end, peduncle up to 14 mm long.

Type: Saudi Arabia, Al-Hail, Migahid, El-Sheikh *et* S. Awad 574/A (*Holotype*: CAI; *Isotype*: KSU!).

Vernacular names: Harm, Rotreyt.

Phenology: February to June.

Distribution: Saudi Arabia: North Saudi Arabia: Nafud Desert, West-central Saudi Arabia: Nejd Desert and Eastern Saudi Arabia (Fig. 13). Worldwide: Iraq.

Habitat: Sandy salt habitats and gravels desert.

Specimens examined: **SAUDI ARABIA:** Alkhasrah, Taif-Riyadh road (23°24'59"N 43°43'27"E), May 2013, *Alzahrani D5* (KAUH); Almuzahmeiah, West of Riyadh (24°25'38"N 45°57'32"E), May 2013, *Alzahrani D6* (KAUH); Khurais Road, 150 km before Al Ahsa (25°11'47"N 48°19'12"E), May 2013, *Alzahrani D7* (KAUH); Al Ahsa-Dammam road (25°37'33"N 49°32'12"E), May 2013, *Alzahrani D27* (KAUH); Shedgum, next to the cement factory, Al Ahsa-Dammam road (25°40'07"N 49°30'31"E), May 2013, *Alzahrani D29* (KAUH); Buqaiq, Al Ahsa-Dammam road (26°54'03"N 49°50'09"E), May 2013, *Alzahrani D31* (KAUH); Riyadh King Khaled International Airport road (24°50'57"N 46°44'14"E), May 2013, *Alzahrani D32* (KAUH); Alsharamiah, Riyadh-Taif road (25°40'07"N 49°30'31"E), May 2013, *Alzahrani D33* (KAUH); before Alhumiat Riyadh-Taif road (23°22'41"N 43°37'29"E), May 2013, *Alzahrani D36* (KAUH); Dhalam, Riyadh-Taif road (22°44'17"N 42°12'48"E), May 2013, *Alzahrani D37* (KAUH); Beirut Street, Hail (27°33'48"N 41°43'47"E), May 2013, *Alzahrani D53* (KAUH); Wadi Tarabah, May 2013, *Aldahan 1* (KAUH); Al-Qaeid road, Hail (27°41'18"N 41°44'38"E), April 2013, *Asiri1* (KAUH); Al-Hail, May 1976; *Migahid et al. 574/A* (*Holotype*: CAI; *Isotype*: KSU!); Riyadh, March 1993, *Thomas 1253* (KSU); Al-Kharj road, April 1981, *Noor 2296* (KSU); Buraidah, May 1983, *Chaudhary H7832* (RIY); Unaizah, May 1978, *Chaudhary s.n.* (RIY); Chara, May 1985, *Heemstra 7428* (RIY); Sulayyil, May 1996, *Chaudhary H14228* (RIY); Riyadh, May 1984, *Chaudhary H8356* (RIY); Aflaj, June 1984, *Jahangir H8398* (RIY); RAWRC, 1984, *Chaudhary H8489* (RIY); Aarqah, May 1984, *Chaudhary 8355* (RIY); 30 km southwest Harad, November 1987, *Mandaville 8696* (CAI); Southern of Riyadh, October 1987, *Collenette 6314* (K, E); 2 km south east of Khurmah, Riyadh road, July 1991, *Collenette 7851* (K). **IRAQ.** Habbanya, June 1966, *Rawi et Alizzi 34453* (K); 40 km south of Baghdad, road to Karbala, November 1958, *Rawi 26883* (K).

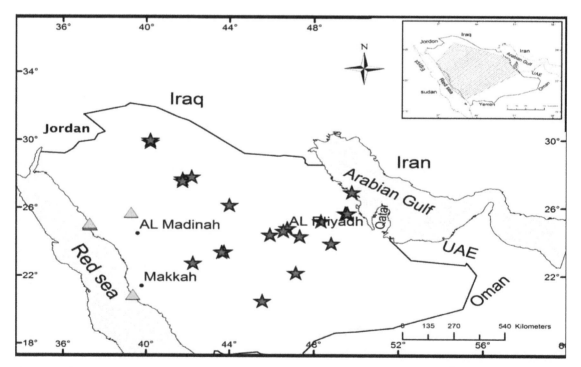

Fig. 13. Distribution map of *Tetraena propinqua* subspecies in Saudi Arabia. △ *T. propinqua* subsp. *propinqua*, ★ *T. propinqua* subsp. *migahidii* (Alzahrani, 2017).

Tetraena simplex (L.) Beier & Thulin, Pl. Syst. Evol. 240: 36 (2003). *Zygophyllum simplex* L., Mant. Pl.: 68 (1767); *Z. portulacoides* Forssk., Fl. Egypt. Arab.: 88 (1775). **(Fig. 14)**.

Diagnosis: *T. simplex* is an annual herb and differs from other *Tetraena* species by its simple, opposite and sessile leaves, yellow flowers, bipartite staminal appendages and 5-lobed obovoid fruits.

Type: Egypt 1762-1763; Forsskål *s.n.* (*Holotype*: C; *Isotype*: LD Herb. Retzius).

Herbs, annual, green, 10–30 cm tall, 50–70 cm wide. Stem glabrous. Leaves simple, opposite, up to 20×2.5 mm, sessile, cylindrical, fleshy; stipules triangular, membranous, 1×1 mm, glabrous. Flowers bisexual, yellow, 2–4×3–5 mm, pedicel 1–2 mm long. Sepals 5, rounded-obtuse at the apex, herbaceous, yellowish green, obovate, 2×1 mm, aestivation imbricate. Petals 5, yellow, spathulate, longer than sepals, 2.5–3.0×1.0–1.5 mm, aestivation valvate. Stamens 10, 2.5–3.0 mm long, appendages bipartite, hyaline, 1.0×0.1–0.3 mm, anthers 2-lobed, yellow, dorsifixed, dehiscent longitudinally; disc smooth. Ovary 5-locular, glabrous; style single, 1–3 mm long. Fruit a schizocarp, obovoid, 5-lobed, glabrous, 2–3 ×1.5–3.0 mm, peduncle 1–2 mm long.

Vernacular names: Harm, Om thoreyb, Hamd, Qarmal.

Phenology: February to June and September to November.

Distribution: Saudi Arabia: Widely distributed throughout the country. Worldwide: Arabian Peninsula, United Arab Emirates, Oman, Yemen, Iran, Jordan, Palestine, Pakistan, India, and tropical Africa (Ghazanfar, 2007).

Habitat: Grows in sandy soils.

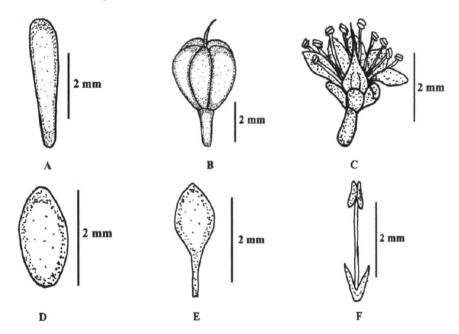

Fig. 14. *Tetraena simplex*: A. Leaf; B. Fruit, C. Flower; D. Sepal; E. Petal; F. Stamen.

Specimens examined: **SAUDI ARABIA**: Dhalam, Taif Riyadh road (22°12'10"N 41°24'19"E), May 2013, *Alzahrani D1* (KAUH); South of Jeddah (22°49'52"N 39°04'27"E), February 2013, *Alzahrani et Albokhari D&E103* (KAUH); 12 km south of Alleith (19°50'10"N 40°30'07"E), February 2013, *Alzahrani D121* (KAUH); Bishah, March 2013, *Al Dahhan 2* (KAUH); Alnuqrah, Prince Abdul Aziz bin Muqrin road, Hail (27°27'27"N 41°38'59"E), April

2013, *Asiri 2* (KAUH); Wady Fatmah, January 1945, *Khattab K 48* (CAIM) Mahazat Al-Said reserve area, July 2002, *Al-Abbasi et Shalhoub SAU-13* (K); Musaymir Wadi Tuban, March 1967, *Smih et Lawranos 25* (K); Wadi Gdeidat 130 km northwest of Mecca near Rabeh Saabar *et* Johfar, May 2004, *Al-Abbasi 0220802* (K); Jizan, 10 km south of Baysh from Sabya just before Wadi Guman, March 1996, *Van Slageren et Al-sa'doon 261* (K); Eastern provinces Al-Hasa lower south slope of Jabal Shaban, February 1965, *Mandaville 375* (K); Madina road 30 km north-east from Yanbu junction, March 1977, *Collenette 16* (K). **QATAR**: Qatar Road to Doha, January 1971, *Wilcox 57* (K); Wadi Al Galaiel toward the southern end of the Qatar peninsula, April 1977, *Boulos H11126* (K); Dukhan road, April 1979, *Batanouny 2462* (K). **BAHRAIN**: Southern Jebeh, May 1979, *Virgo 81* (K); Ras Noma, February 1970, *Gauaeher 50* (K); western plains of Bahrain main Island, April 1984, *Rezk 119* (K). **OMAN**: Matrah in Wadi behind town, March 1969, *Dickson 1095* (K); Sultan Qaboos University campus Seeb, April 1987, *Cope 172* (K); Al Hallaniyah, Kuria Muria Islands, November 1993, *McLeish 3028* (E). **UNITED ARAB EMIRATES**: Abu Dhabi in the vicinity of Umm am Nar near to the old Abu Dhabi airport, May 1982, *Western 292* (CAI); Persian Gulf, March 1937, *Holmes 346* (K); Abu Dhabi, March 1972, *Wilcox 207* (K). **YEMEN**: Aden, Jabal Shamsan tower of Silence and vicinity, June 1987, *Boulos et al. 16531* (K); Socotra, 1898, *Grant et Expedition 73* (E); Socotra, February 1989, *Miller et al. M. 8498* (E). **EGYPT**: Ismailia-Cairo road, November 1979, *Costantin et al. 499* (CAIM). **PAKISTAN**: Baluchistan, Bela to Uthal, April 1965, *Lamond 223* (E); Baluchistan, Makrani; Pasni to Kappan road to Gwadar, April 1965, *Lamond 445* (E); Karachi, October 1949, *Jafri s.n.* (E). **INDIA**: Punjab, January 1886, *Drummond 21622* (E). **NAMIBIA**: 5.3 km east of Goageb along road to Keetmanshoop, May 1993, *Strohbach 2339* (E); Omaruru District, Uis-Barandberg west road 50 km, April 1987, *Long et Rae 757* (E). **KENYA**: August 1938, *Pole Evans et Erens 1608* (E).

Zygophyllum fabago L., Sp. Pl. 1: 385 (1753). (**Fig. 15**).

Diagnosis: *Z. fabago* can be distinguished by its compound leaves, with two flat, obovate-elliptical leaflets, glabrous ovary and oblong loculicidal capsule.

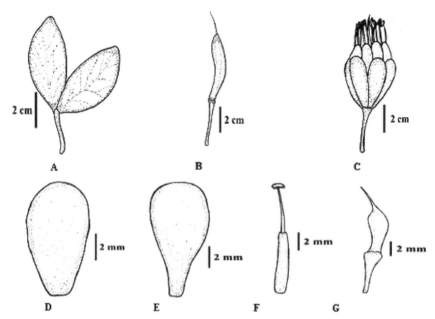

Fig. 15. *Zygophyllum fabago*: A. Leaf; B. Fruit; C. Flower; D. Sepal; E. Petal; F. Stamen; G. Ovary.

Small shrub, perennial, green, up to 75 cm tall. Leaves 2-foliolate, leaflets flat, obovate-elliptic, 40×25 mm, obtuse, petiole 20 mm long. Flowers bisexual, white-creamy, 12×15 mm, pedice 16 mm long. Sepals 5, yellowish green, obovate-elliptic, 10×5 mm, aestivation imbricate. Petals 5, white, 12×5 mm, aestivation valvate. Stamens 10, c. 13 mm long, staminal appendages undivided, 6×1 mm; anthers 2-lobed, yellow, dorsifixed, dehiscent longitudinally. Ovary glabrous; style single, 6.5 mm long. Fruit a loculicidal, oblong-cylindrical capsule, c. 30×(5)10 mm, glabrous, peduncle up to 10 mm long.

Vernacular names: Rotreyt, Qyllab.

Distribution: Saudi Arabia: Northern region (Tabarjal) (Fig. 16). Worldwide: Egypt, Palestine, Syria, Jordan, Iraq, Iran, Pakistan, Turkey, Spain, Georgia, Armenia, Russia, Afghanistan, Azerbaijan, France and Armenia.

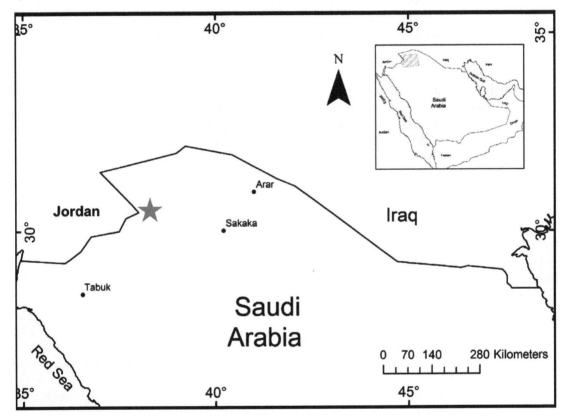

Fig. 16. Distribution map of *Zygophyllum fabago* in Saudi Arabia.

Habitat: Sandy soils.

Specimens examined: **SAUDI ARABIA**: Al Asawia 10 km before Tabarjal, April 1988, *Alaadin, Al Yahya et Al Said H19038* (RIY). **EGYPT**: Wadi Qoseib, December, 1944, *Alhabetai Z6477* (CAIM). **IRAQ**: Abu Ghraib, June 1959, *Rawi 26935* (CAIM); Baghdad, near Saadun State, October 1954, *Haines 72* (E); Baghdad Liwa, May 1956, *Polunin et Duri 68* (E). **IRAN**: Kazvin in Ditione Oppindikeredj, September 1948, *Rechinger 6841* (E); Azerbaijan, Moghan, bank of Aras River, 40 km from Parsabad on road to Aslanduz, May 1971, *Lamond 3182* (E); Azerbaijan south of Khoi, July 1960, *Furse et Synge 799/2* (E). **JORDAN**: Azraq, April 1936, *Dinsmore 11805* (E). **SYRIA**: July 1890, *Post s.n.* (E); June 1910, *Haradjian 3406* (E); Palmyra small dunes at north end of seat lake, April 1943, *Davis 5900* (E). **TURKEY**: Kars, Iğdir, state

Breeding Farm, around Boralar tepesi (hill), July 1956, *Demiris 3301* (E); Vily er Agri, October 1910, *Post 2053* (E); Prov. Sivas, east of Susehri, August 1957, *Davis et Hedge 32706* (E). **GEORGIA**: Davit Gorgeji, at caves, August 2009, *Mitchell et al. 52* (E); Tbilisi, hill sides near Dababane gorge, opposite the Tbilisi Botanical Institute, June 1959, *Davis 33708* (E); URSS-Géorgie-Tbilisi, July 1978, *Leonard 7155* (E). **SPAIN**: Coastal steppe in the hills between La Unión and Cartagena, April 1957, Stud. biol. *Rheno-Trai 57-535* (CAI); Almeria, Velez Blanco, roadside 10 km west of the village (37°38'N 2°6'E), July 1981, *Gardner et Gardner 1501* (E); Champs incultes prés de Cartagena, July 1852, *Bourgeau 2050* (E). **RUSSIA**: 1900, *Kulikowski E* (E); July 1896, *Callier 51* (E). **ARMENIA**: Circa ruinas sanctuarii Zvartnoc ad occidentem urbis Ervan, July 1975, *Gabrielian 12853* (E). **AFGHANISTAN**: Tashkurghan, Septiembre 1937, *Koelz 13182* (E); Samangan Tangi Taschkurgan Streamside, June 1969, *Ekberg 9072* (E); Maymana, halfway between Maymana and Andkhui, June 1962, *Hedge et Wendelbo W3834* (E). **AZERBAIJAN**: Elisabethpol distr. Araeseh Geok-tapa, in ruderatis, May 1908, *Schelkownikow et Woronow 312* (E). **FRANCE**: Héraull, Port de Cette, July 1891, *Bernard s.n.* (E).

Acknowledgements

We thank the authority of the herbaria BM, CAI, CAIM, E, K, KAUH, KSU and RIY for giving us access to their collections. We are thankful to Dr. Shahina Ghazanfar of the Royal Botanic Gardens, Kew, London for her assistance during our visit to this herbarium. We are grateful to Mrs. Mona S. Al Harbi for her help in preparing distribution maps, and to Mr. Mohammed Alnaggar for his assistance during field trips.

References

Al-Arjany, K.M. 2011. Molecular taxonomic perspective and eco-physiological variations of some species of *Tribulus, Zygophyllum* and *Fagonia* genera of family Zygophyllaceae in Saudi Arabia. Master Dissertation, King Saud University, Saudi Arabia.

Alzahrani, D.A. 2017. Systematic studies on the Zygophyllaceae of Saudi Arabia: Two new subspecies combination in *Tetraena*Maxim. Saudi J. Biol. Sci. DOI: 10.1016/j.sjbs.2016.12.022.

Alzahrani, D.A. and Albokhari, E.J. 2017a. Molecular phylogeny of Saudi Arabian *Tetraena* Maxim. and *Zygophyllum* L. (Zygophyllaceae) based on plastid DNA sequences. Bangladesh J. Plant Taxon. **24**(2): 155–164.

Alzahrani, D.A.and Albokhari, E.J. 2017b. Systematic studies on the Zygophyllaceae of Saudi Arabia: a new variety and new variety combination in *Tetraena*. Saudi J. Biol. Sci. **24**: 1574–1579.

Alzahrani, D.A. and Albokhari, E.J. 2017c. Systematic studies on the Zygophyllaceae of Saudi Arabia: new combinations in *Tetraena* Maxim. Türk. J. Bot. **41**: 96–106.

APG (Angiosperm Phylogeny Group). 2009. An update of the Angiosperm Phylogeny Group classification for the orders and families of flowering plants. Bot. J. Linn. Soc. **161**: 105–121.

Azevedo, L.B. 2014. Development and application of stressor–response relationships of nutrients. Ph.D. Thesis, Radboud University Nijmegen, the Netherlands.

Beier, B.A., Chase, M.W. and Thulin, M. 2003. Phylogenetic relationships and taxonomy of subfamily Zygophylloideae (Zygophyllaceae) based on molecular and morphological data. Plant Syst. Evol. **240**: 11–39.

Chaudhary, S.A. 2001. Flora of the Kingdom of Saudi Arabia, Vol. **3**. Ministry of Agriculture and Water, Riyadh, Saudi Arabia.

Collenette, S. 1985. An Illustrated Guide to the Flowers of Saudi Arabia. Scorpion Publishing, London, pp. 506–507.

Collenette, S. 1998. A Checklist of Botanical Species in Saudi Arabia. International Asclepiad Society, Ashford: Headley Brothers Ltd.

Collenette, S. 1999. Wild Flowers of Saudi Arabia. NCWCD, Kingdom of Saudi Arabia, pp. 764–766.

Cronquist, A. 1968. The Evolution and Classification of Flowering Plants. Boston: Houghton Mifflin, 396 pp.

Dahlgren, R. 1980. A revised system of classification of the angiosperms. Bot. J. Linn. Soc. **80**: 91–124.

Endlicher, S.L. 1841. Genera Plantarum Secundum Ordines Naturales Disposita. Part **18**: 1161. Vienna: Fr. Beck.

Engler, A. 1931. Zygophyllaceae. *In*: Engler, A. and Prantl, K. (2nd ed.), Die Naturlichen Pflanzenfamilien **19**: 144–184. Leipzig: Engelmann.

Engler, A. 1964. Syllabus der Pflanzenfamilien. Gebrüder Borntraeger, Berlin Nikolassee, 203 pp.

Ghazanfar, S.A. 2007. Flora of the Sultanate of Oman 2, Crassulaceae – Apiaceae. Scripta Bot. Belg. **36**: 1–220.

Ghazanfar, S.A. and Osborn, J. 2015. Typification of *Zygophyllum propinquum* Decne. and *Z. coccineum* L. (Zygophyllaceae) and a key to *Tetraena* in SW Asia. Kew Bull. **70**: 1–9.

Hadidi, M.N. 1977. Two new *Zygophyllum* species from Arabia. Publications from Cairo University Herbarium **7 & 8**: 327–329.

Hadidi, M.N. 1978. Zygophyllaceae. *In*: Boulos, L. (Ed.), Materials for a flora of Qatar. Webbia **32**: 369–396.

Hosny, A.I. 1988. Genus *Zygophyllum* L. in Arabia. Taeckholmia **11**: 19–32.

Hutchinson, J. 1969. Evolution and Phylogeny of Flowering Plants. Dicotyledons: Facts and theory with over 550 illustrations and maps by the author. London, New York: Academic Press.

IUCN 2014. The IUCN Red List of Threatened Species, version 2014.1 Cambridge UK: IUCN Red List Unit, Available from: http:// www.iucnredlist.org (accessed on 20.12.2014).

Linnaeus, C. 1753. Species Plantarum. Holmiae: Impensis Laurentii Salvii, pp. 385–386.

Louhaichi, M., Salkini, A.K., Estita, H.E. and Belkhir, S. 2011. Initial assessment of medicinal plants across the Libyan Mediterranean coast. Adv. Environ. Biol. **5**: 359–370.

Mandaville, J.P. 1990. Flora of Eastern Saudi Arabia. London and New York: Kegoin Pul International Ltd.

Migahid, A.M. 1978. Flora of Saudi Arabia, Vol. **1**, Seconded. Riyadh University Publications, Riyadh, Saudi Arabia.

Migahid, A.M. 1996. Flora of Saudi Arabia, Vol. **1**, Fourthed. Riyadh University Publications, Riyadh, Saudi Arabia.

Mosti, S., Raffaelli, M. and Tardelli, M. 2012. Contribution to the flora of Central-Southern Dhofar (Sultanate of Oman). Webbia **67**: 65–91.

Norton, J., Abdul Majid, S., Allan, D., AlSafran, M., Böer, B. and Richer, R. 2009. An Illustrated Checklist of the Flora of Qatar. Browndown Publications, Gosport, UK, 67 pp.

Oltmann, O. 1971. Pollenmorphologisch-systematische untersuchungen innerhalb der Geraniales. Dissertation, Berlin.

Sakkir, S., Kabshawi, M. and Mehairbi, M. 2012. Medicinal plants diversity and their conservation status in the United Arab Emirates (UAE). J. Med. Plant Res. **6**: 1304–1322.

Sheahan, M.C. and Chase, M.W. 1996. A phylogenetic analysis of Zygophyllaceae based on morphological, anatomical and *rbc*L DNA sequence data. Bot. J. Linn. Soc. **122**: 279–300.

Sheahan, M.C. and Chase, M.W. 2000. Phylogenetic relationships within Zygophyllaceae based on DNA sequences of three plastid regions, with special emphasis on Zygophylloideae. Syst. Bot. **25**: 371–384.

Soliman, M.S., El-Tarras, A.S. and El-Awady, M.A. 2010. Seed exomorphic characters of some taxa from Saudi Arabia. J. Amer. Sci. **6**: 906–910.

Symanczik, S., Blanzkowski, J., Koegel, S., Boller, T., Wiemken, A. and Al-Yahya'ei, M. 2014. Isolation and identification of desert habituated arbuscular mycorrhizal fungi newly reported from the Arabian Peninsula. J. Arid Land **6**(4): 488–497.

Takhtajan, A.L. 1980. Outline of the classification of flowering plants (Magnoliophyta). Bot Rev. **46**: 225–359.

Takhtajan, A.L. 1987. Systema Magnoliophytorum. Leningrad, Nauka.

Van Huyssteen, D.C. 1937. Morphologisch-systematische Studien über die Gattung *Zygophyllum*. Dissertation. Berlin.

Van Zyl, L. 2000. A systematic revision of *Zygophyllum* in the southern African region. Ph.D. thesis, University of Stellenbosch, Stellenbosch.

Waly, N.M., Al-Ghamdi, F.A. and Al-Shamrani, R.I. 2011. Developing methods for anatomical identification of the genus *Zygophyllum* L. (Zygophyllaceae) in Saudi Arabia. Life Sci. **8**: 451–459.

Zumbruch, H.J. 1931. Über die Bedeutung des Saponins für die systematische Gliederung der Zygophyllaceen-Gattungen. Dissertation, Berlin.

THREE LICHEN TAXA NEW FOR TURKEY

Kenan Yazici[1] and André Aptroot[2]

*Biology Department, Faculty of Science, Karadeniz Technical University,
61080, Trabzon, Turkey*

Keywords: Ascolichen, *Lecanoraceae, Hymenchiaaceae; Verrucariaceae.*

Abstract

Three lichen taxa viz. – *Aspicilia asiatica* (H. Magn.) Yoshim., *Lecanora subcarnea* (Sw.) Ach. var. *soralifera* H. Magn., and *Thelidium minutulum* Körb. were identified as new to Turkey as a result of a lichenological survey in the Bitlis and Muş regions Turkey. In addition, *Lecanora subcarnea* var. *soralifera* is also new to Asia. A detail taxonomic account, notes on known distribution, substrates, and chemistry under each taxon and comparisons with morphologically similar taxa are furnished under each taxon.

Introduction

Recently, a lot of lichen taxa have been recorded for Turkey since the surveys about lichen flora are poor (Aptroot and Yazici, 2012; Arslan *et al.*, 2011; Yazici *et al.*, 2010a, b, c, 2011a, b, 2012, 2013; Karagöz and Aslan, 2012; Karagöz *et al.*, 2011; Kinalioğlu and Aptroo, 2011; Osyczka *et al.*, 2011) but more surveys are still needed of unexplored regions in the country.

Aspicilia A. Massal (*Hymeneliaceae*) contains approximately 230 species (Nordin *et al.*, 2010). *Lecanora* Ach. (*Lecanoraceae*) comprises about than 600 species (McCarth and Mallett, 2004), while *Thelidium* A. Massal (Verrucaraceae) has about 100 lichen taxa (Orange, 1991). From Turkey 42 taxa of *Aspicilia*, 105 taxa of *Lecanora*, and 4 taxa of *Thelidium* have thus far been reported. Of approximately 1650 lichen taxa that have been recorded for the country only 6 lichenized fungi have been reported from Muş Province (Yazici and Aslan, 2016a,b). On the other hand, 31 lichen species were noted from Bitlis region (Çobanoğlu, 2005; Çobanoğlu and Yavuz, 2007; Vondrak *et al.*, 2012). The present study aims at exploring the lichens in the regions of Muş and Bitlis, eastern Turkey. We report here three lichen taxa which are new records for Turkey and Asia.

Materials and Methods

The present study is based on collections from the Bitlis and Muş regions made in 2015-2016. Air-dried samples were examined with a Nikon SMZ1500 stereomicroscope and a Nikon Eclipse 80i compound light microscope. Relevant keys were consulted (Dickhäuser *et al.*, 1995; Ceynowa-Giełdon and Adamska, 2014; Orange, 2008; Thüs and Nascimbene, 2008; Poelt and Wirth, 1968; Poelt and Vĕzda, 1981) for the identifications. Vouchers are stored in the Herbarium of the Biology Department, Karadeniz Technical University, Trabzon, Turkey (KTUB). The diagnosis are based on Turkish specimens.

Study area

Muş: Center, mostly formed by vast areas of meadow and steppe, and high mountains, are mountainous by *Quercus* L. communities locally and *Salix* L. trees are rarely seen in some areas in this region (Baytop and Denizci, 1963). Muş region has a climate characterized by very cold and

---

[1]Corresponding author. Email: kcagri_1997@yahoo.com
[2]ABL Herbarium G.v.d.Veenstraat 107 NL-3762 XK Soest, The Netherlands.

very snowy winters, and hot, dry and short summers, with temperatures ranging from –29 to 41.6°C. Annual rainfall ranges from 350–1000 mm and the average humidity is 60.3% (Akman, 1999).

Bitlis region (Tatvan: Nemrut mountain and Adilcevaz) are mountainous with vast open areas, large plain and sometimes *Quercus*, *Populus* and *Salix* trees are seen in some places. Nemrut mountain is a second large extinct crater of the World. There is a lake, many rocks and trees such as *Quercus* and *Populus* (Baytop and Denizci, 1963). Thence crustose and foliose lichens are predominantly seen. Collecting localities are well-lit, windswept, treeless areas with gently sloping terrain containing streams, grass, and calcareous and siliceous rocks. The climate is characterized by very cold snowy winters and short hot dry short summers, with a temperature range of -21.3°C to 37°C, a mean annual rainfall is around 822.9 mm, and mean annual humidity of 61% (Akman, 1999).

Results

Aspicilia asiatica (H. Magn.) Yoshim., Nov.Sist. Niz. Rast. 9: 286 (1972).   **(Fig. 1).**

Thallus crustose, up to 5 cm diam, ± cycloid or ± elliptic, gray, gray-beige, with deep cracked, and areolate; areoles uneven, blistered, corrugated, areolae up to 800 μm diam; lobes thin and narrow towards the ends, ± contiguous, or with light space, sometimes ± partly overlapping, rarely dichotomic, about 165 μm, bulky, lobe tips black-brown as if burned. Apothecia up to 1.25 mm diam, regular or sometimes irregular and with depressed proper margin, aggregated mostly in the middle, scarce towards the lobes, constricted at the base, one per even fertile areol; thallin exciple more or less distinct, thick, 125 μm diam, gray, concolorous with the thallus, large; disc concav, pruinose, dark red or dark brown-black, to 900 μm diam; epihymenium yellow-brown; hypothecium 50-60 μm, yellow brown-gray; hymenium 90-100 μm; paraphyses contiguous, apices subglobe, upper part filiform. asci 8-spored, clavate, 65-75 × 18-20 μm; ascospores 17 × 10 μm, more or less ellipsoid Thallus and medulla K-, C-, KC-, P-, under upper cortex K more or less yellow-orange.

A detailed description is provided by Oxner (1972).

Aspicilia asiatica grows on calcareous rocks. Previously known from Austria, Afghanistan, Altai-Sayan, China, Mongolia, Tajikistan, Kazakhstan, Kyrgyzstan (Poelt and Wirth, 1968; Abbas *et al.*, 2001; Bredkina and Makarova, 2005; Sedelnikova, 2013). New to Turkey.

Specimen examined: Turkey. Muş: Center, between Üçevler and Muş mainroad, roadside, 38°40'49.85"N 41°25'30.87"E, 2585 m, on calcareous rock, 29.05.2015, leg. K.Yazici. (KTUB–2452).

Accompanying species were: *Aspicilia cinerea* (L. Körb.), *Acarospora fuscat* (Nyl.) Th. Fr., *Acarospora impressula* Th. Fr. var. *hospitans* (H. Magn.) Clauzade & Cl. Roux, *Candelariella vitelline* (Hoffm.) Müll..Arg., *Immersaria athroocarpa* (Ach.) Rambold & Pietschm., *Protoparmeliopsis muralis* (Schreb.) M. Choisy, *Rhizocarpon geographicum* (L.) DC., *Rhizoplaca melanophthalma* (DC.) Räsänen, *Rinodina milvina* (Wahlenb.) Th. Fr. and *Xanthoria elegans* (Link) Th. Fr.

Lecanora subcarnea (Sw.) Ach. var. **soralifera** H. Magn., Bot. Notiser: 433 (1932).   **(Fig. 2).**

Thallus crustose up to 5 cm diam, thick, more or less gray, grayish or yellowish white, epruinose, deeply cracked, areolate; areola blistered, more or less verrucose, surface uneven, margins indistinct. Apothecia up to 1.25 mm diam.; disc light red-brown, light brown or red-brown, slightly pruinose, slightly concave, P+ orange-red, C-; soralia 0.5-0.7 mm, blue-grey, ±

hemisphaerical, occurring on areola, side of apothecia, also on exciple and disc hymenium 90-95 µm high, yellow, yellow-gray, hyaline, clear; paraphyses with thickened upper cells; epihymenium greenish gray-brown; hypothecium hyaline, 150-190 µm, not oil droplets thallin exciple concolous with the thallus, smooth, entire, prominent (Fig. 2a); asci clavate, 8-spored, 40-45 × 8-10 µm; ascospores simple, hyaline, ellipsoid, 9-15 × 6-8 µm (Fig. 2c).Thallus K- or slightly yellow-brown, C-, KC-, P + orange-red. Medulla K-, C-, KC, P-. Disc P+ orange-red. Soralia spot tests are negative.

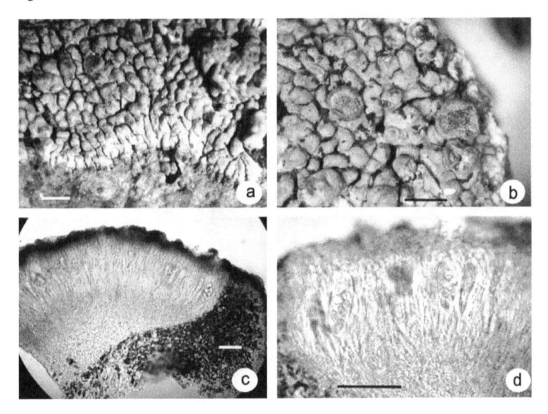

Fig. 1. *Aspicilia asiatica,* a). Thallus with lobes. Scale = 1 mm, b). Apothecia with pruinose disc. Scale = 1 mm, c). Section through apothecium with hymenium, epihymenium, hypothecium, ascus and ascospores. Scale = 50 µm, d). Section of apothecium with hymenuium, ascus and ascospores. Scale = 50 µm.

Lecanora subcarnea var. *soralifera* is a mild-temperate to Mediterranean species, mostly growing on calcareous rock, sometimes on walls. Previously known from Austria, Germany, Sweden, Norway, North America (Berger and Priemetzhofer, 2014; Dickhäuser et al.,1995; Eichler et al., 2010). New to Turkey and Asia.

A detailed descriptions are provided by Dickhäuser et al., (1995), Poelt and Vězda, (1981).

Specimen examined: Turkey, Bitlis: Tatvan, Nemrut mountain, 38°36'08.60"N 42°15' 35.18"E, 2360 m, on calcareous rock, 29.06.2016, leg. K.Yazici. (KTUB–2458).

Thelidium minutulum Körb., Parerga lichenol. (Breslau) 4: 351 (1863). **(Fig. 3).**

Thallus crustose, epilithic, thin to moderately thick 50-100 µm, continuous, grey, partly grey-brown, margin indistinct, up to 5 cm diam, lightly cracked, uneven, rough, corrugated, granular or rimose, also cracked surroundig perithecia; perithecia small, about 150-325 μm diam., 320 μm immersed in the thallus, 160 μm on the thallus, or 0.5 mm immersed 0.35 mm on the thallus, more or less globose to ovate, basal part bounded by algae layer; periphyses present; involucrellum absent or very thin; exciple dark-brown to black, about 100-180 μm diam; asci 8-spored, more or less clavate, 85-90 × 21-23 μm; ascospores colourless, ellipsoid, 17–21 × 6–8 μm, 2-celled (Fig. 3d). All spot tests are negative.

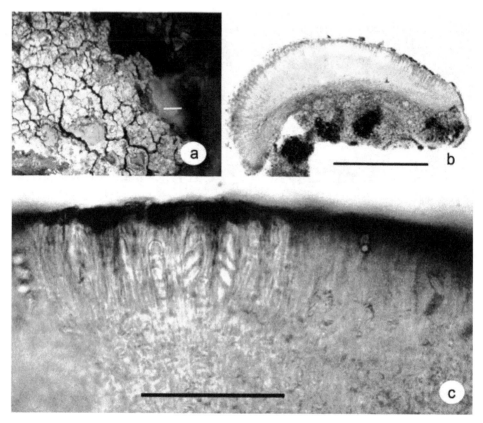

Fig. 2. *Lecanora subcarnea* var. *soralifera*, a). Thallus with apothecia and blue-gray soralia. Scale = 1 mm, b). Cross-section of apothecium with hymenium, epihymenium, hypothecium. Scale = 500 μm, c). Section of apothecium with hymenium, ascus and ascospores. Scale = 500 μm.

A detailed description is provided by Orange (2008).

Thelidium minutulum is a widespread, cool-temperate to arctic-alpine, circumpolar lichen, occuring on calcareous or siliceous rocks, metal-rich, old walls, often vertical faces, limestones, rarely on soil, sterile and grows on steeply inclined faces (Ceynowa-Giełdon and Adamska, 2014; Adamska, 2010; 2012; Ceynowa-Giełdon, 2001). It is known from throughout the Europe. Asia (Taiwan) and North America, (Freire *et al.*, 1999; Thüs and Nascimbene, 2008; Redchenko *et al.*, 2010; Vondrák *et al.*, 2010; Coste ,2011; Pykälä *et. al.*, 2012; Toetenel *et al.*, 2012; Ceynowa-Giełdon and Adamska, 2014). New to Turkey.

Specimen examined: Turkey, Bitlis: Adilcevaz, Karşıyaka village, surrounding Sodalı Lake, 38°49′26.29″N 42°57′16.60″E, 1712 m, on calcareous rock, 17.07.2016, leg. K.Yazici (KTUB–2460).

Notes: Some members of *Thelidium minutulum* can be confused with *Thelidium rehmii* Zschacke, but the thallus in *T. minutulum* is more granular than that of *T. rehmii*. The photobionts in *T. minutulum* are in small aggregated groups, while those of *T. rehmii* distributed irregularly in the thallus. Habitat of these two speceis are Also different (Ceynowa-Giełdon, 2001). Moreover this species is morphologically confused with *Verrucaria bryoctona* (Th. Fr.) Orange. However *T. Minutulum* can be distinguished from *V. bryoctona* in having 2-celled ascospores and structure of excipulum (Aslan and Yazici, 2013). Accompanying species was *Verrucaria nigrescens* Pers.

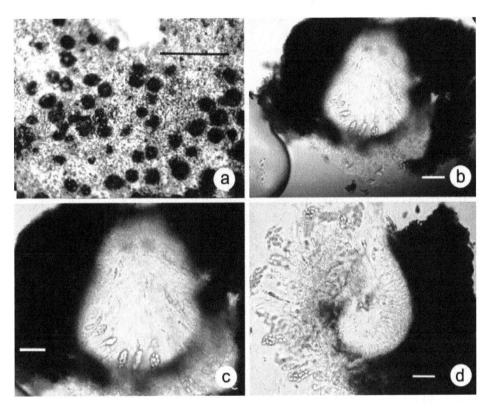

Fig. 3. *Thelidium minutulum,* a). Thallus with perithecia, habitus. Scale = 1 mm. b). Perithecium covered by algae in small group, periphyses, indistinct brown wall of perthecium, ascus and ascospores. Scale = 50 μm, c). Section through perithecium coverd by algae, ligth distict brown wall of perithecium, exciple, periphyces, ascus and ascospores. Scale = 50 μm, d.Section of perithecium with periphyses, ascus and ascospores. Scale = 50 μm.

Acknowledgements

This study was supported by TUBITAK (Project 114Z892).

References

Abbas, A., Mijit, H., Tumur, A. and Jinong, W. 2001. A Checklist of the lichens of Xinjiang, China. Harvard Papers in Botany **5** (2): 359–370.

Adamska, E. 2010. Biota of lichens on the Zadroże Dune and its immediate surroundings, Ecological Questions **12**: 51–28.

Adamska, E. 2012. Protected and threatened lichens in the city of Toruń. In Lipnicki, L. (Ed.) Lichens Protection–Protected Lichen Species, Sonar Literacki, Gorzów Wielkopolski, pp. 313–323.

Akman, Y. 1999. Climate and bioclimate (The methods of bioclimate and climate types of Turkey). 1st Edn., Kariyer Matbaacılık Ltd., Şti, Ankara. 350 pp.

Aptroot, A. and Yazıcı, K. 2012. A new *Placopyrenium* (Verrucariaceae) from Turkey. The Lichenologist **44**: 739–741.

Arslan, B., Öztürk, S. and Oran, S. 2011. *Lecanora, Phaeophyscia* and *Rinodina* species new to Turkey. Mycotaxon **116**: 49–52.

Aslan, A. and Yazici, K. 2013. New *Lecanora, Lecidea, Melaspilea, Placynhium,* and *Verrucaria* records for Turkey and Asia. Mycotaxon **123**: 321–326.

Baytop, A. and Denizci, R. 1963. Türkiye'nin Flora ve Vejetasyonuna Genel Bir Bakış. Ege Üniv. Fen Fak. Monografiler Ser. 1, Ege Üniv. Mat., İzmir. 43 p.

Berger, F. and Priemetzhofer, F. 2014. Erläuterungen und Erstnachweise von Flechten in Oberösterreich, sowie weitere erwähnenswerte Beobachtungen. 1. Update des Flechtenatlas. Stapfia **101**: 53–65.

Bredkina, L .I. and Makarova, I.I. 2005. Checklist of lichens of the central Tian Shan (Kyrgyzstan). Academia Scientarium Rossica **39**: 199–218.

Ceynowa-Giełdon, M 2001. Kalcyfilne porosty naziemne na Kujawach (Calciphilous terricolous lichens in Kujawy), Wydawnictwo Uniwersytetu Mikołaja Kopernika, Toruń.

Ceynowa-Giełdon, M., Adamska, E. 2014. Notes on the genus *Thelidium* (Verrucariaceae, lichenized Ascomycota) in the Kujawy region (north-central Poland). Ecological Questions **19**: 25–33.

Coste, C. 2011. Aperçu de la flore et de la végétation lichéniques de la réserve biologique intégrale du cirque de Madasse (Forêt domaniale du causse Noir, Aveyron). Bull. Soc. Hist. Nat. Toulouse,pp. 1–25.

Çobanoğlu, G. 2005. Lichen collection in the Herbarium of the University of Istanbul (ISTF). Turkish J. Bot. **29**: 69–74.

Çobanoğlu, G.and Yavuz, M. 2007. Muzeul Oltenici Craiova. Oltenia. Studii şi comunicări. Ştiintele Naturii. Tom **23**: 23-26.

Dickhäuser, A., Lumbsch, H.T. and Feige, G.B. 1995. A synopsis of the Lecanora subcarnea group. Mycotaxon **56**: 303–328.

Eichler, M and Cezanne R and Teuber, D 2010. Ergänzungen zur Liste der Flechten und flechtenbewohnenden Pilze Hessens. Zweite Folge Botanik und Naturschutz in Hessen **23**: 89–110.

Freire, M., Dopaza, M.F. and Molares, A.G. 1999. Flora liquenica saxicola y arenícola de la Península de o Grove (Pontevedra, NW de España. Acta Botanica Malacitana **24**: 13–25.

Karagöz, Y. and Aslan, A. 2012. Floristic lichen records from Kemaliye District (Erzincan) and Van Province. Turkish J. Bot. **36**: 558–565.

Karagöz, Y., Aslan, A., Yazıcı, K. and Aptroot, A. 2011. *Diplotomma, Lecanora,* and *Xanthoria* lichen species new to Turkey. Mycotaxon **115**: 115–119.

Kinalioğlu, K. and Aptroot, A. 2011. *Carbonea, Gregorella, Porpidia, Protomicarea, Rinodina, Solenopsora,* and *Thelenella* lichen species new to Turkey. Mycotaxon **115**: 125–129.

McCarth, P.M. and Mallett, K. 2004. Flora of Australia. CSIRO Publishing Vol. **56**. A. Lichens 4, Canberra.

Nordin, A., Savić, S. and Leif Tibell, L. 2010. Phylogeny and taxonomy of *Aspicilia* and *Megasporaceae*. Mycologia **102**(6): 1339–1349.

Orange, A. 1991. *Thelidium pluvium* (*Verrucariaceae*), a new lichenized species from north-west Europe, Lichenologist **23**: 99–106

Orange, A. 2008. British Pyrenocarpous Lichens. 69 pp. Distributed by the author.

Osyczka, P., Yazici, K. and Aslan, A. 2011. Note on *Cladonia* species (lichenized *Ascomycota*) from Ardahan Province (Turkey). Acta Societatis Botanicorum Poloniae **80**: 59–62.

Oxner, A.N. 1972. Conbinationes taxonomicae ac nomina specierum Aspiciliae novae. Novosti Sistematiki Nizshikh Rastenii **9**: 286–292

Poelt, J and Wirth, V. 1968. Flechten aus dem Nordöstlichen Afghanistan. Mitt, Bot. München Band **7**: 219–261.

Poelt, J. and Vězda, A. 1981. Bestimmungsschlüssel europäischer Flechten Ergänzungsheft II. J Cramer, Vaduz. 390 p.

Pykälä, J., Stepanchikova, I.S., Himelbrant, D.E., Kuznetsova, E.S. and Alexeeva, N.M. 2012. The lichen genera *Thelidium* and *Verrucaria* in the Leningrad Region (Russia). Folia Cryptog. Estonica Fasc. **49**: 45–57.

Redchenko, O. and Košnar, J. and Gloser, J. 2010. A contribution to lichen biota of the central part of Spitsbergen, Svalbard Archipelago. Polish Polar Research **31**(2): 159–168.

Sedelnikova, N.V. 2013. Species diversity of lichen biota of the Altai-Sayan ecological region Растительный мир Азиатской России **2**(12с): 12–54

Thüs, H. and Nascimbene J. 2008. Contributions toward a new taxonomy of Central European freshwater species of the lichen genus *Thelidium* (*Verrucariales*, Ascomycota). The Lichenologist **40**(6): 499-521.

Toetenel, H., Aptroot, A. and Sparrius, L. 2012. De licheenflora van de Kop van Schouwen: een vergelijking over vier decennia. Buxbaumiella **93**: 6–21.

Vondrák, J., Halda, J.P., Meliček, J. and Müller, A. 2010. Lichens recorded during the Spring bryo-lichenogical meeting in Chriby Mts (Czech Republic). Bryonora **45**: 36–42.

Vondrák, J., Halici, M.G., Kocakaya, M. and Ondrakova, O.V. 2012. *Teloschistaceae* (lichenized *Ascomycetes*) in Turkey. 1. Some records from Turkey. Nova Hedwigia **94**: 385–396.

Yazici, K., Aptroot, A., Aslan, A., Etayo, Spier, J. and Karagöz, Y. 2010a. Lichenized and lichenicolous fungi from nine different areas in Turkey. Mycotaxon **111**: 113–116.

Yazici, K., Aptroot, A. and Aslan, A. 2010b. Three lichenized fungi new to Turkey and the Middle East. Mycotaxon **111**: 127–130.

Yazici, K., Elix, J.A. and Aslan, A. 2010c. Some parmelioid lichens new to Turkey and Asia. Mycotaxon **111**: 489–494.

Yazici, K., Aptroot, A., Aslan, A., Vitikainen, O. and Piercey-Normore, M.D. 2011a. Lichen biota of Ardahan province (Turkey). Mycotaxon **116**: 480.

Yazici, K., Aptroot, A. and Aslan, A. 2011b. *Lecanora wrightiana* and *Rhizocarpon inimicum*, rare lichens new to Turkey and Middle East. Mycotaxon **117**: 145–148.

Yazici, K., Aptroot, A. and Aslan, A. 2012. *Candelariella*, *Ochrolechia*, *Physcia*, and *Xanthoria* species new to Turkey. Mycotaxon **119**: 149–156.

Yazici, K., Aslan A. and Aptroot, A. 2013. New lichen records from Turkey. Bangladesh J. Plant Taxon. **20**(2): 207–211.

Yazici, K. and Aptroot, A. 2015. *Buellia*, *Lempholemma*, and *Thelidium* species new for Turkey and Asia. Mycotaxon **130**: 701–706.

Yazici, K. and Aslan, A. 2016a. *Aspicilia*, *Lobothallia*, and *Rhizocarpon* species new for Turkey and Asia. Mycotaxon **131**: 227–233.

Yazici, K. And Aslan A 2016b. *Merismatium*, *Porpidia* and *Protoparmelia* spp. new for Turkey and Asia. Mycotaxon **131**: 337–343.

NOTES ON THE GENUS *TYLOPHORA* R. BR. (ASCLEPIADACEAE) OF INDIA

L. Rasingam[1], J. Swamy and S. Nagaraju

Botanical Survey of India, Deccan Regional Centre, Plot. No. 366/1, Attapur, Hyderguda Post, Hyderabad-500048, Telangana, India

Keywords: Tylophora; New subsp. *andamanica*; New combination; Andaman Islands: India.

Abstract

A new subspecies, *Tylophora perakensis* King & Gamble subsp. *andamanica* is described and illustrated from Little Andaman Island, Andaman and Nicobar Islands, India. A new combination, *Tylophora hookeriana* is proposed and the distributional status of *T. indica* Merr. var. *intermedia* M.A. Rahman & Wilcock is also discussed based on the fresh collections from Andaman Islands.

Introduction

The genus *Tylophora* R. Br. (Asclepiadaceae) consists of c. 60 species, and distributed mainly in tropical and subtropical Asia, Africa and Australia (Tseng and Chao, 2011; Murugan and Kamble, 2012). In India, *Tylophora* is represented by 21 species and two varieties (Jagtap and Singh, 1999; Karthikeyan *et al.*, 2009) and recently one more species was described from the Andaman and Nicobar Islands (Murugan and Kamble, 2012). At present *Tylophora* R. Br. is known to be represented in Andaman and Nicobar Islands by four species *viz.*, *T. globifera* Hook. f., *T. indica* (Burm. f.) Merr., *T. flexuosa* R. Br. and *T. nicobarica* Murugan & M.Y. Kamble.

While working on the flora of Andaman Islands, the first author collected an interesting specimen of *Tylophora* R. Br. from the evergreen forests of Little Andaman Island. Critical examination of the specimens and survey of relevant literature revealed that it is morphologically similar to *T. perakensis* King & Gamble but differs from shape and size of corolla which warrants sufficiently to recognize subspecies of it. Hence, it has been described as a new subspecies under *T. perakensis* King & Gamble.

Tylophora perakensis King & Gamble subsp. **andamanica** L. Rasingam & J. Swamy, **subsp. nov.** (Fig. 1).

Diagnosis: The new subspecies *T. perakensis* subsp. *andamanica* is similar to the typical subspecies, *perakensis* by its vegetative characters, but differs in the broadly ovate, acuminate corolla lobes with 7–9 veins (vs. oblong, obtuse corolla lobes with 3–5 veins).

Type: INDIA, Andaman and Nicobar Islands, Little Andaman Island, on the way to Ramkrishnapur dam, 19.5.2008, *L. Rasingam 25991* (*Holotype*: CAL; *Isotypes*: PBL).

Paratype: INDIA, Andaman and Nicobar Islands, South Andamans, Cadell-gunj hill jungle, 25.7. 1891, *Dr. King s.n.* (CAL!).

A climbing shrub, up to 3 m long; branchlets fleshy, striate, twisted, pale brown when dry; internodes 13–15 cm long, glabrous. Leaves simple, opposite-decussate, ovate, 11–15×8–9 cm, apex acuminate; acumen up to 1 cm long, base deeply cordate, margin entire, membranous, glabrous; midrib slender, raised beneath, bearing a small cluster of glands at the base just above the petiole; lateral veins 9 or 10 pairs, curving upwards to anastomose near margin with an obscure looped vein; petiole upto 4.3 cm long, fleshy, glabrous. Inflorescence axillary or lateral

---

[1]Corresponding author. Email: rasingam@gmail.com

between the petioles, up to 16×14 cm, as long as or longer than the leaves, divaricately branched, glabrous; peduncles 1–3 cm long, pubescent when young, later glabrescent. Flowers 5 to 11 in umbellate clusters; bracts minute, c. 0.6 mm; pedicels 6–8 mm long, striate; buds ovoid, c. 3.5×1.5 mm. Calyx lobes ovate, c. 1.3×1.1 mm, apex acute, 5–7-veined, margin ciliate. Corolla campanulate-rotate; tube 0.6–0.7 mm long; lobes broadly ovate, acuminate, c. 3.5×2.2 mm, 7-veined, hairy inside, pubescent outside. Corona processes subglobose, c. 1×2 mm, fleshy, shorter than the anthers, point small, appressed to the anthers. Anthers slender above; appendages lanceolate, up to 1.1 mm long, acuminate; pollen masses globose, very minute, attached by slender, straight caudicles to the minute pollen-carriers. Style apex pentagonal, c. 1.1×1.0 mm, with convex top. Gynoecium bicarpellate, c. 1.2×0.1 mm, pubescent.

Flowering and fruiting: May to August.

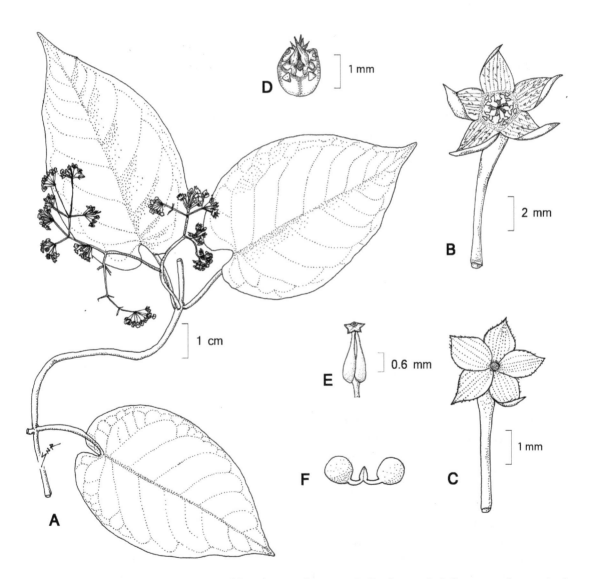

Fig. 1. *Tylophora perakensis* King & Gamble subsp. *andamanica* L. Rasingam & J. Swamy, **subsp. nov**. A. Flowering twig; B. Flower; C. Calyx; D. Corona process; E. Gynoecium; F. Pollinia.

Habitat: Very rare on the edges of inland evergreen forests.

Distribution: India: Andaman and Nicobar Islands, Little Andaman and South Andaman Islands.

Etymology: The infraspecific epithet is named after the type locality, the Andaman Islands.

Nomenclatural notes on *Tylophora macrantha* (Wight) Hook. f.

T. macrantha (Wight) Hook. f. is an endemic species known from Andhra Pradesh, Kerala, Madhya Pradesh and Tamil Nadu (Karthikeyan *et al.*, 2009). It was described by Robert Wight (1834) as a variety under *T. fasciculata* Buch.-Ham. *ex* Wight. Later, Hooker (1883) raised it to species level in his Flora of British India without knowing the name *T. macrantha* Hance (1882), has already been used for a species described from HongKong. Hence, *T. macrantha* (Wight) Hook. f. became an illegitimate later homonym, and for which, Kuntze (1891) proposed a new name, *Vincetoxicum hookerianum*. However, while critically studying the morphological features of the type specimen and other voucher specimens housed in CAL, MH and BSID herbaria, it is strongly felt that this species should be treated under *Tylophora* rather under *Vincetoxicum* as the diagnostic features fall well within the circumscription of *Tylophora*. Therefore, a new combination is proposed here.

Tylophora hookeriana (Kuntze) L. Rasingam & J. Swamy, comb. nov.

Basionym: *Vincetoxicum hookerianum* Kuntze, Revis. Gen. Pl. 2: 424 (1891).

Tylophora fasciculata Buch.-Ham. *ex* Wight var. *macrantha* Wight, Contr. Bot. India: 50 (1834). *T. macrantha* (Wight) Hook. f., Fl. Brit. India 4: 40 (1883), non *T. macrantha* Hance in J. Bot. 20: 79 (1882).

Type: Neelgherry, *Wight Numer. List No.1540* (K, image!).

Distribution: Endemic to India: Andhra Pradesh, Kerala, Madhya Pradesh and Tamil Nadu.

Additional specimens examined: INDIA, Telangana, Amrabad Tiger Reserve, Mallayalodhi, 22.9.2013, *L. Rasingam & M. Sankara Rao 3721* (BSID); Andhra Pradesh, Srisailam Tiger Reserve, Istakameshwaram, 24.9.2014, *L. Rasingam & M. Sankara Rao 5771* (BSID); Tamil Nadu, Nilgiri district, Pykara fall, June 1884, *J.S. Gamble 14239* (CAL), Coimbatore district, Dhimbam, 28.5.1905, *C.E.C. Fischer 63* (CAL); North Arcot district, Vasanthapuram RF, 20.11.1977, *E. Vajravelu 51989* (MH).

Note: *T. hookeriana* resembles *T. fasciculata* by its habit and vegetative characters but differs by its flower length. The flowers of *T. hookeriana* are c. 7 mm long, whereas in *T. fasciculata* the flowers are up to 4 mm long.

Distributional notes on *Tylophora indica* var. *intermedia* M.A. Rahman & Wilcock

T. indica (Burm. f.) Merr. var. *intermedia* M.A. Rahman & Wilcock was described from Bangladesh and reported from India by Rahman and Wilcock (1989) based on the collections of Wight (without any locality, *Wight prop. n. 1548*) and Sedgwick and Bell [Bihar (Kasmar, sandy Sea shore, *Sedgwick* and *Bell 5084, 6746*)] preserved at K. While revising the family Asclepiadaceae for India, Jagtap and Singh (1999) doubted about its distribution and stated "As there is no Sea shore in Bihar state, its distribution is doubtful in India". Further, there is no report on the variety from Indian region after it was described. However, during the documentation of floral wealth of Mount Harriet National Park of Andaman and Nicobar Islands the first author had collected this variety from the Sea shores of the National Park, thus the collection confirms its distribution in India. This variety differs from var. *glabra* (Decne.) H. Huber by its pubescent stems and inflorescences and from its typical variety *indica* by its glabrous corolla lobes and lower surface of leaves.

Tylophora indica (Burm. f.) Merr. var. **intermedia** M.A. Rahman & Wilcock in J. Econ. Taxon. Bot. 13(1): 184 (1989); Rahman & Wilcock in Khan & Rahman, Fl. Bangladesh 48: 55 (1995); Jagtap & Singh, Fasc. Fl. India 24: 161 (1999). **(Fig. 2)**.

A climbing shrub, up to 2 m long; branches minutely ridged, pale brown when dry; internodes 7–11 cm long, pubescent, sparsely hairy at nodes. Leaves simple, opposite-decussate, ovate or ovate-oblong, 2–5×1.0–2.6 cm, base cordate, asymmetric and hairy, margin entire, apex acuminate and mucronate, glabrous; lateral veins 4 or 5 pairs; petioles terete, 5–8×c.1 mm, sparsely hairy. Inflorescence axillary umbels, as long as or slightly longer than the leaves; peduncles angled, 5–9×0.4–0.5 mm, arising between petioles nearer to one of them, hairy; bracts linear, c. 4.0×0.5 mm, apex acute, sparsely hairy; bracteole subulate, 2.0–2.5×c.0.3 mm, margin sparsely hairy; pedicels filiform, 5–27 mm long, hairy. Flowers 6–7×c. 4 mm. Calyx lobes 5, free, slightly attached at base, linear-lanceolate, 2.8–3.0×0.4–0.9 mm, hairy outside, 4 or 5-veined. Corolla rotate, c. 8.2×9.2 mm; lobes 5, united for 2.0–2.8 mm, ovate, 5.2–6.0×2.6–2.7 mm, apex obtuse, glabrous, c. 10-veined. Corona process ovate-oblong, c. 1.2×2.0 mm, uniseriate, 5-lobed, adnate below the staminal column with free points above. Stamens 5, c. 1.0×0.6 mm; pollinia 5, pollen masses solitary in each anther cell, c. 190×130 μm, waxy, yellow, attached by c. 80 μm long, brown caudicles; corpusculum c. 110×70 μm, dark brown. Gynostegium c. 2.2 mm long; carpels 2, c. 1.6×1.0 mm, glabrous; style apex pentagonal, c. 1.1×0.4 mm.

Flowering and fruiting: March to September.

Habitat: Very rare along the littoral forests.

Distribution: India (Andaman Islands), Bangladesh, Myanmar and Sri Lanka.

Specimen examined: INDIA, Andaman and Nicobar Islands, Mount Harriet National Park, near north Bay, 10.3. 2007, *L. Rasingam 2982* (BSID).

Key to the *Tylophora* species in Andaman and Nicobar Islands

1. Leaves deeply cordate at base — *T. perakensis* subsp. *andamanica*
- Leaves truncate, rounded to cordate at base — 2
2. Flowers small, in much branched panicles or corymbose cymes — *T. tetrapetala*
- Flowers large, in simple or rarely branched umbellate cymes and racemes — 3
3. Flowers in simple or branched umbellate racemes — *T. nicobarica*
- Flowers in simple umbellate cymes — 4
4. Plants quite glabrous; sepals ovate-lanceolate; coronal scales globose, very large — *T. globifera*
- Plants pubescent; sepals linear or linear-lanceolate; coronal scales ovoid, small — 5
5. Lower surface of the leaves and corolla lobes glabrous — *T. indica* var. *intermedia*
- Lower surface of the leaves and corolla lobes pubescent — *T. indica* var. *indica*

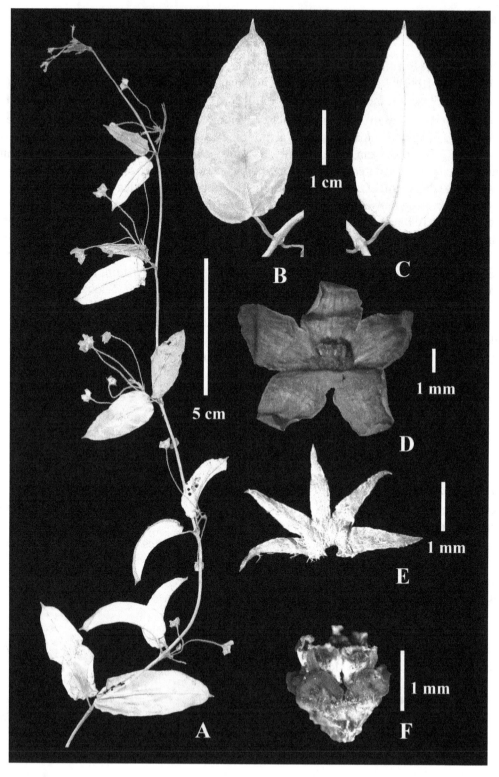

Fig. 2. *Tylophora indica* (Burm. f.) Merr. var. *intermedia* M.A. Rahman & Wilcock A. Flowering twig; B. Leaf – adaxial view; C. Leaf – abaxial view; D. Open flower; E. Calyx; F. Gynostegium.

Acknowledgements

The authors are grateful to Dr. P. Singh, Director, Botanical Survey of India, Kolkata and Dr. M. Ahmedullah, Scientist-E, Botanical Survey of India, Deccan Regional Centre, Hyderabad for facilities and encouragements. We are also thankful to the officials and field assistants of Andaman and Nicobar Forest Plantation & Development Corporation Ltd., and Hut Bay for field support.

References

Hance, H.F. 1882. A decade of new Hong-Kong Plants. J. Bot. **20**: 77–80.

Hooker, J.D. 1883. Asclepiadaceae. *In*: Hooker J.D. (Ed.), The Flora of British India. Vol. **4**. L. Reeve & Co., London, pp. 1–78.

Jagtap, A. and Singh, N.P. 1999. Asclepiadaceae and Periplocaceae. Fascicle of Flora of India. Fascicle **24**: 1–332. Botanical Survey of India, Calcutta.

Karthikeyan, S., Sanjappa, M. and Moorthy, S. 2009. Flowering Plants of India. Dicotyledons. Volume **1** (Acanthaceae–Avicenniaceae). Botanical Survey of India, Kolkata.

Kuntze, O. 1891. Revisio Generum Plantarum: vascular iumomnium at quecellular iummultarum secund umleges nomeclaturae internationales cum enumerationeplantarumexoticarum in itinere mundi collectarum, Pars II. Dulau & Co, 37, Soho Square, London, p. 424.

Murugan, C. and Kamble, M.Y. 2012. A new species of *Tylophora* (Apocynaceae–Asclepiadoideae – Asclepiadeae) from the Nicobar Islands, India. Rheedea **22**(2): 83–87.

Rahman, M.A. and Wilcock, C.C. 1989. Notes on tropical Asian Asclepiadaceae – II. J. Econ. Taxon. Bot. **13**(1): 181–185.

Tseng, Y.H. and Chao, C.T. 2011. *Tylophora lui* (Apocynaceae), a new species from Taiwan. Ann. Bot. Fennici **48**: 515–518.

Wight, R. 1834. Asclepiadeae Indicae. Contributions to the Botany of India, Parbury, Allen & Co., London, pp. 29–67.

PHYLOGENY OF *GALIUM* L. (RUBIACEAE) FROM KOREA AND JAPAN BASED ON CHLOROPLAST DNA SEQUENCE

KEUM SEON JEONG, JAE KWON SHIN[1], MASAYUKI MAKI[2] AND JAE-HONG PAK[3]

Division of Forest Biodiversity, Korea National Arboretum, Pocheon, Gyeonggi-do 487-821, Korea

Keywords: Chloroplast DNA; Korean-Japan *Galium;* Molecular data; Phylogeny.

Abstract

The present paper deals with the phylogeny and inter-and intragenic relationships using four chloroplast DNA sequences within 19 *Galium* L. species from Korea and Japan. Maximum parsimony and Bayesian analyses were conducted to clarify the relationships among the section and species. The strict consensus tree had three main clades. Clade I comprises of the only individuals of *G. paradoxum* Maximowicz (sect. *Cymogalia*), which is distinguished by opposite leaves in the genus, supported by the 100% bootstrap value (PP: 0.98); Clade II consists of members of eight sections (sect. *Galium*, sect. *Hylaea*, sect. *Kolgyda*, sect. *Trachygalium,* sect. *Leptogalium*, sect. *Orientigalium*, sect. *Aparine*, and sect. *Leiogalium*); Clade III comprises members of eight sections (sect. *Baccogalium*, sect. *Lophogalium*, sect. *Platygalium*, sect. *Relbunium*, sect. *Depauperata*, sect. *Aparinoides,* sect. *Leiogalium* and *Trachygalium*). The sect. *Leptogalium* which includes two taxa namely *G. tokyoense* Makino and *G. dahuricum* var. *lasiocarpum* (Makino) Nakai is paraphyletic. Four taxa of *Trachygalium* group (*G. trachyspermum* A. Gray, *G. gracilens* (A. Gray) Makino, *G. pogonanthum* Franch. & Sav., *G. koreanum* Nakai) were placed from sect. *Cymogalia* to sect. *Platygalium* based on molecular and morphological data.

Introduction

Galium L., the largest genus of the tribe Rubieae in the family Rubiaceae (Robbrecht and Manen, 2006), is taxonomically diverse and comprises over 650 species (Govaerts, 2006). *Galium* is divided into 16 sections based on characters of leaf and fruit by Ehrendorfer *et al.* (2005). The species of *Galium* are distributed centrally in temperate regions and are mostly annual and perennial herbaceous plants. The genus is characterized by more than two leaf-like whorls, number of divided petal, rudimentary calyx and a two locular ovary.

Phylogenetic relationships among species of tribe Rubieae including eleven genera have been studied by many researchers (Ehrendorfer *et al.*, 1994, 2014; Manen *et al.*, 1994; Manen and Natali, 1995; Natali *et al.*, 1995, 1996; Soza and Olmstead, 2010). Molecular phylogenetic studies using chloroplast DNA *atp*B-*rbc*L intergenic region have shown monophyly of the tribe Rubieae with seven major clades, and confirmed that genera *Asperula* and *Galium* is not a monophyletic group (Manen *et al.*, 1994; Natali *et al.*, 1995, 1996). Soza and Olmstead (2010) conducted more clearly molecular phylogenetic analysis of tribe Rubieae using three chloroplast DNA makers and their results indicated that *Galium* is polyphyletic, and species of *Galium* occur in three major clades (Clades III, V, VII). Recently, phylogenetic relationships study of tribe Rubieae including

---

[1] Division of Forest Resource Conservation, Korea National Arboretum, Pocheon, Gyeonggi-do 487-821, Korea.
[2] Division of Ecology and Evolutionary Biology, Graduate School of Life Sciences, Tohoku University, Aoba, Sendai 980-8578, Japan.
[3] Research Institute for Dok-do and Ulleung-do Island, Kyungpook National University, Daegu 702-701, Korea. Corresponding author. Email: jhpak@knu.ac.kr

some *Galium* species by Ehrendorfer *et al.* (2014) has evaluated that genus *Galium* is paraphyletic. Although there have been several phylogenetic study to investigate relationships of tribe Rubieae, very little is known about phylogenetic relationships among Korean species of *Galium*. Soza and Olmstead (2010) determined the phylogenetic relationships among Rubieae including members of *Galium* but this study included only three common species distributed in Korea and Japan. In Korea, twenty taxa of seven sections are currently recognized (Lee, 1995; Lee, 1979; Lee, 2004). *G. koreanum* Nakai, *G. verum* var. *asiaticum* for. *pusillum* (Nakai) M. Park are endemic to Korea and latter species is restrictedly distributed in Mt. Halla of Jeju Island. *G. kikumugura* Ohwi is broadly expanded to Japan. Jeong and Pak (2009, 2012) conducted morphological and somatic chromosome number counts of Korean *Galium*. These studies however, provided very little phylogenetic relationships among the species. Therefore, further studies are needed to understand their phylogenetic relationships among Korean *Galium* species and taxonomic position of Korean and Japan taxa within the *Galium* spp. occurring worldwide. This study aims to clarify inter-and intragenic relationships within Korean and 10 Japanese *Galium* species, and to determine the taxonomic position of Korean endemic taxa within the closely related *Galium* spp. using the chloroplast DNA sequences.

Materials and Methods
Plant materials

Total 19 species of *Galium* distributed in Korea and Japan were collected (Table 1). We selected two outgroup taxa [*Didymaea alsinoides* (Cham. and Schltdl.) Standl., and *Rubia cordifolia* L.] based on the results of the analyses of Soza and Olmstead (2010). The sequences of *Galium* and outgroups obtained from National Center for Biotechnology Information (NCBI) database with the exception of sequences of sample from Korea-Japan. All sources and voucher specimens of materials were deposited at the Herbarium of Kyungpook National University (KNU).

DNA extraction, amplification and sequencing

Total genomic DNA was extracted from fresh leaf tissues and field-collected silica-gel dries tissue using the 2 % hexa decyltrimethyl ammonium bromide (CTAB) procedure (Doyle and Doyle, 1987). We amplified the *rpo*B-*trn*C region and *trn*C-*ycf*6 region with primers designed by Demesure *et al.* (1995). The *trn*L-*trn*F-*ndh*J region was amplified using primers published in Taberlet *et al.* (1991) and Shaw *et al.* (2007) (Table 2). Polymerase chain reaction (PCR) conditions were an initial denaturation of 94°C for 5 min, 35 cycles of 94°C denaturation for 30 s, 48°C-57°C annealing for 30 s extension for 1m, and final extension at 72°C for 10 min. PCR products were purified using the QIAquick PCR purification kit following the instructions of the manufacturer. Sequencing reactions were carried out for the purified PCR products using Big Dye Terminator Cycle Sequencing reagents (Applied Biosystem, Foster city, CA, USA). For sequencing, we used the same primers as those used for PCR. All sequences have been deposited in GenBank (Table 1).

Data analysis

The DNA sequences were aligned with Clustal X (Thompson *et al.*, 1997). All chloroplast regions were combined and analyzed using Maximum Parsimony (MP) and the Bayesian analyses. Gaps introduced from the alignment were treated as missing characters in subsequent analyses. MP analyses were conducted in a PAUP* (version 4.0b 10; Swofford, 2003) using a heuristic searches with TBR branch swapping and MULTREES option. Relative support of various monophyletic groups revealed in the most parsimonious trees was examined with the bootstrap

Table 1. Sampling sites of plant materials used for phylogenetic analyses.

Taxon	Locality	Voucher	GenBamk acc. No.			
			trnC-ycf6	trnF-ndhJ	TrnL	rpoB-trnC
Sect. *Aparine*						
Galium spurium var. *echinospermon*	Chilgok-gun, Korea	J20050310	KC339150	KC339020	KC339085	LC062539
Sect. *Aparinoides*						
G. trifidum	Jeju-si, Korea	J20060807	KC339148	KC339018	KC339083	LC062537
	Tokyo metro, Japan	M20100501	KC339149	KC339019	KC339084	LC062538
Sect. *Cymogalia*						
G. paradoxum	Pyeongchang-gun, Korea	J20090814	KC339164	KC339034	KC339099	LC062552
	Jeongseon-gun, Korea	J20050618	KC339163	KC339033	KC339098	LC062551
	Muju-gun, Korea	J20100844	KC339162	KC339032	KC339097	LC062550
Sect. *Hylaea*						
G. trifloriforme	Ulleung-gun, Nari, Korea	J20080621	KC339204	KC339074	KC339139	LC062581
	Ulleung-gun, Korea	J20080635	KC339203	KC339073	KC339138	LC062580
	Ulleung-gun, Taehwa, Korea	J20080603	KC339205	KC339075	KC339140	LC062582
	Miyagi, Japan	J20100748	KC339206	KC339076	KC339141	LC062583
G. japonicum	Ulleung-gun, Nari, Korea	J20080611	KC339207	KC339077	KC339142	LC062584
	Ulleung-gun, Nari, Korea	J20080612	KC339151	KC339021	KC339086	LC062540
	Jeongeup-si, Korea	J20100845	KC339209	KC339079	KC339144	LC062585
	Jeju-si, Korea	J20070901	KC339210	KC339080	KC339145	LC062586
	Miyagi, Japan	J20100758	KC339211	KC339081	KC339146	LC062587
	Yamagata, Japan	J20100759	KC339212	KC339082	KC339147	LC062588
Sect. *Leptogalium*						
G. dahuricum var. *lasiocarpum*	Namyangju-si, Korea	J20100897	KC339189	KC339059	KC339124	LC062569
	Pyeongchang-gun, Korea	J20090807	KC339192	KC339062	KC339127	LC062571
	Yeongwol-gun, Korea	J20080926	KC339155	KC339025	KC339090	LC062543
	Seongju-si, Korea	J20100658	KC339188	KC339058	KC339123	LC062568
	Jecheon-si, Korea	J20091021	KC339190	KC339060	KC339125	LC062570
	Yamagata, Japan	J20100708	KC339194	KC339064	KC339129	LC062573
G. kikumugura	Mt. Zao, Japan	J20100765	KC339200	KC339070	KC339135	LC062577
G. pseudoasprellum	Miyagi, Japan	J20100789	KC339202	KC339072	KC339137	LC062579
G. tokyoense	Pocheon-si, Korea	J20070938	KC339195	KC339065	KC339130	LC062572
	Pocheon-si, Korea	J20090808	KC339193	KC339063	KC339128	LC062574
	Tokyo metro, Japan.	M20090503	KC339197	KC339067	KC339132	LC062575
Sect. *Platygalium*						
G. boreale	Yeongwol-gun, Korea	J20050625	KC339152	KC339022	KC339087	LC062541
	Mongolia	L20090830	KC339153	KC339023	KC339088	LC062542
G. gracilens	Sunchen-si, Korea	J20090801	KC339181	KC339051	KC339116	LC062566
	Hwasun-gun, Korea	J20090830	KC339180	KC339050	KC339115	LC062565
G. kamtschaticum var. *yakusimense*	Jeju-si, Korea	J20070907	KC339166	KC339036	KC339101	LC062553
G. koreanum	Sancheong-gun, Korea	J20100808	KC339186	KC339056	KC339121	LC062567
G. kinuta	Yeongwol-gun, Korea	J20050626	KC339167	KC339037	KC339102	LC062554
G. pogonanthum	Hamyang-gun, Korea	J20090504	KC339172	KC339042	KC339107	LC062559
	Jeju-si, Korea	J20050706	KC339171	KC339041	KC339106	LC062558
G. trachyspermum	Inje-gun, Korea	J20080906	KC339170	KC339040	KC339105	LC062546
	Andong-si, Korea	J20070751	KC339157	KC339027	KC339092	LC062545
	Gyeongju-si, Korea	J20100913	KC339159	KC339029	KC339094	LC062547
	Geoje-si, Korea	J20090327	KC339156	KC339026	KC339091	LC062544
	Miyagi, Japan	J20100723	KC339160	KC339030	KC339095	LC062548
	Yamagata, Japan	J20100747	KC339161	KC339031	KC339096	LC062549
Sect. *Galium*						
G. verum var. *asiaticum*	Geoje-si, Korea	J20100524	KC339173	KC339043	KC339108	LC062563
	Jeju-si, Korea	J20090685	KC339174	KC339044	KC339109	LC062562
	Fukui, Japan	M20100503	KC339176	KC339046	KC339111	LC062561
G. verum var. *trachycarpum* f. *nikkoense*	Ulsan metro., Korea	J20050830	KC339198	KC339068	KC339133	LC062576
	Tokushima, Japan	J20100732	KC339177	KC339047	KC339112	LC062564
G. verum var. *asiaticum* f. *pusillum*	Jeju-si, Korea	J20050807	KC339175	KC339045	KC339110	LC062560

method (Felsenstein, 1985). Bootstrap values were calculated from 1,000 replicates with the random addition and heuristic search options. The Bayesian phylogenetic analyses were conducted with MrBayesver 3.1.2 (Ronquist and Huelsenbeck, 2003). The suitable model was determined to be GTR+I+G for combined sequence data by MrModeltest 2.3 (Nylander, 2004). Each Morkov chain was started from a random tree and run for 1,000,000 generations, sampling a tree every 100 generations. Burn-in time was estimated from the plot of likelihoods generated using the 'sump' command in MrBayes. Posterior probabilities (pp) were based on analysis of post-burn-in tree. Nodes were considered highly supported when pp values were higher than 0.95 (Felesenstein, 1985).

Results and Discussion
Sequence characteristics

The total of 4,341 lengths of the aligned sequences was used for phylogenetic analysis. Of a total of investigated character sites, 2,793 characters were constant and 824 characters were parsimony informative including out groups. The parsimony analyses generated 10,620most parsimonious trees with a total length of 2,970 steps, a consistency index of 0.65 and a retention index of 0.88.The MP tree with bootstrap values(BP) and PP are shown in Fig. 1.

Phylogenetic analyses

The strict consensus tree had three main clades (clade I, clade II and clade III). Clade IV is highly supported by the 100% bootstrap value (PP: 0.98) and was sister to the rest of the species, which were grouped in two other clades. This clade was only composed of the individuals of *G. paradoxum* Maxim. Clade IIa is supported 99% bootstrap value (PP<0.95). Clade IIb consists of two highly supported subclades (subclade IIa and IIb). Subclade IIa included three taxa: *G. dahuricum* var. *lasiocarpum* (Makino) Nakai., *G. pseudoasprellum* Makino and *G. triflorum* Michx. comprising of Group B. *G. triflorum* (sect. *Trachygalium*) was sister to *G. dahuricum* var. *lasiocarpum* from Korea-Japan and *G. pseudoasprellum* from Japan (99% bootstrap value). Subclade IIb is supported by 91% bootstrap value (PP<0.95). This subclade contained 8 taxa from Korea-Japan. It was further divided into Group C and D. Group C contained members of three sections (sect. *Galium*, sect. *Leiogalium* and sect. *Leptogalium*) which are identified by Soza and Olmstead (2010), *G. tokyoense* Makino, *G. kikumugura*, and three species belonging to sect. *Galium* from Korea-Japan. But the *G. verum* group from Korean and Japanese were not well resolved. In the Group D, *G. japonicum* (Maxim.) Makino & Nakai from Korea and Japan is monophyletic, although the individuals of *G. trifloriforme* Kom. did not form monophyletic group. These two taxa share its most recent common ancestor with *G. spurium* var. *echinospermum* (Wallr.) Hayekand *G. odoratum* (L.) Scop (61% bootstrap value (PP: 0.96)). Clade III is supported by 91% bootstrap value (PP: 0.97), comprising eight sections; sect .*Baccogalium*, sect. *Lophogalium*, sect. *Platygalium*, sect. *Leiogalium*, sect. *Trachygalium*, sect. *Relbunium*, sect. *Depauperata*, sect. *Aparinoides*. The members of sect. *Depauperata*, and sect. *Aparinoides* are sister to the rest of the species within this Clade. *G. trifidum* L. is paraphyletic and unresolved within the clade. Group A in Clade III included four taxa from *G. trachygalium* group (*G. gracilens* (A. Gray) Makino, *G. koreanum*, *G. pogonanthum* Franch. & Sav. and *G. trachyspermum* A. Gray) and members of sect. *Platygalium* (BS: 80%, PP<0.95). The previous classification based on morphological study of the four taxa of the *G. trachygalium* group was not resolved (Jeong and Pak, 2009). The individuals from the same taxa did not even form the monophyletic. *G. kinuta* Nakai & Hara belonging to sect. *Platygalium* with *G. boreale* L. was resolved as paraphyletic.

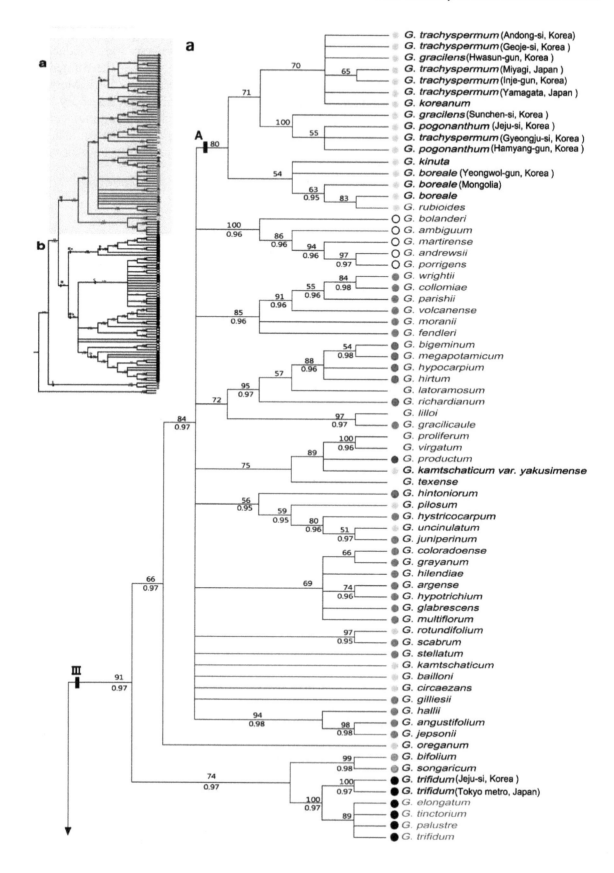

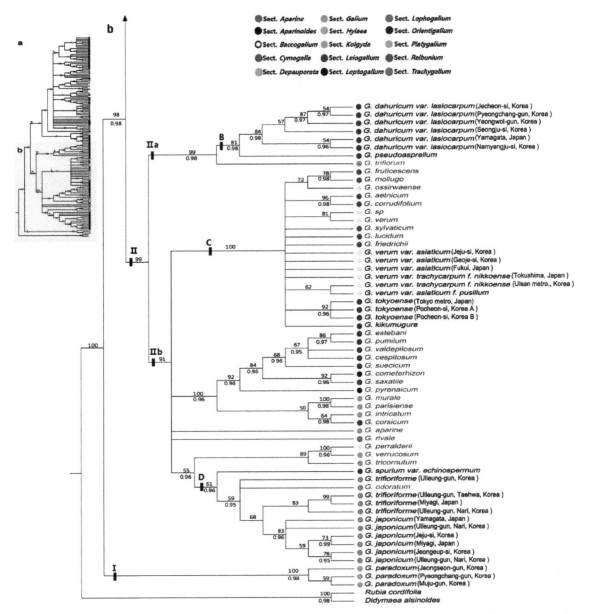

Fig. 1. Strict consensus tree of genus *Galium* based on Chloroplast DNA data, Bootstrap values and posterior probabilities are shown above and below branches, respectively. Different shapes were used for sectional treatments (taxon without shape "represents not classified"). Species in black represent the taxa sampled in this study.

Phylogenetic relationships of Korean-Japanese Galium

The phylogenetic relationships among Korean *Galium* and some of Japanese *Galium* were, for the first time, assessed in this study. We confirmed that the cpDNA phylogeny has significantly higher resolution and better support than previous study in Korean-Japanese *Galium* using morphological and chromosome number data by Jeong and Pak (2009, 2012). In some of taxa, our data were incongruent with previous classifications of *Korean-Japanese Galium* based on morphological data.

G. *paradoxum* was sister to the group consisting of the rest of the *Galium* species (Fig 1). It also support the study of Ehrendorfer *et al.* (2014) using the plastid DNA sequences. The species is a perennial herb with opposite leaves, a pair of scale-like small stipules, one vein, white petiole and corolla, and rotate flowers. *G. paradoxum* was placed into a sect. *Cymogalia* based on the characters of inflorescence and hairs of a fruit (Pobedimova *et al.*, 2000; Ehrendorfer *et al.*, 2005). Its main distributions is in eastern Asia (Ehrendorfer *et al.*, 2014), and mainly occurs in moist high elevations in mountain forests.

The taxa in the clade II have whorls of six or eight leaf-like organs. The five taxa from Korea-Japan are contained in Group C. The taxa of *G.verum* group (sect. *Galium*; *G. verum* var. *asiaticum* Nakai, *G. verum* f. *nikkoense* var. *trachycarpum* (Nakai) Ohwi and *G. verum* var. *asiaticum* f. *pusillum*) showed polytomies in the MPtree with weak PP. *G. verum* var. *asiaticum* is widely distributed throughout Korea and Japan. In our study, G. *verum* var. *asiaticum* have five chloroplast types from five individuals. But we cannot find morphological variation among the individuals. The three taxa are erect and have whorls of six or more than leaf-like organs, inflorescences of branched panicles with white or yellow flowers, and glabrous fruits. These three taxa don't exhibit significant morphological differences. But the plant and leaves size of *G. verum* var. *asiaticum* f. *pusillum* are smaller than those of other two taxa, and Korean endemic species in Mt. Halla on Jeju Island (Lee, 2004). It formed a clade with *G. verum* f. *nikkoense* var. *trachycarpum* from Ullsan-si (eastern part of Korea) with weak BS. It could provide crucial information for origin of Korean endemic, *G. verum* var. *asiaticum* f. *pusillum*. It needs additional study to investigate the origin and in these evolutionary relationship among these taxa. The four taxa of *G. dahuricum* group from Korea-Japan; *G. dahuricum* var. *lasiocarpum*, *G. kikumugura*, *G. tokyoense*, and *G. pseudoasprellum*, are have been included into sect. *Trachygalium* (Ehrendorfer *et al.*, 2005). There is no study of phylogenetic using molecular makers before. The four taxa of *G. dahuricum* group occur in East Asia, and have serious identification problems and taxon delimitation due to severe variations in the morphology of leaves, seed hairs and flower and inflorescences (Chen and Enrendorfer, 2011). We confirmed the phylogenetic relationship among these taxa, for the first time. *G. kikumugura* and *G. tokyoense* were included in Group C. *G. kikumugura* having whorls of four leaf-like organs and fruit with generally hooked hairs were closely related to *G. tokyoense*, morphologically (Yamazaki, 1993). Lee (1995) reported the distribution of *G. kikumugura* in Korea but we could not find the distibution during the this study although the species is widely distributed in Japan. We also could not confirm *G. kikumugura* specimens collected from Korea at Korean and Japan herbria. Therefore we assumed that the distribution report of this taxa by Lee (1995) was based on misclassification. *G. pseudoasprellum* was treated as synonyms of *G. dahuricum* by Ehrendorfer *et al.* (2005), but in our results did not support his opinion. *G. pseudoasprellum* is similar to *G. dahuricum* var. *lasiocarpum*, morphologically but it can be distinguished from *G. dahuricum* based on leaf shapes, which whorl of 6 elliptic or lanceolate leaves. *G. tokyoense* has glabrous fruit and white flower compare with *G. dahuricum* var. *lasiocarpum*. Previous studies based on morphology (Yamazaki, 1993; Pobedimova *et al.*, 2000; Chen and Ehrendorfer, 2011) were argument for classification of *G. tokyoense*. We confirmed that the *G. tokyoense* and *G. dahuricum* var. *lasiocarpum* were polyphyletic. Also our result is supported that previous classification that *G. tokyoense* be regarded as a species. *G. kamchaticum* Steller *ex* Schultes & J. H. Schultes and *G. kamchaticum* var. *yakusimense* (Masamune) Yamazakiwere place to clade□ with polytomy at MP tree with weak PP value. *G .kamchaticum* is distributed in an alpine meadow of worldwide with centers of the diversity in eastern Asia and eastern North America (Ehrendorfer *et al.*, 2005). *G. kamchaticum* var. *yakusimense* is smaller leave and tall than *G. kamchaticum*. This species is erect, with round leaves, one vein, whorls of four leaf-like organs, 4-parted white, and a fruit with generally hooked hairs.

We confirmed that *G. kinuta* is closer to *G. boreale*. Two taxa usually occur in northern part of Korean peninsula, especially in the mountain forests in lower elevation. The somatic chromosome number of *G. kinuta* and *G. boreale* were 4X (2*n*=44) and/or 2X (2*n*=11), respectively (Jeong and Pak, 2009). *G. kinuta* is erect, four leaf-like organs, three veins, branched panicles of inflorescences, and white flowers. *G. kinuta* and *G. boreale* are generally very similar in morphology and can be distinguished by the characters of leaf-shape.

Table 2. Primers used for amplification of cpDNA regions in this study.

Region	Primer	Sequence (5'-3')	Annealing temperature (°C)	References
*trn*C-*ycf*6	trnC^{GCA}F	CCAGTTCRAATCYGGGTG	52	Demesure *et al.* (1995)
	ycf6R	GCCCAAGCRAGACTTACTATATCCAT		Demesure *et al.* (1995)
*trn*F-*ndh*J	ndhJ	ATGCCYGAAAGTTGGATAGG	57	Shaw *et al.* (2007)
	TabE	GGTTCAAGTCCCTCTATCCC		Taberlet *et al.* (1991)
*Trn*L intron	c	CGAAATCGGTAGACGCTACG	55	Taberlet *et al.* (1991)
	d	GGGGATAGAGGGACTTGAAC		Taberlet *et al.* (1991)
*rpo*B-*trn*C	rpoBb	CGGATATTAATAKMTACATACG	55	Soza and Olmstead (2010)
	rpoBd	GTTGGGGTTTACATATACT		Soza and Olmstead (2010)

The *G. trachygalium* group consisted of four species; *G. trachygalium*, *G. pogonanthum*, *G. gracilens*, which occur in both Korea and Japan, and *G. koreanum* endemic to Korea. Although, the four taxa placed into Group A, our data did not provide insights into the specific phylogenetic relationships among *G. trachygalium* group species. These taxa are characterized by whorls of four leaf-like organs, cymose inflorescences with several terminal flowers, 4-parted rotate flowers and tuberculate fruit. The identification and delimitation of these species are usually difficult because they are very similar in morphology. The four species are distinguished by the differences in leaf size, shape, and fruit hairs (Jeong and Pak, 2012). These taxa usually occur in the near or same population, and share a common habitat. The somatic chromosome number of these species are 2X (2*n*=22) and/or 4X (2*n*=44) (Jeong and Pak, 2009). This inconsistencies phylogeny can be explained the speciation processes of the *G. trachygalium*group. But it is yet to be determined whether incomplete lineage sorting of ancestral polymorphisms in the population, or chloroplast capture by hybridization and introgression. It needs additional study to understand origin and clear relationship among these taxa. *G. trachyspermum*, *G. pogonanthum* and *G. gracilens* previously been placed into a sect. *Cymogalia* by Yamazaki (1993) but our data showed that these four taxa including *G. koreanum*, are more closely related to members of sect. *Platygalium* (Table 1). We suggest that the four taxa have to be transferred to sect. *Platygalium* based on molecular and morphological data.

Acknowledgement

This research was supported by Basic Science Research Program through the National Research Foundation of Korea (NRF) funded by the Ministry of Education (2016R1A 6A1A05011910).

References

Chen, T. and Ehrendorfer, F. 2011. Rubia.Vol 19. In: Wu ZY, Raven PH, Hong DY(Eds). Flora of china. Beijing: Science Press; St. Louis: Missouri Botanical Garden Press. pp. 104-141.

Demesure, B., Sodzi, N., Petit, R.J. 1995. A set of universal primers for amplification of polymorphic noncoding regions of mitochondrial and chloroplast DNA in plants. Molec. Ecol. **4**: 129-131.

Dolye, J.J. and Dolye, J.L.1987. A rapid DNA isolation procedure for small quantities of fresh leaf tissue.Phytochem. Bul. Bot. Soc. Amer. **19**:11-15.

Ehrendorfer, F., Manen, J.-F.and Natali, A. 1994. Cp DNA intergene sequences corroborate restriction site data for reconstructing Rubiaceae phylogeny. Pl. Syst. Evol. **190**: 195-211.

Ehrendorfer, F., Schönbeck-Temesy, E., Puff, C. and Rechinger, W. 2005. Rubiaceae.eds. K. H. Rechinger, Flora Iranica. no. 176. Verlag des Naturhistorischen Museums Wien, Vienna, Austria.

Ehrendorfer, F., Vladimirov, V. and Barfuss, M.H.J. 2014. Paraphyly and polyphyly in the worldwide tribe Rubieae (Rubiaceae): Challenges for genetic delimitation. Ann. Missouri Bot. Gard. **100**: 79-88.

Felsensteijn, J. 1985. Confidence limits on phylogenies: an approach using the bootstrap. Evolution. **39**: 783-791.

Govaerts, R. 2006. World checklist of selected plant families. In: F. A. Bisby, Y.R. Roskov, M. A. Ruggiero, T. M. Orrell, L. E. Paglinawan, P. W. Brewer, N.Bailly, and J. van Hertum, (eds.), Species 2000 & ITIS catalogue of life: www.catalogueoflife.org/annualchecklist/ 2007/. Species 2000, Reading, U. K.

Jeong K.S. and Pak, J. H. 2009. A cytotaxonomic study of *Galium* (Rubiaceae) in Korea. Korean J. Pl. Tax. **39**: 42-47.

Jeong K.S. and Pak, J. H. 2012. The morphological study of *Galium* L. (Rubiaceae) in Korea. Korean J. Pl. Tax. **42**: 1-12.

Lee, T.B. 1979. Illustrated Flora of Korea. Hyangmunsa, Seoul (in Korean).

Lee, W.T. 1995. Lineamenta Florae Koreae. Academy Press, Seoul (in Korean).

Lee, Y.N. 2004. Flora of Korea. Kyohaksa, Seoul (in Korean).

Manen, J.-F., Natali, A. and Ehrendorfer, F. 1994. Phylogeny of Rubiaceae-Rubieae inferred from the sequence of a cpDNA intergene region.Pl. Syst. Evol.**190**: 195-211.

Manen, J.-F.and Natali, A. 1995. Comparison of the evolution of ribulose-1,5-biphosphate carboxylase (*rbcL*)and *atb-rbcL* noncoding spacer sequences in a recent plant group, the tribe Rubieae (Rubiaceae). J. Molec. Evol. **41**: 920-927.

Natali, A., Manen, J.-F.andEhrendorfer, F. 1995. Phylogeny of the Rubiaceae-Rubioideae, in particular the tribe Rubieae: Evidence from a non-coding chloroplast DNA sequence. Ann. Missouri Bot. Gard. **82**: 428-439.

Natali, A., Manen, J.-F., Kiehn, M. and Ehrendorfer, F. 1996. Tribal, generic, and specific relationships in the Rubioideae-Rubieae (Rubiaceae) based on sequence data of a cpDNA intergene region. Opera Bot. Belg. **7**: 193–203.

Nylander, J.A.A. 2004. Mrmodeltest 2.3. Program distributed by the author. Evolutionary biology Centre, Uppsala University.

Pobedimova, E.G. 2000. *Galium* L. in: B. K. Schischkin (ed.), Flora of the U.S.S.R, vol. 23. Bishen Singh Mahendra Pal Singh, Dehra Dun. India, and Koeltz Scientific Books, Koenigstein, Germany. pp. 345–459.

Robbrecht, E. and Manen, J.F. 2006. The major evolutionary lineages of the coffee family (Rubiaceae, angiosperms). Combined analysis (nDNA and cpDNA) to infer the position of Coptosapelta and Luculia, and supertree construction based on *rbcL*, *rps*16, *trnL-trnF* and *atpB-rbcL* data. A new classification in two subfamilies, Cinchonoideae and Rubioideae. Syst. Geogr. Plants. **76**: 85-146.

Ronquist, F. and Huelsenbeck, J.P. 2003. Mrbayes 3: Bayesian phylogenetic inference under mixed models. Bioinfomatics. **19**: 1572-1574.

Shaw, J., E. B. Lickey, J. T. Beck, S. B. Farmer, W. Liu, J. Miller, K. C. Siripun, C. T. Winder, E. E. Schilling, and R. L. Small. 2005. The tortoise and the hare II: Relative utility of 21 noncoding chloroplast DNA sequences for phylogenetic analysis. Am. J. Bot. **92**: 142-166.

Shaw, J., Lickey, E.B., Schililling, E.E. and Small, R.L. 2007. Comparison of whole chloroplast genome sequences to choose noncoding regions for phylogenetic studies in angiosperms; the tortoise and the hare. Am. J. Bot. **94**: 275-288.

Soza, V.L. and Olmstead, R.G. 2010. Molecular systematics of tribe Rubieae (Rubiaceae): Evolution of major clades, development of leaf-like whorls, and biogeography. Taxon **59**: 755-771.

Swofford, D.L. 2003. PAUP: Phylogenetic analysis using parsimony (and other method). Ver.4.0b10. Sinauer Associates, Sunderlad, Massachusetts.

Taberlet, P., Gielly, L. Pautou, G. and Bouvet, J.1991. Universal primer for amplification of three non-coding regions of chloroplast DNA.Pl.Molec. Biol. **17**: 1105-1109.

Thompson, J.D., Gibson, T.J., Plewniak, F., Jeanmougin, F. and Higins, D.G. 1997. The CLUSTAL X windows interface: flexible strategies for multiple sequence alignment aided by quality analysis tools. Nucleic Acids Res. 22: 4676-4882.

Yamazaki, T. 1993. Angiospermae. Vol☐a. *In:* Iwatsuki K., David E.B., Ohba H. (Eds.) Flora of Japan. Tokyo: Kodansa. pp. 233-240.

TYPIFICATION OF FOURTEEN NAMES OF TWELVE RECOGNIZED TAXA IN *LEUCAS* R. BR. (LAMIACEAE) AND ONE NEW COMBINATION

RAJEEV KUMAR SINGH[1]

Botanical Survey of India (BSI), Southern Regional Centre (SRC), TNAU Campus, Lawley Road, Coimbatore 641 003, Tamil Nadu, India

Keywords: Isolectotype; Isoneotype; Lectotype; Neotype; Syntype; *Leucas*.

Abstract

Eight binomials of six recognized species of Indian *Leucas* R. Br. are lectotypified, namely, *Leucas beddomei* (Hook. f.) Sunojk. & P. Mathew, *L. diffusa* Benth., *L. helianthemifolia* Desf., *L. nepetifolia* Benth., *L. pilosa* Benth., *L. pilosa* Benth. var. *pubescens* Benth., *L. ternifolia* Desf. and *L. vestita* Benth. Two recognized taxa are neotypified, namely, *L. angularis* Benth. and *L. lanata* Benth. var. *candida* Haines. *L. lanata* Benth. var. *candida* Haines is raised to species rank as *L. candida* (Haines) R.Kr. Singh. *L. pilosa* Benth. is added to the flora of India. Additionally the following four recognized endemic species of *Leucas* of Myanmar are also lectotypified, *Leucas collettii* Prain, *L. helferi* Hook. f., *L. ovata* Benth. and *L. teres* Benth.

Introduction

During the present study on the systematics of *Leucas* in India, a total of 22 taxa have already been lectotypified (Singh, 2015). In the present communication, eight names of six recognized species of Indian *Leucas* are lectotypified and two names of two recognized taxa are neotypified here to avoid any ambiguity in the application of these names, because no specific herbarium sheet was cited as holotype in protologue of these taxa and also not lectotypified in earlier works (Singh, 2001; Sunojkumar and Mathew, 2002, 2008; Sunojkumar, 2008; Singh, 2015). The variety *candida* Haines of *L. lanata* Benth. is raised to species rank and *L. pilosa* Benth., which was earlier considered as endemic to Myanmar, is now added to the Flora of India. Further, during the present study on *Leucas* in India, author studied type specimens of *Leucas* (held at CAL and K), which are specimens of species that are endemic to Myanmar. These are lectotypified here. While designating lectotypes and neotypes, the guidelines of Art. 9.2, 9.23, 9.3(c) and 9.6 and recommendations 9A, 9B, 9C and 9D of the Melbourne Code (McNeill *et al.*, 2012) were followed.

Typification of Indian Leucas

1. **Leucas angularis** Benth., Pl. Asiat. Rar. (Wallich) 1: 62 (1830).

   Type citation: "ex Ceylona. (*Herb. Lindley.*)"

   Neotype (here designated): India, Tamil Nadu, Glen Fall, Kodaikanal Hills, 15 Oct 1919, *Jacob 16135* (MH41559!); isoneotype: MH41558!. (Fig. 1)

   Distribution: India (Kerala and Tamil Nadu) and Sri Lanka.

   Notes: The above neotype is required as the original collection or gathering on the basis of which *Leucas angularis* was described is not known to exist. Within the protologue, Bentham (1830) cited only 'ex Ceylona. (*Herb. Lindley.*)' but did not provide any further information. Bentham's types are held at K and Lindley's at BM, CGE and K. However, attempts to locate type

---
[1]Email: rksbsiadsingh@yahoo.co.in

specimens in these herbaria were unsuccessful. Since no original material of the species appears to be extant, the specimen from MH41559 is chosen here as the neotype. The specimen selected is well preserved, has mature leaves and well developed flowers.

2. **Leucas beddomei** (Hook. f.) Sunojk. & P. Mathew, Rheedea 12(2): 170 (2002).

Leucas hirta (B. Heyne ex Roth) Spreng. var. *beddomei* Hook. f., Fl. Brit. India 4: 687 (1885).

Type citation: "Chambra Peek, Wynaad, alt. 5000 ft., *Beddome*"

Lectotype (here designated): India, Kerala, Wynaad [Wayanad], Chambra Peak, 5000 ft., March 1880, *Beddome s.n.* (K000929538!); isolectotype: BM000950511!. (Fig. 2)

Distribution: India, endemic and rare (Kerala, restricted to Wayanad district).

Notes: J.D. Hooker (1885) described *Leucas hirta* var. *beddomei* on the basis of specimens collected by Beddome from Chambra Peak, Wynaad, but no specific herbarium sheet was designated as the holotype nor did he mention the name of herbarium where the specimens were housed. Two herbarium sheets, collected by Beddome from Chambra Peak, Wynaad, with J.D. Hooker's annotation 'L. hirta var. beddomei Hf' were traced (BM000950511 and K000929538). Of these two, the better preserved K000929538, is designated here as the lectotype as it agrees well with the protologue and also in having dissected flower parts pasted on the sheet.

Singh (2001) cited the type information as "*Holotype* : India, Chambrapeek, Wynaad, 5000 ft., *Beddome s.n.* (BM)" and Sunojkumar and Mathew (2002) as "Type: India, Kerala, Wayanad, Chembra peak, 5000 ft., *Beddome s.n.* – type of *Leucas hirta* var. *beddomei* Hook. f. (holotype – K, Cibachrome photo!)". Although, they cited BM and K as housing the holotype, but their citation of holotype cannot be corrected to lectotype as per Article 9.23 of ICN 2012, which state that 'On or after 1 January 2001, lectotypification or neotypification of a name of a species or infraspecific taxon is not effected unless indicated by use of the term "lectotypus" or "neotypus", its abbreviation, or its equivalent in a modern language'. They also did not mention the phrase, "designated here" or its equivalent according to Article 7.10.

3. **Leucas candida** (Haines) R.Kr. Singh, **comb. et stat. nov.**

Leucas lanata Benth. var. *candida* Haines, Bot. Bihar Orissa 4: 747 (1922).

Type citation: "Var. *candida* occurs on the hills of the Central Provinces", "It possibly occurs on the higher Sirguja mountains."

Neotype (here designated): India, Tamil Nadu, Nilgiris district, Marappalam–Burliar road, 1225 m, 29 Apr 1971, Rathakrishnan 38130 (MH73387!); isoneotype: MH73388!. (Fig. 3)

Distribution: India, endemic (Madhya Pradesh, Odisha and Tamil Nadu).

Notes: The above neotype is required as the original collection or gathering on the basis of which *Leucas lanata* var. *candida* was described is not known to exist. Within the protologue, Haines (1922) cited only the locality but did not provide the date of collection, number of collection/gathering and the name of herbarium where the specimens were housed. Haines's types are known to exist at K and some at CAL, I tried to trace the type specimens in these two herbaria but no specimen was found extant. Since no original material of the species appears to be extant, the specimen from MH (MH73387) is chosen here as the neotype. The specimen selected is well preserved, has mature leaves and well developed flowers.

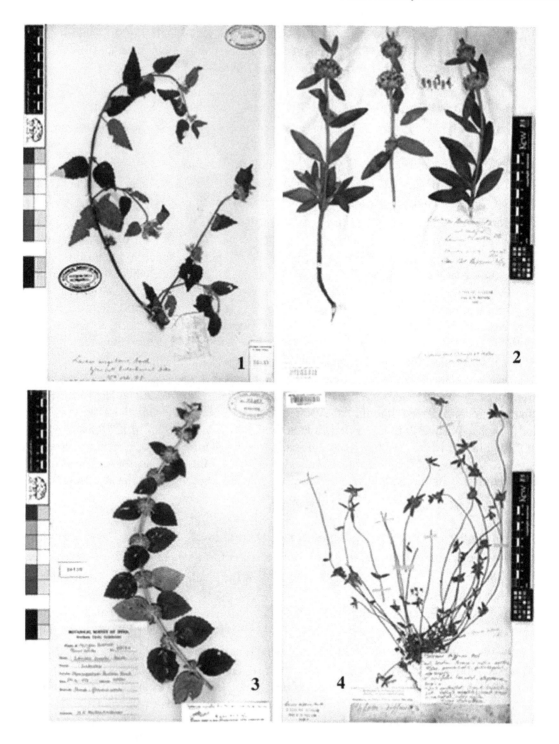

Figs 1-4: 1. Neotype of *Leucas angularis* Benth. (MH, Accesssion no. 41559, © Botanical Survey of India, SRC, Coimbatore). 2. Lectotype of *Leucas beddomei* Sunojk. & P. Mathew (K000929538, © the Board of Trustees of the Royal Botanic Gardens, Kew). 3. Neotype of *Leucas candida* R.Kr. Singh (MH, Accession no. 73387, © Botanical Survey of India, SRC, Coimbatore). 4. Lectotype of *Leucas diffusa* Benth. (K000929557, © the Board of Trustees of the Royal Botanic Gardens, Kew).

Key to distinguish *Leucas candida* from *L. lanata*

1. Leaves broadly ovate-rounded, 1.5–7 × 1.3–5.5 cm, veins not impressed above, pubescent above, tomentose beneath, dark above on maturity; petioles 0.8–2 cm long; calyx pubescent, teeth 0.6–0.9 mm long; corolla tube included within calyx, lower lip 1.2–1.3 cm long; nutlets smooth, rounded at apex *L. candida*

– Leaves ovate-lanceolate or ovate-oblong, 1–4 × 0.5–1.6 cm, veins distinctly impressed above, tomentose above, silky beneath, grey on maturity; petioles absent in upper leaves, or short (0.3–0.9 cm long) in lower ones; calyx tomentose, teeth 1.3–1.6 mm long; corolla tube usually exserted from calyx, lower lip < 0.8–0.9 cm long; nutlets tuberculate, truncate at apex *L. lanata*

4. **Leucas diffusa** Benth., Labiat. Gen. Spec. : 615 (1834).

Leucas dimidiata sensu Benth, Pl. Asiat. Rar. (Wallich) 1: 61 (1830), *non* (Roth) Spreng. (1825).

Type citation: "L. dimidiata. Benth. in Wall. Pl. As. Rar. non Roth.", "Hab. in Indiae Orientalis Peninsula Herb. Madr. (*h. s. sp. e Mus. Angl. Ind.*)"

Lectotype (here designated): India, Penins. Indiae Orientalis [Peninsular India], Madras, without date, *Rottler s.n.* (K000929557!). (Fig. 4)

Residual syntypes: Without locality, 1829, Herb. Madr., *Wallich s.n.* (K000929558!); Without locality, without date, Herb. Madr., *Wallich cat. n. 2528 E* (CAL362830!).

Distribution: India, endemic (Andhra Pradesh, Delhi, Karnataka, Kerala and Tamil Nadu).

Notes: Bentham (1834) described *Leucas diffusa* based on the specimens from Peninsular India and Herb. Madr., but no type was indicated nor did he cite the name of the collector(s), date of collection, collection number and the name of herbarium where the specimens were housed. In the protologue, he mentioned that his earlier *L. dimidiata* in *Plantae Asiaticae Rariores (Wallich)* is *L. diffusa* now. As per the specifications given in protologues of *L. diffusa* and *L. dimidiata sensu* Benth. (1830) in *Plantae Asiaticae Rariores*, it is clear that Peninsular India specimen belongs to Herbarium Rottlerianum and specimens of Herb. Madr. belongs to *Wallich cat. n. 2528 E*. Two specimens of Herb. Madr. (K000929558 and CAL362830) and one of Penins. Indiae Orientalis was traced (K000929557). Since Bentham worked at K, only the two specimens at K have been considered for lectotypification. The collection K000929557 is better preserved and more complete than the other, it agrees well with the protologue and also has dissected flower parts and short descriptive notes pasted on the sheet. Therefore, this collection is here designated as the lectotype.

5. **Leucas helianthemifolia** Desf., Mém. Mus. Hist. Nat. 11: 2 (1824).

Type citation: "M. Lechenault", "Cette jolie espèce est indigène des la presqu'île de l'Inde; elle croît sur la base des montagnes de Nelligerry."

Lectotype (here designated): India, montagnes de Nelliggerry [Nilgiri Mountains], without date, *Leschenault 34* (P00738007!). (Fig. 5)

Residual syntype: India, montagnes de Nelliggerry [Nilgiri Mountains], without date, *Leschenault s.n.* (P00215013!).

Leucas ternifolia Desf., Mém. Mus. Hist. Nat. 11: 4 (1824).

Type citation: "M. Lechenault", "elle croît également sur la base des montagnes de Nelligerry."

Lectotype (here designated): India, montagnes de Nelliggerry [Nilgiri Mountains], without date, *Leschenault 206* (P00738006!); isolectotype: P00215014!. (Fig. 6)

Distribution: India, endemic (Kerala and Tamil Nadu).

Notes: Desfontaines (1824) described *Leucas helianthemifolia* based on a gathering by Leschenault from Nilgiri hills, India but no specific herbarium sheet was designated as the holotype nor did he mention the name of herbarium where the specimens were housed. Within the protologue, Desfontaines gave the precise locality and collector name but did not provide the number and date of collection. Two herbarium sheets, collected by Leschenault from Nilgiri Mountains, India are held at P (P00215013 and P00738007). The better preserved sheet, P00738007, is chosen here as the lectotype because the illustration in the protologue is based on this and it agrees well with the protologue.

Leucas ternifolia was described by Desfontaines (1824) on the basis of specimens collected by Leschenault from Nilgiri Mountains, India but no specific herbarium sheet was designated as the holotype nor did he mention the name of herbarium where the specimens were housed. In the protologue, Desfontaines gave the precise locality and collector name but did not provide the number and date of collection. Two herbarium sheets at P (P00215014 and P00738006), collected by Leschenault from Nilgiri Mountains, India were traced. Of these, the best one, P00738006, is chosen here as the lectotype because the illustration in the protologue is based on this and it agrees well with the protologue.

6. **Leucas nepetifolia** Benth., Pl. Asiat. Rar. (Wallich) 1: 62 (1830).

Type citation: "Hab. . . . . (Herb. Madr.)"

Lectotype (here designated): Without locality, without date, Herb. Madr., *Wallich cat. n. 2526* (K001116355!); isolectotype: CAL362295!. (Fig. 7)

Distribution: India, endemic and rare (Andhra Pradesh, Karnataka and Tamil Nadu).

Notes: In the protologue of *Leucas nepetifolia*, Bentham (1830) indicated only 'Hab. . . . . (Herb. Madr.)' as type citation but did not provide the name of collector, date of collection, locality, number of collection/gathering and the name of herbarium where the specimens were housed. Pertaining to the specification given in protologue, two specimens of *L. nepetifolia* of Herb. Madr. (CAL362295 and K001116355), belonging to *Wallich cat. n. 2526* were traced and better preserved sheet, K001116355, is designated here as the lectotype as it agrees well with the protologue.

7. **Leucas pilosa** Benth., Pl. Asiat. Rar. (Wallich) 1: 62 (1830).

Type citation: "Hab. α. ad ripas Irawaddi."

Lectotype (here designated): Myanmar, Irawaddi, 1829, *Wallich cat. n. 2058* [1] (K000929509!). (Fig. 8)

Residual syntypes: Myanmar, Irawaddi, 1826, *Wallich cat. n. 2058* [1] (K001115020! and CAL362482!).

Leucas pilosa Benth. var. **pubescens** Benth., Pl. Asiat. Rar. (Wallich) 1: 62 (1830) *et* Labiat. Gen. Spec. 609 (1834) *et* Prodr. (A. P. de Candolle) 12: 526 (1848).

Type citation: "β Rajemahl."

Lectotype (here designated): India, Jharkhand, Rajemahl [Rajmahal], 7 Aug 1820, *Wallich cat. n. 2058* [β] (K001115022!). (Fig. 9)

Figs 5-8: 5. Lectotype of *Leucas helianthemifolia* Desf. (P00738007, © Muséum National D'Histoire Naturelle, Paris). 6. Lectotype of *Leucas ternifolia* Desf. (P00738006, © Muséum National D'Histoire Naturelle, Paris). 7. Lectotype of *Leucas nepetifolia* Benth. (K001116355, © the Board of Trustees of the Royal Botanic Gardens, Kew). 8. Lectotype of *Leucas pilosa* Benth. (K000929509, © the Board of Trustees of the Royal Botanic Gardens, Kew).

Residual syntypes: India, Jharkhand, Rajemahl [Rajmahal], 6 Aug 1820, *Wallich cat. n. 2058* [β] (CAL362483!); Without locality, without date, *Wallich cat. n. 2058* [β] (K001115021!).

Distribution: India (Jharkhand, Madhya Pradesh and Uttarakhand) and Myanmar.

Notes: Bentham (1830) described *Leucas pilosa* based on the specimens from Irawaddi, but no specific herbarium sheet was designated as the holotype nor did he cite the name of the collector, date of collection, collection number and the name of herbarium where the specimens were housed. Pertaining to the specification given in protologue, three specimens from Irawaddi of *Wallich cat. n. 2058* [1] were known (CAL362482, K000929509 and K001115020). Only the two sheets at K have been considered here to choose the lectotype for this name because Bentham worked at K. The herbarium specimen, K000929509 belongs to herbarium Benthamianum and is designated here as the lectotype as it agrees well with the protologue.

In the protologue of *Leucas pilosa* var. *pubescens*, Bentham (1830) indicated only 'β Rajemahl.' as type citation but did not provide the name of collector, date of collection, number of collection/gathering and the name of herbarium where the specimens were housed. Pertaining to the specification given in protologue, three specimens from Rajemahl [Rajmahal] of *Wallich cat. n. 2058* [β] were traced (CAL362483, K001115021 and K001115022), which should be considered as original material. Only the two K specimens are considered as suitable lectotypes specimens for this name. The best one and better preserved sheet, K001115022, is designated here as the lectotype as it agrees well with the protologue.

J.D. Hooker (1885) did not include *Leucas pilosa* Benth. var. *pubescens* Benth. in the Flora of British India, furthermore, he did not mention the place of occurrence of this variety. He mentioned the occurrence of *L. pilosa*, only in Burma [Myanmar]. Singh (2001) in Monograph of Indian *Leucas* did not clearly conclude the identity of var. *pubescens* and simply wrote that this variety was considered conspecific to *L. pilosa* by earlier workers and it may probably be a form of *L. decemdentata* (Willd.) Sm., but he mention the species *L. pilosa* Benth. is endemic to Myanmar. Bentham (1830) described var. *pubescens* on the basis of specimens from Rajmahal, Jharkhand state, India. In *Labiatarum Genera et Species* and in *Prodromus Systematis Naturalis Regni Vegetabilis (DC.)*, he cited Royle's collection from Deyra Dhoun [Dehra Dun] and Wallich's collection from Rajemahl [Rajmahal] for var. *pubescens*. After study of type specimens of var. *pubescens* from Rajmahal (K and CAL) and Royle's collection from Dehra Dun (DD), it is now concluded that this variety *pubescens* is conspecific with *L. pilosa* Benth. Hence, *Leucas pilosa* Benth. is now added to Indian flora.

8. **Leucas vestita** Benth., Pl. Asiat. Rar. (Wallich) 1: 61 (1830), *p.p. et* Labiat. Gen. Spec. 613 (1834), *p.p. et* Prodr. (A. P. de Candolle) 12: 530 (1848).

Type citation: "Hab. in Sillet", "Hab. in Indiae orientalis provincia Sillet *Wallich !* et Peninsulae montibus Madurensibus *Wight !* (*h. s. sp. e Mus. Angl. Ind. et comm. a cl. Wight.*)", "In Indiae orientalis Peninsulae montibus Madurensibus (Wight ! n. 2530).— Wight ic. 2, t. 338."

Lectotype (here designated): India, Peninsula Ind. orientalis [Peninsular India], without date, *Wight 2530* (K000929531!). (Fig. 10)

Residual syntype: Without locality, without date, 1829, Herb. Wight, *Wallich s.n.* (K000929529, *p.p.!*).

Distribution: India, endemic (Andhra Pradesh, Karnataka, Kerala and Tamil Nadu).

Notes: In *Plantae Asiaticae Rariores (Wallich)*, Bentham (1830) indicated only 'Hab. in Sillet' as type citation for *Leucas vestita*, but did not provide the name of collector, date of collection, number of collection and the name of herbarium where the specimens were housed.

Later in 1834 (*Labiatarum Genera et Species*), he amended the description and cited 'Hab. in Indiae orientalis provincia Sillet *Wallich !* et Peninsulae montibus Madurensibus *Wight !* (*h. s. sp. e Mus. Angl. Ind. et comm. a cl. Wight.*)' as type. Again in 1848 (*Prodromus Systematis Naturalis Regni Vegetabilis*), he amended the description and cited type as 'In Indiae orientalis Peninsulae montibus Madurensibus (Wight ! n. 2530).— Wight ic. 2, t. 338.' So, finally *L. vestita* was correctly described by Bentham in *Prodromus Systematis Naturalis Regni Vegetabilis (DC.)* and it is endemic to south India. Pertaining to these specifications, two specimens (K000929529, *p.p.* and K000929531) from herbarium Benthamianum were traced. Of these two, the best one, K000929531, is designated here as the lectotype, as it agrees well with the protologue. The herbarium sheet K000929529 is a mixed collection, the upper half plant specimen belongs to *Herb. Wight* (syntype of *L. vestita* Benth. var. *vestita*), whereas the lower half is of Beddome from Anamallay hills (syntype of *L. vestita* Benth. var. *sericostoma* Hook. f.). Two herbarium sheet of *Wallich cat. n. 2039* (K001114953! and CAL362672!) collected from Sillet [Sylhet] though written as *L. vestita* Benth. by Wallich are actually *L. ciliata* Benth. The herbarium sheet K000929542 of *Wallich cat. n. 2046 [β]* collected from Sillet [Sylhet] was identified as *L. vestita* by Bentham (1834) in *Labiatarum Genera et Species* and later in 1848 treated as *L. ciliata* var. *hirsuta* in *Prodromus Systematis Naturalis Regni Vegetabilis (DC.)*.

Typification of Leucas species endemic to Myanmar

1. **Leucas collettii** Prain, J. Asiat. Soc. Bengal, Pt. 2, Nat. Hist. 59(4): 313 (1891).

   Type citation: "UPPER BURMA; Popah hill, 5000, *Collett* n. 29."

   Lectotype (here designated): Myanmar, Popah hill, 5000 ft., Dec 1887, *Collett 29* (CAL0000020543!); isolectotype: K000929569!. (Fig. 11)

   Notes: Prain (1891) described *Leucas collettii* based on the specimens collected from Popah hill, Myanmar, but no specific herbarium sheet was designated as the holotype nor did he mention the name of herbarium where the specimens were housed. Only two herbarium specimens of *Collett 29* are now extant, CAL0000020543 and K000929569. Of these two, CAL0000020543, is designated here as the lectotype as it agrees well with the protologue and also includes short descriptive notes and drawing of flower on the sheet by Prain.

2. **Leucas helferi** Hook. f., Fl. Brit. India 4: 681 (1885).

   Type citation: "TENASSERIM; *Helfer*."

   Lectotype (here designated): Myanmar, Tenasserim, without date, *Helfer 4046* (CAL0000020544!). (Fig. 12)

   Notes: J.D. Hooker (1885) described *Leucas helferi* based on the specimens collected by Helfer from Tenasserim but no specific herbarium sheet was designated as the holotype nor did he cited the date of collection, collection number and the name of herbarium where the specimens were housed. Pertaining to the specification given in protologue only one specimen of *L. helferi* collected by Helfer from Tenasserim is extant now at CAL (CAL0000020544). The types of J.D. Hooker's are known to be at K, sometimes at BM, E and P, but no original materials are found there. Although, the specimen CAL0000020544 was not examined by J.D. Hooker, but was collected by Helfer from Tenasserim and it is a part of original gathering. So, this should be considered as original material according to Art. 9.3(c) of Melbourne Code (McNeill *et al.* 2012) and is chosen here as the lectotype as it agrees well with the protologue.

3. **Leucas ovata** Benth., Pl. Asiat. Rar. (Wallich) 1: 61 (1830).

Type citation: "Wall. Cat. Herb. Ind. n. 2057.", "Hab. ad ripas Irawaddi."

Lectotype (here designated): Myanmar, Irawaddi, 1829, *Wallich cat. n. 2057* (K000929508!). (Fig. 13)

Residual syntypes: Myanmar, Irawaddi, 1826, *Wallich cat. n. 2057* (K001115019); Myanmar, Irawaddi, without date, *Wallich cat. n. 2057* (CAL362481!).

Figs 9-14: 9. Lectotype of *Leucas pilosa* Benth. var. *pubescens* Benth. (K001115022, © the Board of Trustees of the Royal Botanic Gardens, Kew). 10. Lectotype of *Leucas vestita* Benth. (K000929531, © the Board of Trustees of the Royal Botanic Gardens, Kew). 11. Lectotype of *Leucas collettii* Prain (CAL0000020543, © Central National Herbarium, Howrah). 12. Lectotype of *Leucas helferi* Hook. f. (CAL0000020544, © Central National Herbarium, Howrah). 13. Lectotype of *Leucas ovata* Benth. (K000929508, © the Board of Trustees of the Royal Botanic Gardens, Kew). 14. Lectotype of *Leucas teres* Benth. (K000929483, © the Board of Trustees of the Royal Botanic Gardens, Kew).

Notes: Bentham (1830) described *Leucas ovata* based on the gathering from Irawaddi of *Wallich cat. n. 2057*, but no specific herbarium sheet was designated as the holotype nor did he mention the date of collection and name of herbarium where the specimens were housed. Pertaining to the specification given in protologue, three specimens from Irawaddi of *Wallich cat.*

n. 2057 were known (CAL362481, K000929508 and K001115019). Only the two specimens at K have been considered here to choose the lectotype specimen for this name because Bentham worked at K. The herbarium sheet K000929508 belongs to herbarium Benthamianum and is designated here as the lectotype as it agrees well with the protologue.

4. **Leucas teres** Benth., Pl. Asiat. Rar. (Wallich) 1: 62 (1830).

   Type citation: "Wall. Cat. Herb. Ind. n. 2060", "Hab. ad ripas Irawaddi."

   Lectotype (here designated): Myanmar, Irawaddi, 1829, *Wallich cat. n. 2060* (K000929483!).

   Residual syntypes: Myanmar, Irawaddi, 1826, *Wallich cat. n. 2060* (K001115024!); Myanmar, Irawaddi, without date, *Wallich cat. n. 2060* (K000929482!). (Fig. 14)

   Notes: Bentham (1830) described *Leucas teres* based on the gathering from Irawaddi of *Wallich cat. n. 2060*, but no specific herbarium sheet was designated as the holotype nor did he mention the date of collection and name of herbarium where the specimens were housed. Pertaining to the specification given in protologue, three specimens from Irawaddi of *Wallich cat. n. 2060* were known (K000929482, K000929483 and K001115024). Of these, the herbarium sheet K000929483, from the Benthamianum herbarium, is designated here as the lectotype as it agrees well with the protologue.

Acknowledgements

The author is thankful to Dr. P. Singh, Director, Botanical Survey of India (BSI), Kolkata and Dr. G.V.S. Murthy, Head of Office, Botanical Survey of India, Southern Regional Centre, Coimbatore for facilities. I am also grateful to the Curators of BM, CAL, CGE, E, DD, K, MH and P for information and images of type specimens.

References

Bentham, G. 1830. *Leucas*. In: Wallich N (ed.), Plantae Asiaticae Rariores: or, Descriptions and figures of select number of unpublished East Indian plants 1: 60–62. Treuttel and Würtz, London.

Bentham, G. 1834. Labiatarum Genera et Species **6**: 609, 613 & 615. James Ridgway & Sons, London.

Bentham, G. 1848. *Leucas*. In: de Candolle AP (ed.), Prodromus Systematis Naturalis Regni Vegetabilis **12**: 523–533. Sumptibus Victoris Masson, Paris.

Desfontaines, R.L. 1824. Observations Sur les genres *Leucas* et *Phlomis*. Description de plusieurs Espèces. Mémoires du Muséum d'Histoire Naturelle, Paris **11**: 1–10.

Haines, H.H. 1922. The Botany of Bihar and Orissa **4**: 747–748. Adlard & Son & West Newman Ltd., London.

Hooker, J.D. 1885. The Flora of British India **4**: 680–691. L. Reeve & Co., London.

Mabberley, D.J. 2008. Mabberley's Plant-Book: A portable dictionary of plants, their classification and uses. Third Edition. Cambridge University Press, Cambridge, pp. 485.

McNeill, J., Barrie, F.R., Buck, W.R., Demoulin, V., Greuter, W., Hawksworth, D.L., Herendeen, P.S., Knapp, S., Marhold, K., Prado, J., Proud'homme van Reine, W.F., Smith, G.F., Wiersema, J.H. and Turland, N.J. (eds.). 2012. International Code of Nomenclature for algae, fungi and plants (Melbourne Code): Adopted by the Eighteenth International Botanical Congress, Melbourne, Australia, July 2011. Regnum Vegetabile **154**: 1–274.

Prain, D. 1891. Noviciae Indicae. III. Some additional species of Labiatae. J. Asiat. Soc. Bengal, Pt. 2, Nat. Hist. **59**(4): 294–318.

Singh, R.Kr. 2015. Lectotypification of Indian taxa of *Leucas* (Lamiaceae). Telopea **18**: 395–424.

Singh, V. 2001. Monograph on Indian *Leucas* R. Br. (Dronapushpi) Lamiaceae. Scientific Publishers, Jodhpur, India.

Sunojkumar, P. 2008. Taxonomical change in *Leucas ciliata* Benth. and *Leucas vestita* Benth. (Lamiaceae: Lamioideae). Candollea **63**(1): 81-83.

Sunojkumar, P. and Mathew, P. 2002. *Leucas beddomei* (Hook. f.) Sunojkumar & P. Mathew (Lamiaceae), a new status and name for *Leucas hirta* var. *beddomei* Hook. f. – a little known endemic from India. Rheedea **12** (2): 169-174.

Sunojkumar, P. and Mathew, P. 2008. South Indian *Leucas*: A Taxonomic Monograph. Centre for Research in Indigenous Knowledge Science and Culture, Calicut.

TAXONOMY AND REPRODUCTIVE BIOLOGY OF THE GENUS *ZEPHYRANTHES* HERB. (LILIACEAE) IN BANGLADESH

Sumona Afroz, M. Oliur Rahman[1] and Md. Abul Hassan

Department of Botany, University of Dhaka, Dhaka 1000, Bangladesh

Keywords: Zephyranthes Herb.; Pollination; Seed germination; Pseudovivipary; Pollen viability.

Abstract

The genus *Zephyranthes* Herb. is revised along with its pollination mechanism, seed germination and vegetative propagation. Detailed taxonomy of four *Zephyranthes* species occurring in Bangladesh, namely, *Z. atamasco* (L.) Herb., *Z. candida* (Lindl.) Herb., *Z. carinata* Herb. and *Z. tubispatha* (L'Her.) Herb. *ex* Traub. was studied with their updated nomenclature, important synonyms, phenology, specimens examined, habitat, distribution, economic value and mode of propagation. A dichotomous bracketed key is provided for easy identification of the species. Pollination investigation reveals that all studied species of *Zephyranthes* are self-pollinated. Minimum five days were required for germination of seeds in *Z. atamasco*, and three days each in *Z. candida*, *Z. carinata* and *Z. tubispatha*. Pseudovivipary type of germination has been reported in *Z. candida* and *Z. carinata* for the first time. The maximum number of seeds (30) per fruit are produced in *Z. tubispatha*, whereas the minimum seeds (2) per fruit are found in *Z. atamasco*. Vegetative propagation through bulb was found more suitable than seeds in *Z. atamasco*, *Z. candida* and *Z. carinata*. Pollen viability has been found 100% in *Z. candida*, and *Z. tubispatha*, whereas, *Z. atamasco* and *Z. carinata* have shown 80% and 98% viability, respectively.

Introduction

The genus *Zephyranthes* Herb. (Liliaceae) comprises about 70 species and native to diverse areas of the New World including Argentina, the Caribbean, Mexico and North America (Chowdhury and Hubstenberger, 2006; Spurrier *et al.*, 2015). In Bangladesh, this genus is represented by four species and found under cultivation. *Zephyranthes* are characterized by linear or lorate leaves, solitary flower, funnel shaped perianth, three carpels and sub-globose or depressed fruits. Pharmacological studies of *Zephyranthes* have revealed that the genus has anticancer, antifungal, and antibacterial activities (Katoch and Singh, 2015). Leaf decoction of *Z. candida* is used in South Africa as a remedy for diabetes mellitus (Pettit *et al.*, 1984). Several studies on the genus *Zephyranthes* were carried out based on morphology (Spencer, 1973; Flagg and Flory, 1976; Flagg *et al.*, 2002). Recently, Flagg and Smith (2008) studied three closely related species of *Zephyranthes*, i.e. *Z. atamasca* (the correct specific epithet is *atamasco*), *Z. treatiae*, and *Z. simpsonii* from southern United States, all of which have linear stigmatic lobes, green perianth tubes and white perianth segments. Based on cytological, herbarium, and field studies, and on Principal component analysis (PCA) and scatter diagram analysis, they concluded that all three taxa are distinct at the species level. Raina and Khoshoo (1972) studied cytogenetics of *Z. candida* and *Z. sulphurea*, while the breeding system of *Z. atamasco* was investigated by Broyles and Wyatt (1991).

Studies on reproductive biology disclose the nature of species, adaptation, speciation, hybridization, and systematics (Anderson *et al.*, 2002; Neal and Anderson, 2005). Several studies

---

[1]Corresponding author. Email: prof.oliurrahman@gmail.com; oliur.bot@du.ac.bd

on reproductive biology and pollination mechanism have been conceded in different group of plants (Cox, 1990; Wyatt and Broyles, 1990; Singer and Sazima, 1999; Liza *et al.*, 2010). Studies on seed germination in different plants are also well documented, and factors affecting seed germination have been recognized in different species (Yang *et al.*, 1999; Hassan and Fardous, 2003; Chauhan and Johnson, 2008; Rahman *et al.*, 2012; Ferdousi *et al.*, 2014).

Despite the systematic studies of *Zephyranthes* were carried out in different countries (Flagg *et al.*, 2002; Flagg and Smith, 2008; Arroyo-Leuenberger and Leuenberger, 2009) there has been no detailed study on taxonomy of this genus occurring in Bangladesh. Pollination, seed germination and propagation of *Zephyranthes* have never been investigated. Because of medicinal and ornamental value, the members of this genus need to be brought under cultivation, and prior to bring them under cultivation their reproductive biology need to be investigated in detail. The objectives of the present study are to revising the genus *Zephyranthes* and to investigate reproductive biological characteristics including mode of pollination and seed germination of these ornamental and economically important species, which might help in conveying the plants under rapid cultivation.

Materials and Methods

Plant materials

Four species of *Zephyranthes* namely, *Z. atamasco*, *Z. candida*, *Z. carinata* and *Z. tubispatha* were collected from different places and planted in the Dhaka University Botanical Garden for further study. The collected specimens were critically examined, and the study was supplemented by the herbarium specimens preserved at the Dhaka University Salar Khan Herbarium (DUSH) and Bangladesh National Herbarium (DACB). Identification of the *Zephyranthes* species were confirmed in consultation with standard literature (Karthikeyan *et al.*, 1989; Noltie, 1994; Hajra and Verma, 1996; Raven and Zhengyi, 2000; Utech, 2002; Siddiqui *et al.*, 2007) and by comparing with herbarium specimens deposited in DUSH and DACB. A dichotomous key to the species has been constructed for easy identification of the taxa. The voucher specimens have been deposited at DUSH.

Pollination

To study pollination the species were meticulously observed after flowering. Bagging experiment was conducted to understand the mechanism of pollination (Hassan and Khan, 1996). Bags of 15x12 cm made from fine cloth were used for bagging the flowers of individual plant at a stage, when all flowers were unopened. During bagging if open flowers exist, they were removed manually so that only unopened buds remained in the bags. Bagged inflorescence or flowers were kept under continuous observation for fruit formation which was compared with that of the plants kept under control.

Seed germination

Seeds were collected from mature fruits and sown in earthen pots of 10 inch in diameter filled up with a mixture of soil and compost (2:1). Seeds were sown at regular intervals in the earthen pot and the pots were kept in the semi-shaded position and watered everyday. Seeds were sown in different pots in different times of the year to record dormancy (if any), suitable period for germination, percentage and nature of germination.

Vegetative propagation

Vegetative propagation was performed through bulbs, separated from main plants.

Pollen viability

Pollen was taken from recently opened flower anthers. A drop of acetocarmine was taken onto the slide. After removing the anther from the flower pollen was touched into the acetocarmine placed on a slide and observed under light microscope. Viable pollen takes acetocarmine and the shapes are regular, whereas non-viable pollen remains non-coloured with acetocarmine and were very irregular in shape.

Results

Taxonomic treatment

Zephyranthes Herb., App. [Bot. Reg.]: 36 (1821).

Argyropsis M. Roem., Syn. Ensat. : 125 (1847); *Arviela* Salisb., Gen. Pl. Fragm.: 135 (1866); *Habranthus* Herb., Bot. Mag.: t. 2464 (1824); *Mesochloa, Plectronema* and *Pogonema* Rafin., Fl. Tell. 4: 10 (1836); *Pyrolirion* Herb., App. [Bot. Reg.]: 37 (1821).

Small herbs with tunicate bulbs. Leaves simple, linear or lorate, appearing with or after the flowers. Flowers solitary, usually at the top of the long scape. Perianth funnel-shaped, tube short or long, dilated upward; tepals 6, rarely up to 8, in 2 series of 3 each, united at the base. Stamens 6, rarely up to 8, adnate to the perianth base; filaments long; anthers linear, dorsifixed. Carpels 3, united, ovary 3-celled, ovules many; style filiform; stigma 3-lobed. Fruit a capsule, sub-globose or depressed, loculicidally 3-valved. Seeds oblong, black.

Key to species of *Zephyranthes*

1.	Leaves terete; spathe covering the ovary	*Z. candida*
-	Leaves flat; spathe not covering the ovary	2
2.	Spathe not 2-fid; flowers yellow	*Z. tubispatha*
-	Spathe 2-fid; flowers pink or white	3
3.	Outer 3 tepals obtuse, pink in colour	*Z. carinata*
-	Outer 3 tepals acute, white but turn pink at maturity	*Z. atamasco*

Zephyranthes atamasco (L.) Herb., App. Reg. 36 (1821); Utech, Fl. North. America 26: 298 (2002). *Amaryllis atamasco* L., Sp. Pl. 1: 292 (1753); *A. atamasco* Blanco, Fl. Filip. : 254 (1837).

(Figs 1 & 5A).

English names: Atamasco Lily, Fairy Lily, Rain Lily, Easter Lily, Zephyr Lily.

Local name: *Sada Ghashphul*.

A perennial bulbous herb, bulb ovoid, c. 2.5 cm in diam., neck 2.5–5.0 cm long. Leaves linear, up to 15 cm long, bright green. Flowers solitary, terminal, bisexual; peduncle c. 21 cm long, hollow. Spathe simple, c. 3.0×0.6 cm, hyaline, tubular, 2-notched. Perianth segments 6, c. 5×2 cm, arranged in two rows, inner 3 smaller than the outer 3, white but lower 2 green in colour, changes from pure white to pink at maturity. Stamens 6, outer 3 large, c. 2.5 cm long, inner 3 small, c. 1.7 cm long; anthers linear, yellowish-orange. Carpels 3, united; ovary inferior, c. 0.4 cm long, 3-celled, placentation axile; style slender; stigma 3-notched, c. 4 cm long. Fruit a capsule, subglobose.

Flowering and fruiting: April to May.

Specimens examined: Dhaka: Dhaka University Botanical Garden, 11.4.2007, Sumona 21 (DUSH); 10.9.2013, Sumona 85 (DUSH).

Chromosome number: 2n = 12, 24 (Kumar and Subramaniam, 1986).

Habitat: Cultivated in gardens.

Distribution: Native to south-east America, naturalized in southern North America (Wade *et. al.*, 2014),widely cultivated in many countries including Bangladesh.

Uses: Cultivated as an ornamental plant in gardens. All parts are toxic especially bulb, may be fatal if eaten (Kates *et al.*, 1980).

Propagation: Through bulbs and seeds.

Zephyranthes candida (Lindl.) Herb., Bot. Mag. 53: t. 2607 (1826); Hajra & Verma, Fl. Sik. M. 1: 138 (1996); Raven & Zhengyi, Fl. China 24: 265 (2000); Utech, Fl. North America 26: 302 (2002). Hassan, Encycl. Flora & Fauna of Bangladesh 11: 351 (2007). *Amaryllis candida* Lindl., Bot. Reg. 9: t. 724 (1823). **(Figs 2 & 5B)**.

English name: Fairy Lily.

Local name: *Sada Ghashphul*.

A perennial clump-forming bulbous herb, bulb tunicated, ovoid, c. 2.5 cm in diam., neck 2.5–5.0 cm long. Leaves simple, terete, linear, up to 35 cm long and 0.5 cm in diam., hollow, obtuse, glabrous, dark green. Inflorescence solitary on terminal leafless scape. Flowers bisexual, incomplete, actinomorphic, epigynous, white; peduncle c. 26 cm long; spathe like bract present at the top of a long scape covered the ovary; bract c. 3×1 cm, brown in colour, lanceolate, glabrous. Tepals 6, c. 3.7×1.5 cm, free, ovate-lanceolate, white. Stamens 6, free, about half as long as the perianth; anthers c. 0.9 cm long, oblong, dorsifixed, yellow; filaments white, glabrous, more or less as long as anthers. Carpels 3, syncarpous; ovary inferior, c. 0.5×0.3 cm, 3-celled, ovules many; style slender, c. 1.7 cm long with stigma; stigma 3-notched; placentation axile. Fruit a capsule, subglobose, c. 0.8×1.2 cm, yellowish-green, 16–25 seeded. Seeds angular, flattened; testa black.

Flowering and fruiting: August to November.

Specimens examined: Dhaka: Science Library compound, University of Dhaka, 20.9.2007, Sumona 46 (DUSH); Nazrul Institute compound, University of Dhaka, 20.8.2011, Sumona 68 (DUSH).

Chromosome number: 2n = 19, 20, 36, 38, 40, 41, 48, 50 (Kumar and Subramaniam, 1986).

Habitat: Gardens, where it is widely cultivated.

Distribution: Originated from Argentina and Uruguay (Bateman *et al.*, 2004). Native to South America, naturalized in South China, cultivated in many countries including Bangladesh (Siddiqui *et al.*, 2007).

Uses: Used as an ornamental plant in gardens, containers or as a landscape plant. Bulb contains cytostatic constituents which can be used in the treatment of cancer (Pettit *et al.*, 1990).

Propagation: Through clumps of bulbs and seeds.

Zephyranthes carinata Herb., Bot. Mag. : t. 2594 (1825); Hajra & Verma, Fl. Sik. M. 1: 138 (1996); Raven & Zhengyi, Fl. China 24: 265 (2000); Utech, Fl. North America 26: 299 (2002). *Z. grandiflora* Lindl., Bot. Reg.: t. 902 (1825); Hassan, Encycl. Flora & Fauna of Bangladesh 11: 351 (2007). **(Figs 3 & 5C)**.

English names: Pink Rain Lily, Fairy Lily, Zephyr Lily, Pink Storm Lily.

Local names: *Golapi Ghashphul, Peyazphul*.

A bulbous, clump forming perennial herb, bulb tunicated, up to 2 cm in diam. Leaves simple, exstipulate, linear, obtuse, entire, glabrous, green, up to 35.0×0.8 cm, appearing with flowers. Inflorescence solitary on terminal leafless scape. Scape c. 18 cm long, light green, produce a single maroon lipstick-like bud on a top. Flowers c. 7.5 cm long, c. 7.5 cm across, spreading, last a few days, closing up at night. Spathe simple, c. 2.0×0.6 cm, hyaline, tubular, 2-notched. Perianth segments 6, rarely up to 8, funnel-shaped, rose or pink, 2–4 cm long, sub-elliptic to oblong–lanceolate. Stamens 6, sometimes 7–8, adnate to the throat of the perianth; anthers linear, yellow, narrow, dorsifixed; filament up to 2 cm long, white,. Carpels 3, syncarpous, ovary 3–celled, ovules many in each cell; style filiform, c. 2.5 cm long, placentation axile; stigma deeply 3–4 fid. Fruit a capsule, c. 0.5×0.5 cm, dark green, 6–10 seeded. Seeds black.

Flowering and fruiting: June to October. Blooming soon after a heavy rainfall.

Specimens examined: Dhaka: Dhaka University Botanical Garden, 19.9.2007, Sumona 45 (DUSH); Dhaka University Botanical Garden, 24.8.2014, Sumona 92 (DUSH); Nazrul Institute compound, University of Dhaka, 20.8.2012, Sumona 76 (DUSH).

Chromosome number: 2n = 24, 36, 48 (Kumar and Subramaniam, 1986).

Habitat: Well-drained soils.

Distribution: Native of Central America and Mexico, distributed in warmer parts of America, widely cultivated in many countries with a warm climate (Siddiqui *et al.*, 2007). In Bangladesh, it is widely grown in many gardens.

Uses: The species is valued as an ornamental plant, along walkway or at the front of a sunny border. In China, bulbs are used to break fever and a paste of the bulb is used for boils. Bulbs possess alkaloids which might be used in the treatment of cancer (Wiart, 2012).

Propagation: Propagated by bulbs or seeds.

Zephyranthes tubispatha Herb., App. Reg.: 36 (1821); Hook. f., Fl. Brit. Ind. 6: 277 (1892); Prain, Beng. Pl. 2: 797 (1903); (L'Her.) Herb. *ex* Traub, Taxon 7: 110 (1958); Hassan, Encycl. Flora & Fauna of Bangladesh 11: 352 (2007). *Amaryllis tubispatha* L'Her., Sert. Angl.: 9 (1789). *Z. nervosa* Herb., Amaryll. : 172 (1837). **(Figs 4 & 5D)**.

English names: Zephyr Lily, Fairy Lily, Rain Lily.

Local name: *Holde Ghashphul*.

A small perennial herb with underground tunicated bulb, bulb c. 1.5×1.0 cm, grows singly. Leaves simple, linear, c. 30 cm long and 3 mm broad, green, entire, obtuse, appearing along with the flowers. Flowers solitary, pedunculate; bracts spathe-like, c. 2.3 cm long, situated at the top of a fistular scape, scape up to 28 cm long. Perianth segments 6, connate below, free above, funnel-shaped, c. 3.7 cm long, yellow. Stamens 6; anthers linear, dorsifixed, c. 0.7 cm long, orange, burst longitudinally; filament c. 1.4 cm long. Carpels 3, syncarpous, ovary 3-celled, c. 0.5 cm long, ovules many; placentation axile; style 1, c. 2 cm long, white; stigma 3–lobed, short. Fruit a subglobose capsule, loculicidally 3-valved, yellowish-green, c. 0.8×1.0 cm, 16–20 seeded. Seeds oblong, black, angled.

Flowering and fruiting: June to September.

Specimens examined: Dhaka: Dhaka University Botanical Garden, 26.5.2007, Sumona 37 (DUSH); Dhaka University Botanical Garden, 10.4.1968, Mozahar 101 (DUSH); Uttara, Sector-3, 18.8.1998, M. Salar Khan K 10115 (DACB); Dhaka University Botanical Garden, 30.6.1970, A.M. Huq 78 (DACB).

Chromosome number: 2n = 24 (Kumar and Subramaniam, 1986).

Habitat: Well-drained soils and grassy ground of hilly areas.

Distribution: Native of Peru, tropical America and the West Indies (Siddiqui *et al.*, 2007). This species is planted in gardens and has been naturalized in many countries including Bangladesh.

Uses: Used as an ornamental plant.
Propagation: By bulbs and seeds.

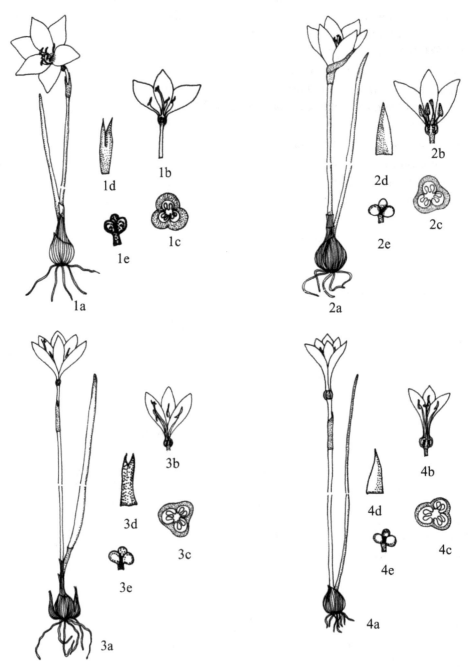

Figs 1-4. Habit sketch of four *Zephyrnthes* species. 1. *Z. atamasco*: 1a. Habit (×0.3); 1b. L.S. of flower (×0.2); 1c. T.S. of ovary (×2); 1d. Bract (×0.5); 1e. Fruit (×1). 2. *Z. candida*: 2a. Habit (×0.3); 2b. L.S. of flower (×0.2); 2c. T.S. of ovary (×3); 2d. Bract (×0.5); 2e. Fruit (×0.5). 3. *Z. carinata*: 3a. Habit (×0.3); 3b. L.S. of flower (×0.2); 3c. T.S. of ovary (×5); 3d. Bract (×0.5); 3e. Fruit (×0.5). 4. *Z. tubispatha*: 4a. Habit (×0.3); 4b. L.S. of flower (×0.2); 4c. T.S. of ovary (×2); 4d. Bract (×0.5); 4e. Fruit (×0.5).

Fig. 5. Habit of four *Zephyranthes* species: A. *Z. atamasco*; B. *Z. candida*; C. *Z. carinata*; D. *Z. tubispatha*.

Reproductive biology

Reproductive biology study on four *Zephyranthes* species revealed that *Z. atamasco, Z. candida, Z. carinata* and *Z tubispatha* all are self-pollinated. In *Z. atamasco* and *Z. carinata* single fruit sets each under both bagged and un-bagged condition after five days of bagging and no fruit formation occurs from emasculated flowers. In *Z. candida*, the emasculated flowers do not produce any fruit setting, while both bagged and un-bagged plants have single fruit setting after four days of bagging. Fruit setting starts after three days of bagging in *Z. tubispatha*, and a single fruit setting has been observed both in bagged and un-bagged plants. Seeds were germinated after five days of sowing in *Z. atamasco* and the rate of germination is very low (20%). There is no dormancy period. In *Z. candida*, three to four days were taken for germination of seeds. The germination rate was 100% when they were sown in August through October indicating that these months are most suitable for seed sowing in this species. Seeds lost complete viability after three months in *Z. candida*. The minimum three days were taken for germination of seeds in *Z. carinata* and the germination rate was 100% when sown in July. After three months of seeds sowing they were not germinated and seeds lost their viability. The study also revealed that after maturation, viability of seeds decreased gradually. The results showed that three to four days were required for germination of seeds in *Z. tubispatha* and the rate of germination was very high when sown just after seed collection. There is no dormancy period in *Z. tubispatha*. Seeds lost their viability after three months. The optimum period of seed germination, minimum days taken for germination and percentage of germination in four *Zephyranthes* species are depicted in Table 1. Different stages of seed germination in *Z. atmasco, Z. candida, Z. carinata* and *Z. tubispatha* are shown in Figure 6.

Table 1. Data on seed germination of four *Zephyranthes* species.

Species	Optimum period of seed germination	Minimum days taken for germination	Percentage of germination	Remark
Z. atamasco	May	5	20	Hypogeal
Z. candida	August-October	3	100	Pseudovivipary
Z. carinata	July	3	100	Pseudovivipary
Z. tubispatha	July-August	3	100	Hypogeal

In the present study, vegetative propagation in *Z. atamasco* and *Z. candida* through bulb has been found effective and more suitable than seeds. Plants propagated from seeds took about three years to bloom, whereas it took around two years from bulb separation. The study revealed that no bulblet was formed in *Z. tubispatha* indicating that seeds are the only means of regeneration in this species. The results showed that propagation through bulb separation took less time than that of seeds. Flower initiation took place through bulb is usually one year earlier than blooming through seeds. A comparative account of reproductive characters of *Zephyranthes* species i.e. time taken for seed germination, scape initiation to first flower, fruit formation after flowering, fruit maturation, number of flowers per scape, number of fruits per scape, number of seeds per fruit, time taken for flowering from seed germinated plants and time taken for flowering from bulb transferred plants are presented in Table 2.

Table 2. Reproductive characteristics of studied four species of *Zephyranthes*.

Species	Time taken for SG	Time taken for SIFF	Time taken for FFAF	Time taken for FMAF	No. of flowers/ scape	No. of fruits/ scape	No. of seeds/ fruit	Time taken for FFSG	Time taken for FFBT
Z. atamasco	5 days	8 days	4 days	7 days	1	1	2-8	3 years	2 years
Z. candida	3-4 days	7 days	3 days	6 days	1	1	6-20	3 years	2 years
Z. carinata	3-4 days	8 days	4 days	7 days	1	1	6-20	3 years	2 years
Z.tubispatha	3-4 days	7 days	3 days	6 days	1	1	8-30	3 years	Not possible

SG=Seed germination; SIFF= Scape initiation to first flowering; FFAF= Fruit formation after flowering; FMAF= Fruit maturation after formation; FFSG= Flowering from seed germination; FFBT= Flowering from bulb transfer.

Z. candida and *Z. carinata* showed pseudovivipary (Fig. 7). Seeds of these species are germinated inside the capsules after a heavy rainfall. In this process of germination, the hypocotyle elongated and came out of the seed forming a loop and developed narrow, straight

Fig. 6. Different stages of seed germination in four *Zephyraanthes* species: A-D. *Z. atamasco*; E-H. *Z. candida*; I-L. *Z. carinata*; M-P. *Z. tubispatha*; A,E,I&M: Flower; B,F,J&N: Fruits; C,G,K&O: Seeds; D,H,L&P: Seedling.

epicotyle. Hypogeal type of germination has been noticed in these species. Further investigation is needed to explore the mechanism of pseudovivipary in these species.

Fig. 7. Pseudovivipary in *Zephyranthes*: A. *Z. carinata*; B&C. *Z. candida*.

Pollen viability was tested in this study because of its importance in reproductive biology, and no seed formation takes place without viable pollen. The present investigation revealed that pollen viability ranged from 80 to 100% among the *Zephyranthes* species. The percentage of pollen viability was found 100% in *Z. candida* and *Z. tubispatha*, whereas, *Z. atamasco* and *Z. carinata* exhibited 80% and 98% viability, respectively. The viable pollens of these species are shown in Figure 8.

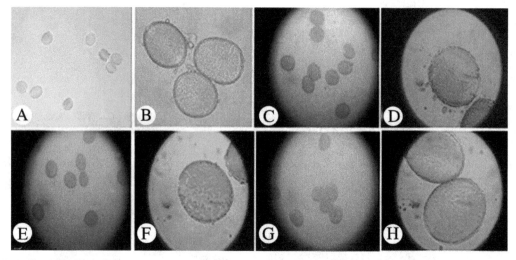

Fig. 8. Pollen viability of four *Zephyranthes* species: A&B. *Z. atamasco* (×10, ×40); C&D. *Z. candida* (×10, ×40); E&F. *Z. carinata* (×10, ×40); G&H. *Z. tubispatha* (×10, ×40).

Discussion

Zephyranthes are remarkable for the wide ecological niche they occupy, from xeric to temporarily flooded conditions, having many coveted ornamental characteristics. Flowers of

Zephyranthes appear in spring through fall after the first rains and they last one to two days, depending on sunlight and temperature, however, new flowers continuously develop for several days (Knox, 2009). In the present investigation we studied floral morphology, phenology pollination, seed germination and vegetative propagation of four *Zephyranthes* species occurring in Bangladesh, *viz.*, *Z. atamasco*, *Z. candida*, *Z. carinata* and *Z. tubispatha*. Among them *Z. candida* can easily be distinguished from the remaining species by its terete leaves and spathe covering the ovary.

Previous investigations in *Zephyranthes* have indicated that species with styles that are long relative to the stamens are self-incompatible, whereas species with short styles are self-compatible. Species with styles as long as the stamens may be either self-compatible or self-incompatible (Broyles and Wyatt, 1991). Studies on breeding system of a long-styled *Z. atamasco* revealed that 5%, 78%, and 92% fruit-set occurred in flowers which were bagged, self-pollinated, or cross-pollinated, respectively (Broyles and Wyatt, 1991). In our study, we found that *Z. atamasco* produced fruit setting in bagged and un-bagged plants, while no fruit formation occurred in emasculated flowers. Although it took very short time for germination of seeds, the germination rate is only 20% as observed in *Z. atamasco*. In *Z. candida*, all seeds sown during August to October were germinated only after three to four days of sowing which indicates this period as the most suitable time for seed sowing. Seeds lost their viability after three months in *Z. candida*. All seeds of *Z. carinata* sown in July were germinated, while the rate of seed germination was 100% when they were sown in July and August. In case of *Z. tubispatha* the rate of seed germination was 100% when sown in July and August, while this rate decreased to 80% when seeds were sown in September. None of the seeds was germinated in any of the four species of *Zephyranthes* employed in this study when they were sown after October. This indicates that seeds are not germinated after rainy seasons and they lost their viability in this period.

Pseudovivipary describes plants that produce apomictic or asexual propagules such as bulbils or plantlets in the place of sexual reproductive structures. Species with true vivipary tend to inhabit shallow marine habitats, either in mangrove or in seagrass communities, while pseudovivipary is most prevalent among terrestrial plants occurring in strongly seasonal environments, either growing at high altitudes and latitudes, or in semi-arid to arid areas (Elmqvist and Cox, 1996). All of these habitats are characterized by extraordinarily coarse-grained environments for seedling establishment, even though with major differences in patch size.

Vegetative propagation through pseudovivipary is known from over 100 species of grasses (Poaceae) and this can be caused by genetic factors, injury or unfavourable environmental conditions (Milton *et al.*, 2008). This phenomenon has also been reported in members of some dicotyledonous families including Crassulaceae (Mabberly, 1987), Oxalidaceae (van der Pijl, 1983) and Saxifragaceae (Lid and Lid, 1994). In Liliaceae, pseudovivipary was documented in *Allium* (Stebbins, 1950) and *Crinum viviparum* (Ansari and Nair, 1987). However, this phenomenon has never been described in *Zephyranthes*. Our study is the first of its nature reporting pseudovivipary in *Z. carinata* and *Z. candida*. Several authors have argued that pseudovivipary has evolved in response to a short growing season, enabling plants to complete the cycle of offspring production, germination and establishment during the few weeks of an arctic or alpine growing season (Lee and Harmer, 1980). Molau (1993) has pointed out that pseudovivipary among tundra plants is mostly prevalent among late-flowering species. The present investigation reveals diagnostic feature of pseudovivipary as noticed in *Z. candida* and *Z. carinata*. In order to understand the mechanism of pseudovivipary in *Zephyranthes* species further detailed study is needed. Based on the present investigation, it could be concluded that reproductive biological characters somehow can be used for delimiting *Zephyranthes* species; however, further studies including more taxa are needed for better understanding the taxonomy and interspecific relationships of the genus *Zephyranthes*.

Acknowledgement

The authors would like to thank anonymous reviewers for their suggestions and comments on the manuscript.

References

Anderson, G.J., Johnson, S.D., Neal, P.R. and Bernardello, G. 2002. Reproductive biology and plant systematics: the growth of a symbiotic association. Taxon **51**: 637–653.

Ansari, R. and Nair, V.J. 1987. Nomenclatural notes on some South Indian plants. J. Econ. Taxon. Bot. **11**: 205–206.

Arroyo-Leuenberger, S. and Leuenberger, B.E. 2009. Revision of *Zephyranthes andina* (Amaryllidaceae) including five new synonyms. Willdenowia **39**: 145–159.

Bateman, H., Kelpac, A., Moody, M., Sweeney, B. and Tomnay, S. 2004. The Essential Flower Gardening Encyclopedia. Fog City Press, USA, 608 pp.

Broyles, S. and Wyatt, R. 1991. The breeding system of *Zephyranthes atamasco* (Amaryllidaceae). Bull. Torrey Bot. Club **118**: 137–140.

Chauhan, B.S. and Johnson, D.E. 2008. Influence of environmental factors on seed germination and seedling emergence of *Eclipta* (*Eclipta prostrata*) in a tropical environment. Weed Science **56**: 383–388.

Chowdhury, M.R. and Hubstenberger, J. 2006. Evaluation of cross pollination of *Zephyranthes* and *Habranthus* species and hybrids. J. Arkansas Acad. Sci. **60**: 113–118.

Cox, P.A. 1990. Pollination and the evolution of breeding systems in Pandanaceae. Ann. Miss. Bot. Gard. **77**: 816–840.

Elmqvist, T. and Cox, P.A. 1996. The evolution of vivipary in flowering plants. Oikos **77**: 3–9.

Ferdousi, A., Rahman, M.O. and Hassan, M.A. 2014. Seed germination behaviour of six medicinal plants from Bangladesh. Bangladesh J. Plant Taxon. **21**(1): 71–76.

Flagg, R.O. and Flory, W.S. 1976. Origin of three Texas species of *Zephyranthes*. Pl. Life **32**: 67–80.

Flagg, R.O. and Smith, G.L. 2008. Delineation and distribution of *Zephyranthes* species (Amaryllidaceae) endemic to the southeastern United States. Castanea **73**(3): 216–227.

Flagg, R.O., Smith, G.L. and Flory, W.S. 2002. *Zephyranthes*. In: Flora of North America Editorial Committee (Eds), Flora of North America: North of Mexico. Magnoliophyta: Liliidae: Liliales and Orchidales. New York and Oxford. Vol. **26**, pp. 296–303.

Hajra, P.K. and Verma, D.M. 1996. Flora of Sikkim: Monocotyledons. Vol.1. Botanical Survey of India, New Delhi, India, 336 pp.

Hassan, M.A. and Fardous, Z. 2003. Seed germination, pollination and phenology of *Gloriosa superba* L. (Liliaceae). Bangladesh J. Plant Taxon. **10**(1): 95–97.

Hassan, M.A. and Khan, M.S. 1996. Pollination mechanism in *Polygonum* L. (*s.l.*). Bangladesh J. Plant Taxon. **3**(1): 67–70.

Karthikeyan, S., Jain, S.K., Nayar, M.P. and Sanjappa, M. 1989. Flora of India. Ser. **4**. Botanical Survey of India, Pune, India, 435 pp.

Kates, A.H., Davis, D.E., McCormack, J. and Miller, J.F. 1980. Poisonous plants of the southern United States. Cooperative Extension Service. University of Georgia, Athens, 30 pp.

Knox, G.W. 2009. Rainlily, *Zephyranthes* and *Habranthus* spp.: Low maintenance flowering bulbs for Florida gardens. Environmental Horticulture Department, Florida Cooperative Extension Service, Institute of Food and Agricultural Sciences, University of Florida, Florida, USA, 12 pp.

Kumar, V. and Subramaniam, B. 1986. Chromosome Atlas of Flowering Plants of the Indian Subcontinent. Vol. **II**, Botanical Survey of India, Calcutta, 464 pp.

Lee, J.A. and Harmer, R. 1980. Vivipary, a reproductive strategy in response to environmental stress? Oikos **35**: 254–265.

Lid, J. and Lid, D.T. 1994. Norsk Flora. 6th ed. with Reidar Elven. Detnorskesamlaget, Oslo.

Liza, S.A., Rahman, M.O., Uddin, M.Z., Hassan, M.A. and Begum, M. 2010. Reproductive biology of three medicinal plants. Bangladesh J. Plant Taxon. **17**(1): 69–78.

Mabberley, D.J. 1987. The Plant Book. Cambridge University Press, Cambridge.

Milton, S.J., Dean, W.R.J. and Rahlao, S.J. 2008. Evidence for induced pseudo-vivipary in *Pennisetum setaceum* (Fountain grass) invading a dry river, arid Karoo, South Africa. South African J. Bot.**74**: 348–349.

Molau, U. 1993. Relationships between flowering phenology and life history strategies in tundra plants. Arct. Alp. Res. **25**: 391–402.

Neal, P. and Anderson, G.J. 2005. Are 'mating systems' 'breeding systems' of inconsistent and confusing terminology in plant reproductive biology? Or is it the other way around? Plant Syst. Evol. **250**: 173–185.

Noltie, H.J. 1994. Flora of Bhutan, Vol. **3**. Part 1. Royal Botanic Garden, Edinburgh, 456 pp.

Pettit, G.R., Gaddamidi, V. and Cragg, G.M. 1984. Antineoplastic agents, 105. *Zephyranthes grandiflora.* J. Nat. Prod. **47**: 1018–1020.

Pettit, G.R., Cragg, G.M., Singh, S.B., Duke, J.A. and Doubek, D.L. 1990. Antineoplastic Agents 162. *Zephyranthes candida*. J. Nat. Prod. **53**(1): 176–178.

Rahman, M.Z., Rahman, M.O. and Hassan, M.A. 2012. Seed germination of two medicinal plants: *Desmodium pulchellum* (L.) Benth. and *D. triflorum* (L.) DC. Bangladesh J. Plant Taxon. **19**(2): 209–212.

Raina, N.S. and Khoshoo, T.N. 1972. Cytogenetics of tropical bulbous ornamentals. IX. Breeding systems in *Zephyranthes*. Euphytica **21**: 317–323.

Raven, P. and Zhengyi, W. (Eds). 2000. Flora of China. Flagillariaceae through Marantaceae. Vol. **24**, Beijing Science Press, 431 pp.

Siddiqui, K.U., Islam, M.A., Ahmed, Z.U., Begum, Z.N.T., Hassan, M.A., Khondker, M, Rahman, M.M., Kabir, S.M.H., Ahmad, A.T.A., Rahman, A.K.A. and Haque, E.U. (Eds). 2007. Encyclopedia of Flora and Fauna of Bangladesh. Vol. **11**, Angiosperms: Monocotyledons (Agavaceae–Najadaceae). Asiatic Society of Bangladesh, Dhaka, 399 pp.

Katoch, D. and Singh, B. 2015. Phytochemistry and pharmacology of genus *Zephyranthes*. Med. & Aromat. Plants **4**: 212.

Singer, R.B. and Sazima, M. 1999. The pollination mechanism in the '*Pelexia* alliance' (Orchidaceae: Spiranthinae). Bot. J. Linn. Soc. **131**: 249–262.

Spencer, L.B. 1973. A monograph of the Genus *Zephyranthes* (Amaryllidaceae) in North and Central America. Winston-Salem, North Carolina Ph.D. dissertation, Wake Forest University.

Spurrier, M.A., Smith, G.L., Flagg, R.O. and Serna, A.E. 2015. A new species of *Zephyranthes* (Amaryllidaceae) from Mexico. Novon **24**: 289–295.

Stebbins, G.L. 1950. Variation and evolution in plants. Columbia Univ. Press, New York, 643 pp.

Utech, F.H. 2002. Flora of North America: North of Mexico. Vol. **26**, Flora of North America Editorial Committee (Eds), Oxford University Press, New York, 752 pp.

Wade, G., Nash, E., McDowell, E., Beckham, B. and Crisafulli, S. 2014. Native Plants for Georgia, Part III: Wildflowers. UGA Extension Bulletin 987-3, 151 pp.

Wiart, C. 2012. Medicinal Plants of China, Korea and Japan: Bioresources for Tomorrow's Drugs and Cosmetics. CRC Press, Taylor & Francis Group, New York, London, 454 pp.

Wyatt, R. and Broyles, S.B. 1990. Reproductive biology of milkweeds (*Asclepias*): Recent advances. *In*: Kawano, S. (Ed.), Biological Approaches and Evolutionary Trends in Plants. Academic Press, London, pp. 255–272.

Yang, J., Lovett-Doust, J. and Lovett-Doust, L. 1999. Seed germination patterns in green dragon (*Arisaema dracontium*, Araceae). Am. J. Bot. **86**(8): 1160–1167.

van der Pijl, L. 1983. Principles of Dispersal in Higher Plants. 3rd ed. Springer, Berlin.

TYPE SPECIMENS OF NAMES IN *BAUHINIA* AND *PHANERA* (FABACEAE: CAESALPINIOIDEAE) AT CENTRAL NATIONAL HERBARIUM, HOWRAH (CAL)

S. BANDYOPADHYAY AND P.P. GHOSHAL[1]

Central National Herbarium, Botanical Survey of India, P.O. Botanic Garden, Howrah 711 103, West Bengal, India

Keywords: Fabaceae; Caesalpinioideae; *Bauhinia*; *Phanera*; Types; CAL.

Abstract

The types of the names in *Bauhinia* L. and *Phanera* Lour. at Central National Herbarium, Howrah (CAL) have been enumerated.

Introduction

In course of detailed taxonomic studies, types of the names have to be examined in order to confirm their identities. The types are deposited in many herbaria worldwide. It becomes easy to locate the types if databases of type specimens in different herbaria are made available. The purpose of the present paper is to inform the botanical community the 38 type specimens of names in *Bauhinia* L. and *Phanera* Lour. at CAL. Earlier, in the two volumes of the book entitled, 'Type collections in the Central National Herbarium', Datta *et al.* (1985) included two type specimens of *Bauhinia viz. B. hallieriana* Elmer and *B. whitfordii* Elmer as part of Elmer's collections and types at CAL, and Sammaddar (1991), in the second volume of the said book, included the types of *Phanera nicobarica* N.P. Balakr. & Thoth.

Materials and Methods

The protologues of all the names have been studied and the current status of the names has been given after scrutiny of relevant literature, *viz.,* de Wit (1956); Thothathri (1965); Larsen and Larsen (1996); Govaerts (1996); Sinou *et al.* (2009); Bandyopadhyay *et al.* (2012); Bandyopadhyay (2013a); Mackinder and Clark (2014); Bandyopadhyay (2014) and Bandyopadhyay and Ghoshal (2015).

The names appearing in bold are presently the accepted names. If the name, whose type citation has been provided is presently not the accepted name, then after the type citation and notes (if any) the correct name has been provided either with the symbol '=' or '≡' to denote whether it is a heterotypic or homotypic synonym, respectively. The terms associated with the types have been given as far as possible. In case of lectotype and neotype, the name of the author(s) who has/have designated it has also been given. If the types are isotype or isosyntype, then the location where the holotype or lectotype is deposited has been mentioned. The 13 character alpha-numeric barcode starting with the acronym CAL for each type specimen has been given if they are digitized; otherwise the accession number (6 digits) of CAL has been provided.

The majority of these type specimens at CAL have been digitized and barcoded. The images of the desired type specimens may be sent on request for study.

---

[1]Corresponding author. Email: pp_ghoshal@rediffmail.com

Systematic enumeration

1. *Bauhinia chalcophylla* H.Y. Chen, J. Arnold Arbor. 19: 130 (1938).
   Type: China, Yunnan, Talang, 3500 ft, 1901, *A. Henry* 13240 (isotype 137840).
   Notes: The holotype is at NY.
   ≡ **Phanera chalcophylla** (L. Chen) Mackinder & R. Clark, Phytotaxa 166: 54 (2014).

2. *Bauhinia curtisii* Prain, J. Asiat. Soc. Bengal, Pt. 2, Nat. Hist. 66: 195 (1897).
   Type: Malaysia, Kedah, Polo Langkawi, at the Lake Suni(?), 1890, *Curtis* 2619 (syntype CAL0000011272).
   ≡ **Phanera curtisii** (Prain) Bandyop. & Ghoshal, Telopea 18: 141 (2015).

3. *Bauhinia diptera* Collett & Hemsl., J. Linn. Soc., Bot. 28: 52 (1890).
   Type: Burma, Upper Burma: Shan Hills, Ywangan, 4000 ft, May 1888, *H. Collett* 727 (lectotype CAL0000011313; isolectotypes CAL0000011311, CAL0000011312); Shan Hills, Koni, 4000 ft, Apr. 1888, *H. Collett* 586 (syntypes CAL0000011314, CAL0000011315).
   Notes: Thothathri (1965) designated the lectotype.
   = **Phanera yunnanensis** (Franch.) Wunderlin, Phytoneuron 19: 1 (2011).

4. *Bauhinia enigmatica* Prain, J. Asiat. Soc. Bengal, Pt. 2, Nat. Hist. 66: 496 (1897).
   Type: Burma, Upper Burma, Southern Shan State, Fort Stedman, 1893, *Abdul Khalil s.n.* (lectotype CAL0000011306); Fort Stedman, 1894, *Abdul Khalil s.n.* (syntype CAL0000011304); Maymyo, June 1888, *Badal Khan* 89 (syntype CAL0000011305).
   Notes: Thothathri (1965) designated the lectotype.
   = **Bauhinia brachycarpa** Wall. *ex* Benth. in Miq., Pl. Jungh. 2: 261 (1852).

5. *Bauhinia ferruginea* Roxb., Fl. Ind. (ed. Carey) 2: 331 (1832).
   Type: Penang, *Wall. Cat. num. list no.* 5776 (isoneotype 137421).
   Notes: The neotype, designated by de Wit (1956: 454), is at K.
   ≡ **Phanera ferruginea** (Roxb.) Benth. in Miq., Pl. Jungh. 2: 262 (1852).

6. *Bauhinia foveolata* Dalzell, J. Linn. Soc., Bot. 13: 188 (1872).
   Type: India, North Canara, near Yellapore, 18 [no year given], *W.A. Talbot* 20 (neotype CAL0000007180).
   Notes: Bandyopadhyay (2011) designated the neotype.
   ≡ **Piliostigma foveolatum** (Dalzell) Thoth., Bull. Bot. Soc. Bengal 19: 131, 1967 (1965).

7. *Bauhinia glabrifolia* (Benth.) Baker var. *maritima* K. Larsen & S.S. Larsen, Thai Forest Bull., Bot. 25: 14 (1997).
   Type: Burma, Tenasserim & Andamans, *Helfer* 1880 (isotype CAL0000011309).
   Notes: The holotype is at K.
   ≡ **Phanera glabrifolia** Benth. var. **maritima** (K. Larsen & S.S. Larsen) Bandyop., Edinburgh J. Bot. 70: 363 (2013).

8. *Bauhinia hallieriana* Elmer, Leafl. Philipp. Bot. 2: 691 (1910).
   Type: Philippine, Romblon, March 1910, *A.D.E. Elmer* 12172 (isolectotype CAL0000011332).

Notes: The lectotype, designated by Larsen and Larsen (1996: 461), is at NY.

= **Phanera aherniana** (Perkins) de Wit, Reinwardtia 3: 448 (1956).

9. *Bauhinia hullettii* Prain, J. Asiat. Soc. Bengal, Pt. 2, Nat. Hist. 66: 183 (1897).

   Type: Malaysia, Perak, Tapa, *L. Wray (Jr.)* 177 (syntype CAL0000011337); Malay Archipelago, Penang, 100–300 ft, Feb. 1881, *Dr. King's Collector* 1347 (syntype CAL0000011336).

   = **Phanera ferruginea** (Roxb.) Benth. in Miq., Pl. Jungh. 2: 262 (1852).

10. *Bauhinia integrifolia* Roxb., Fl. Ind. (Ed. Carey) 2: 331 (1832).

    Type: Penang, *Wall. Cat. num. list no.* 5780 (isolectotypes 137126, 137127).

    Notes: The lectotype, designated by de Wit (1956: 478), is at K.

    ≡ **Phanera integrifolia** (Roxb.) Benth. in Miq., Pl. Jungh. 2: 263 (1852).

11. *Bauhinia khasiana* Baker in Hook.f., Fl. Brit. India 2(5): 281 (1878).

    Type: Meghalaya, Khasia, 1000–3000 ft, *J.D. Hooker & T. Thomson s.n.* (isolectotype CAL0000011245).

    Notes: The first-step designated by Larsen and Larsen (1980a) and second-step lectotype, designated by Bandyopadhyay (2013b), are at K.

    ≡ **Phanera khasiana** (Baker) Thoth., Bull. Bot. Soc. Bengal 19: 131, 1967 (1965).

12. *Bauhinia kingii* Prain, J. Asiat. Soc. Bengal, Pt. 2, Nat. Hist. 66: 189 (1897).

    Type: Malay Peninsula, Perak, *B. Scortechini* 320 (isolectotypes CAL0000011274, CAL0000011276, CAL0000011277, CAL0000011278); Malay Peninsula, Perak, Gunong Batu Pateh, 4500 ft, 1887, *L. Wray (Jr.)* 392 (syntypes CAL0000011273, CAL0000011275, CAL0000011280); Selangor, Bukit Etam, *H. Kellsall* 2001 (syntype CAL0000011279).

    Notes: The lectotype, designated by de Wit (1956: 499), is at K.

    ≡ **Phanera kingii** (Prain) Bandyop. *et al.*, Bangladesh J. Pl. Taxon. 19: 57 (2012).

13. *Bauhinia meeboldii* Craib, Repert. Spec. Nov. Regni Veg. 12: 392 (1913).

    Type: Burma, Lower Burma, Maunglow, Mergui, March 1911, *A. Meebold* 14280 (CAL0000011296, CAL0000011297, CAL0000011298, CAL0000011299).

    ≡ **Phanera meeboldii** (Craib) Thoth., Bull. Bot. Soc. Bengal 19: 133. 1967 (1965).

14. *Bauhinia mirabilis* Merr., Univ. Calif. Publ. Bot. 15: 103 (1929).

    Type: Borneo, Elphinstone Province, Tawao, *A.D.E. Elmer* 21432 (isolectotype CAL0000011333).

    Notes: The lectotype, designated by Larsen and Larsen (1996: 503), is at A.

    = **Lysiphyllum dipterum** (Blume *ex* Miq.) Bandyop. & Ghoshal, Phytotaxa 178: 288 (2014).

15. *Bauhinia mollissima* Wall. *ex* Prain, J. Asiat. Soc. Bengal, Pt. 2, Nat. Hist. 66: 176 (1897).

    Type: Wall. Cat. num. list no. 5782 (isolectotype 137282).

    Notes: de Wit (1956: 403) cited *Wallich num. list no.* 5782 (K) as the 'type'. This should be considered as effective lectotypification according to Art. 7.10 of ICN (McNeill *et al.*, 2012). Larsen and Larsen (1979: 8) cited *Wallich num. list no.* 5782 (K) as the lectotype but in Flora Malesiana (Larsen and Larsen, 1996: 451), *Wallich num. list no.* 5782 (K) as the holotype.

≡ **Bauhinia pottsii** G.Don var. **mollissima** (Wall. *ex* Prain) Govaerts, World Checkl. Seed Pl. 2: 10 (1996).

Notes: B. *pottsii* G. Don var. *mollissima* (Wall. *ex* Prain) K. Larsen & S.S. Larsen, Bot. Tidsskr. 74: 8 (1979) is not a valid combination because the basionym page was not cited.

16. *Bauhinia ornata* Kurz, J. Asiat. Soc. Bengal, Pt. 2, Nat. Hist. 42: 72 (1873).

    Type: Pegu, 11.2 (1871), *S. Kurz* 2579 (lectotype CAL0000011252); (isolectotypes CAL0000011248, CAL0000011249, CAL0000011251); Pegu, 8.4 (1871), *S. Kurz* 2579 (syntype CAL0000011247); Burma, Pegu, E. and W. Slopes, Choungmenah Chg, 8.4.71, *S. Kurz* 2579 (syntype CAL0000011250).

    Notes: Thothathri (1965) designated the lectotype. Bandyopadhyay (2012a) discussed in details about the typification of the name.

    ≡ **Phanera ornata** (Kurz) Thoth., Bull. Bot. Soc. Bengal 19: 134, 1967 (1965).

17. *Bauhinia ornata* Kurz subsp. *mizoramensis* Bandyop., B.D. Sharma & Thoth., Nordic J. Bot. 12: 223 (1992).

    Type: Mizoram, Mizo hills, *R. Dutta* 33793 (holotype CAL0000011253).

    Notes: The type specimen was collected from Saiha, now in East Chimtuipui district, sometimes in July or August, 1963 (see Bandyopadhyay, 2001a).

    ≡ **Phanera ornata** (Kurz) Thoth. subsp. **mizoramensis** (Bandyop. *et al.*) Bandyop. *et al.*, Bangladesh J. Pl. Taxon. 19: 58 (2012).

18. *Bauhinia polycarpa* Wall. *ex* Benth. in Miq., Pl. Jungh. 2: 261 (1852).

    Type: 1827, *Wall. Cat. num. list no.* 5787 (isolectotypes 136699, 136706).

    Notes: The lectotype, designated by Larsen and Larsen (1996: 455) is at K.

    = **Bauhinia viridescens** Desv., Ann. Sci. Nat. (Paris) 9: 429 (1826).

19. *Bauhinia pottingeri* Prain, J. Asiat. Soc. Bengal, Pt. 2, Nat. Hist. 67: 289 (1898).

    Type: N.E. of Burma, Namlao to Bansparao, 500–2000 ft, 23.3. ?, *R.A. Pottinger s.n.* (lectotype CAL0000011318).

    Notes: Thothathri (1965) cited *R.A. Pottinger s.n.* (CAL) as 'type'. This has to be considered as an effective lectotypification according to Art. 7.10 of ICN (McNeill *et al.*, 2012).

    ≡ **Phanera pottingeri** (Prain) Thoth., Bull. Bot. Soc. Bengal 19: 133 (1967).

20. *Bauhinia ridleyi* Prain, J. Asiat. Soc. Bengal, Pt. 2, Nat. Hist. 66: 185 (1897).

    Type: Malay Peninsula, Perak: Pangkalin Bahra, Dec. 1880, *Dr. King's Collector* 1096 (syntypes CAL0000011321, CAL0000011326); Thwbing, Aug. 1884, *B. Scortechini* 519 (syntype CAL0000011322); Batu Gajch?, Feb. 1885, *B. Scortechini* 140a (syntype CAL0000011323); *B. Scortechini* 140 (syntype CAL0000011324).

    ≡ **Phanera ridleyi** (Prain) A. Schmitz,, Bull. Bot. Soc. Roy. Bot. Belgique 110: 15 (1977).

21. *Bauhinia rosea* Kurz, J. Asiat. Soc. Bengal, Pt. 2, Nat. Hist. 42: 72 (1873).

    Type: Burma, Pegu, Ein Forest Palween, between Kwaymapyoo chg., May 54, *D. Brandis s.n.* (lectotype CAL0000011302); *D. Brandis s.n.* (isolectotype CAL0000011301).

    Notes: Thothathri (1965) cited *D. Brandis s.n.* (CAL) as 'type'. This has to be considered as an effective lectotypification according to Art. 7.10 of ICN (McNeill *et al.*, 2012).

    = **Phanera kurzii** (Prain) Thoth., Bull. Bot. Soc. Bengal 19: 133 (1967).

22. *Bauhinia scortechinii* Prain, J. Asiat. Soc. Bengal, Pt. 2, Nat. Hist. 66: 188 (1897).

    Type: Perak, G. Haram, May 84, *B. Scortechini* 698 (isolectotypes CAL0000011319, CAL0000011320).

    Notes: The lectotype, designated by Larsen and Larsen (1996: 507), is at K.

    ≡ **Phanera bidentata** (Jack) Benth. in Miq., Pl. Jungh. 2: 263 (1852).

23. *Bauhinia sulphurea* C.E.C.Fisch., Bull. Misc. Inform. Kew 1927: 85 (1927).

    Type: Burma, Tenasserim, Tenasserim river, 28.2.1926, *C.E. Parkinson* 1951 (paratype CAL0000011316).

    ≡ **Phanera sulphurea** (C.E.C. Fisch.) Thoth., Bull. Bot. Soc. Bengal 19: 133, 1967 (1965).

24. *Bauhinia strychnoidea* Prain, J. Asiat. Soc. Bengal, Pt. 2, Nat. Hist. 66: 195 (1897).

    Type: Perak, Larut, Goping, 300–500 ft, April 1884, *Dr. King's Collector* 5914 (isolectotypes CAL0000011267, CAL0000011268); Malaya Peninsula, Perak, Kinta, 300–800 ft, January 1885, *Dr. King's Collector* 7054 (syntypes CAL0000011265, CAL0000011269, CAL0000011270); Selangor, Caves Kuwala Lumpur, *H. Kelsall* 1971 (syntype CAL0000011266); Perak, *Scortechini s.n.* (syntype CAL0000011271).

    Notes: The lectotype, designated by de Wit (1956: 429), is at K.

    ≡ **Phanera strychnoidea** (Prain) Bandyop. & Ghoshal, Telopea 18: 142 (2015).

25. *Bauhinia tenuiflora* Watt. *ex* C.B. Clarke, J. Linn. Soc., Bot. 25: 18 (1889).

    Type: India, Muneypore, Nongjaibang, 1700 ft, 30 Nov. 1885, *C.B. Clarke* 42342 B (syntype CAL0000011243); *C.B. Clarke* 42304 D (syntype CAL0000011244).

    Notes: A loose pod (CAL0000011243) mounted on the same herbarium sheet bearing the flowering specimen *C.B. Clarke* 42304 D (CAL0000011244).

    ≡ **Phanera glauca** Benth. subsp. **tenuiflora** (Watt *ex* C.B. Clarke) A.Schmitz, Bull. Soc. Roy. Bot. Belgique 110: 14 (1977).

26. **Bauhinia tortuosa** Collett & Hemsl., J. Linn. Soc., Bot. 28: 52, t. 8 (1890).

    Type: Burma, Upper Burma, Shan Hills, Koni, 5000 ft, May 1888, *H. Collet* 561 (lectotype CAL0000011303; isolectotype CAL0000011317)

27. *Bauhinia viridiflora* Backer, Bull. Jard. Bot. Buitenzorg Ser. 3, 2: 323 (1920).

    Type: Indonesia, Java, Birak Dense (Fiji Pasoedja), *Backer* 8801 (isolectotype CAL0000011307).

    Notes: The lectotype, designated by de Wit (1956: 473), is at BO.

    ≡ **Phanera bassacensis** (Pierre *ex* Gagnep.) de Wit var. **backeri** de Wit, Reinwardtia 3: 473 (1956).

28. **Bauhinia whitfordii** Elmer, Leafl. Philipp. Bot. 1: 229 (1907).

    Type: Philippine, Luzon Island, Province Benguet, Baguio, March 1907, *A.D.E. Elmer* 8897 (CAL0000011331).

    Notes: de Wit (1956: 483) cited *A.D.E. Elmer* 8897 (A) as the holotype. Larsen and Larsen (1996: 480) designated *A.D.E. Elmer* 8897(NY) as the lectotype.

    = **Phanera integrifolia** Benth. var. **nymphaeifolia** (Perkins) Mackinder & R.Clark, Phytotaxa 166: 57 (2014).

29. *Bauhinia wrayi* Prain, J. Asiat. Soc. Bengal, Pt. 2, Nat. Hist. 66: 191 (1897).

Type: Malay Peninsula, within 300 ft, Dec. 1883, *Dr. King's Collector* 5243 (isolectotypes CAL0000011290, CAL0000011291, CAL0000011292); Malay Peninsula, Perak, Larut, within 300 ft, 1881, *H. Kunstler* 2446 (syntype CAL0000011281); Perak, *B. Scortechini* 1652 (syntype CAL0000011282); Perak, Assan Kumbong, May 1888, *L. Wray (Jr.)* 1934 (syntype CAL0000011283); Assan Kumbong, *L. Wray (Jr.)* 2782 (syntypes CAL0000011284, CAL0000011285); Malay Peninsula, near ? Selangore, 1000–1200 ft, April 1886, *H. Kunstler* 8758 (syntype CAL0000011286); Malay Peninsula, Perak, Larut, within 300 ft, Aug. 1881, *Dr. King's Collector* 2466 (syntype CAL0000011287); Malay Peninsula, Perak, Larut, within 300 ft, March 1883, *H. Kunstler* 4049 (syntype CAL0000011288); Malay Peninsula, Perak, Larut, within 300 ft, Aug. 1881, *Dr. King's Collector* 2238 (syntype CAL0000011289).

Notes: The lectotype, designated by de Wit (1956: 518), is at K (also see Bandyopadhyay, 2001b).

≡ **Phanera wrayi** (Prain) de Wit, Reinwardtia 3: 517 (1956).

30. **Phanera glabrifolia** Benth. in Miq., Pl. Jungh. 2: 263 (1852).

Type: Cultivated at Hort. Bot. Calc. (labelled as '*Bauhinia piperifolia* Roxb.'), (isolectotypes CAL0000011293, CAL0000011294, CAL0000011295, CAL0000011308).

Note: The lectotype, designated by Larsen and Larsen (1980b), is at K.

31. **Phanera glauca** Benth. in Miq., Pl. Jungh. 2: 265 (1852).

Type: Amherst, 18 Feb. 1827, *Wall. Cat. num. list no.* 5785 (137381).

Notes: The lectotype, designated by Larsen and Larsen (1996: 478), is at K.

32. **Phanera glauca** Benth. subsp. **tenuiflora** (Watt *ex* C.B. Clarke) A. Schmitz var. **gandhiana** Gogoi & Bandyop., J. Bot. Res. Inst. Texas 8: 71 (2014).

Type: Arunachal Pradesh, Anjaw district, in between Changwanti and Walong, 800 m, 20 May 2011, *R. Gogoi* 24374A (holotype CAL0000025065).

33. **Phanera glauca** Benth. subsp. **tenuiflora** (Watt *ex* C.B. Clarke) A. Schmitz var. **murlenensis** Ram Kumar *et al.*, Phytotaxa 166: 155 (2014).

Type: Mizoram, Murlen National Park, in the buffer region of the Park between Vapar to Ngur, ca. 1400 m, 11.4.2013, *Ramesh Kumar & party* 128363 (holotype CAL0000026092).

34. *Phanera griffithiana* Benth. in Miq., Pl. Jungh. 2: 263 (1852).

Type: Malay Peninsula, Malacca, *Griffith* 1867 (CAL0000011330).

≡**Phanera ferruginea** (Roxb.) Benth. var. **griffithiana** (Benth.) Bandyop. *et al.*, Bangladesh J. Pl. Taxon. 19: 57 (2012).

35. *Phanera jampuiensis* Darlong & D. Bhattach., Kew Bull. 69: 9534 (2014).

Type: Tripura, North district, Jampui hill range, Tlangsang, 770 m, 16 April 2013, *L. Darlong* 10397 (holotype CAL000025210).

= **Phanera glabrifolia** Benth., Pl. Jungh. 2: 263 (1852).

36. **Phanera lucida** Benth. in Miq., Pl. Jungh. 2: 262 (1852).

Type: *Wall. Cat. num. list no.* 5779A (isolectotype 137039).

Notes: The lectotype, designated by de Wit (1956: 511), is at K.

37. **Phanera nervosa** Benth. in Miq., Pl. Jungh. 2: 262 (1852).

Type: Mt. Sillhet, *Wall. Cat. num. list no.* 5777 (isolectotype 137449).

Notes: The lectotype, designated by Bandyopadhyay (2012b), is at K.

38. *Phanera nicobarica* N.P. Balakr. & Thoth., Bull. Bot. Surv. India 17: 201 1978 (1975).

Type: Andaman & Nicobar Islands, Great Nicobar: 15 km on East-West Road, ± 100 m, 23 Aug. 1975, *N.P. Balakrishnan* 3043 A (holotype CAL0000011257); 3043 B (isotype CAL0000011258), 3043 C (isotype CAL0000011259), 18 km on North-South Road, ± 25 m, 17 July 1976, *N.P. Balakrishnan* 3824 A (paratype CAL0000011260, CAL0000011261), 3824 B (paratype CAL0000011262, CAL0000011263); on the way from Galathea Bay to Pulobaha Bay, ± 125 m, 26.3.1966, *K. Thothathri & S.P.Banerjee* 11661, '10661' typo. Error in protologue (paratype CAL0000011264).

= **Phanera stipularis** (Korth.) Benth. in Miq., Pl. Jungh. 2: 263 (1852).

Acknowledgements

We thank Dr. Paramjit Singh, Director, Botanical Survey of India and Dr. P.V. Prasanna, Scientist "F' & Head of the Office, Central National Herbarium, Botanical Survey of India for the facilities. We also thank Dr. R. Govaerts (K) for providing a relevant page from World Checklist of Seed Plants, and the anonymous reviewers for improving the manuscript.

References

Bandyopadhyay, S. 2001a. Miscellaneous notes on *Bauhinia* L. (Leguminosae: Caesalpinioideae) – II. J. Econ. Taxon. Bot. **25**: 10–12.

Bandyopadhyay, S. 2001b. On the type of *Bauhinia wrayi* Prain (Leguminosae: Caesalpinioideae). J. Bombay Nat. Hist. Soc. **98**: 490–491.

Bandyopadhyay, S. 2011. Neotypification of *Bauhinia foveolata* (Leguminosae: Caesalpinioideae). J. Jap. Bot. **86**: 169.

Bandyopadhyay, S. 2012a. Typification of *Bauhinia ornata* (Leguminosae: Caesalpinioideae) – One last time. Nelumbo **54**: 263–264.

Bandyopadhyay, S. 2012b. Lectotypification of *Bauhinia nervosa* (Leguminosae: Caesalpinioideae). J. Bot. Res. Inst. Texas **6**: 109–111.

Bandyopadhyay, S. 2013a. Two new varietal combinations in *Phanera* (Leguminosae: Caesalpinioideae). Edinburgh J. Bot. **70**: 363–365.

Bandyopadhyay, S. 2013b. Second-step lectotypification of *Bauhinia khasiana* Baker (Leguminosae: Caesalpinioideae). Candollea **68**: 99–103.

Bandyopadhyay, S. 2014. Tribe *Cercideae (Fabaceae: Caesalpinioideae)*. *In*: Singh, P. and Bandyopadhyay, S. (Eds), *Fasc. Fl. India* 26. Botanical Survey of India, Kolkata.

Bandyopadhyay, S. and Ghoshal, P.P. 2015. Seven new combinations in *Phanera* (Fabaceae: Caesalpinioideae: Cercideae). Telopea **18**: 141–144.

Bandyopadhyay, S., Ghoshal, P.P. and Pathak, M.K. 2012. Fifty new combinations in *Phanera* Lour. (Leguminosae: Caesalpinioideae) from paleotropical region Bangladesh J. Pl. Taxon. **19**: 55–61.

Datta, A., Pramanick, B.B. and Nayar, M.P. 1985. Elmer's Philippine and Bornean collections and their type material at Central National Herbarium (CAL). *In*: Type collections in the Central National Herbarium. Botanical Survey of India, Howrah.

Govaerts, R. 1996. World Checklist of Seed Plants **2**: 1–492. [page 10] Continental Publishing, Deurne.

Larsen, K. and Larsen, S.S. 1979. Taxonomic note on *Bauhinia pottsii* complex. Bot. Tidsskr. **74**: 7–11.

Larsen, K. and Larsen, S.S. 1980a. *Bauhinia.* In: Aubréville, A. and Leroy, J.-F. (Eds), Fl. Cambodge Laos Viêtnam 18: 146–210. Paris.

Larsen, K., and Larsen, S.S. 1980b. Notes on the genus *Bauhinia* in Thailand. Thai Forest Bull., Bot. **13**: 37–46.

Larsen, K. and Larsen, S.S. 1996. *Bauhinia. In:* Kalkman, C., Kirkup, D.W., Nooteboom, H.P., Stevens, P.F. and Wilde, W.J.J.O. de (Eds), Flora Malesiana **12**: 442–535. Rijksherbarium/Hortus Botanicus, Leiden University, The Netherlands.

Mackinder, B.A. and Clark, R. 2014. A synopsis of the Asian and Australasian genus *Phanera* Lour. (Cercideae: Caesalpinioideae: Leguminosae) including 19 new combinations. Phytotaxa **166**: 49–68.

McNeill, J., Barrie, F.R., Buck, W.R., Demoulin, V., Greuter, W., Hawksworth, D.L., Herendeen, P.S., Knapp, S., Marhold, K., Prado, J., Prud'homme, van Reine W.F., Smith, G.F., Wiersema, J.H., Turland, N.J. (Eds) 2012. International Code of Nomenclature for algae, fungi, and plants (Melbourne Code). Adopted by the Eighteenth International Botanical Congress Melbourne, Australia, July 2011, A.R.G. Gantner Verlag KG. [*Regnum Veg.* 154].

Sammaddar, U.P. 1991. Type collections in the Central National Herbarium, Vol. **2**. Botanical Survey of India, Howrah.

Sinou, C., Forest, F., Lewis, G.P. and Bruneau, A. 2009. The genus *Bauhinia s.l.* (Leguminosae): a phylogeny based on the plastid *trn*L-*trn*F region. Botany **87**: 947–960.

Thothathri, K . 1965 publ 1967. Studies in Leguminosae 5. Taxonomic and nomenclatural notes on the Indo-Burmese species of *Bauhinia* Linn. Bull. Bot. Soc. Bengal **19**: 130–134.

de Wit, H.C.D. 1956. A revision of Malaysian Bauhinieae. Reinwardtia **3**: 381–539.

INDIAN CHEILANTHOID FERN - A NUMERICAL TAXONOMIC APPROACH

KAKALI SEN[1] AND RADHANATH MUKHOPADHYAY[2]

Department of Botany, University of Kalyani, Pin-741235, Kalyani, Nadia, West Bengal, India

Keywords: Cheilanthoid fern; Cluster analysis; Micromorphology; Numerical taxonomy.

Abstract

Twenty one species belonging to five genera (*viz. Aleuritopteris* Fēe, *Cheilanthes* Sw., *Doryopteris* J. Sm., *Notholaena* R. Brown, *Pellaea* Link.) of the Indian cheilanthoid ferns were studied to develop the new data set of micromorphological details *viz.* epidermal cells, stomatal morphotypes, venation pattern and spore ultrastructre. Cluster analysis was performed by using the two- state of multiple characters that separate the genus *Aleuritopteris* from *Cheilanthes* at the Eucladian distance of 5.1, though completely linked with other closely related genera, *viz. Doryopteris, Notholaena* and *Pellaea*. The taxonomic conundrum lies within these genera was resolved with numerical taxonomic study.

Introduction

Cheilanthoid ferns form an evolutionary group with the strong tendency to be confined in the three large continental/archipelago land areas of America, Africa and Asia-Malaysia. The centre of diversity of the genera is in America and especially in Mexico, where about 100 species form the richest xeric fern flora in the world (Tryon and Tryon, 1982). In India, the distribution range is very wide from the altitudinal variations of plains (100m) to the slopes and small pockets of Himalaya (3000 m.), Nilgiri and Palni hills of South (Nayar, 1962; Dixit, 1984; Pande and Pande, 2003; Sen and Mukhopadhyay, 2011). The group is characterized by the sporangia on the abaxial side of the lamina, covered or not by a marginal pseudoindusium without veins, sporangia approximate in sori or soral lines, stipes at the base with one vascular bundle, sometimes with two, lamina farinose or efarinose, stems with scales, rarely with hairs, base chromosome no. n=29 or 30 (Nayar, 1962; Tryon and Tryon, 1990).

Cheilanthes and *Aleuritopteris* are old and phylogenetically problematical genera (genus 'arduum' of Fee) generally included among the gymnogrammeoid or placed in the Cheilanthaceae (family *nov.*) by Nayar (1962) or Pteridaceae (Tryon and Tryon, 1982; Smith *et al.*, 2006). Difficulties in identifying discrete generic boundaries among the cheilanthoids have long been attributed to convergent evolution driven by adaptation to arid environment (Tryon and Tryon, 1973). The workers frequently echo the comment on that "there is an obvious need for the development of new data which will give a better insight into the evolutionary lines within the group" (Tryon and Tryon, 1973). And the workers of molecular systematics also inferred for the genus *Cheilanthes* that "it needs redefinition" (Smith *et al.*, 2006). Among cheilanthoid ferns the depositions of farina are used widely to delineate section *Aleuritopteris* (presence of farina) from section-*Cheilanthes* (absence of farina) though it breaks at the wider geographical scales (Nayar, 1962; Khullar, 1994). But the potential adaptive significance of the farina has made it a trait of evolutionary interest.

---

[1] Corresponding author. Email: itskakali@gmail.com
[2] CAS, Department of Botany, University of Burdwan, Pin-713104, Burdwan, West Bengal, India.
 Present Address: 8/3, Dinabandhu Mukherjee Lane, Shibpur, Howrah, Pin-711102, West Bengal, India.

To redefine *Aleuritopteris* and *Cheilanthes* and also to regenerate new character sets to resolve the ambiguity of their generic status our present study is attempted to focus mainly the micromorphological characters. On the basis of the multiple dataset of two character-state, taxa were clustered to establish the interrelationships that exist among them.

Materials and Methods

Detail list of specimens studied are mentioned in the Table 1. For cluster analysis of the 21 taxa studied as many as 9-two state characters (i.e. characters which exist in two alternative forms or states i.e. either present or absent for generic segregation (Table 2); for 12 spp. of *Aleuritopteris*, 43-two state characters (Table 3) and for 6 spp. of *Cheilanthes* 42-two state characters (Table 4), were taken into account to prepare the data matrix. Responses of each taxon to each of these characters were coded in a data matrix as '1' and '0' respectively for two alternative states i.e. presence or absence. The data thus recorded were further utilized in finding the overall similarities or rather the distance between taxa and putting them in clusters using the concept of 'Euclidean Distance' for measuring distance and 'Complete Linkage' for amalgamation or linkage. Statistical analysis was performed using Statistica-6 (Sneath and Sokal, 1973).

Images of the standard character-sets used for the analysis were taken in Leica QWIN 80 microscope and Scanning Electron Microscope, Model No. Japan Hitachi 530.

Results and Discussion

The morphometric study was performed to delineate the taxa of the cheilanthoid fern at the generic and infrageneric level. For numerical taxonomic study the suitable characters used at the generic level are mentioned in the Table 2. For clustering at the infrageneric level used characters are mentioned in Table 3 (Genus *Aleuritopteris*) and Table 4 (Genus *Cheilanthes*) respectively. Figure 4(A-I) and Fig. 5(A-K) shows the contrasting character states as stated in Tables 2-4. Previously, all the works performed by various workers (Nayar, 1962; Tryon and Tryon, 1982; Khullar, 1994) have given much importance to the farina character, which is a potential synapomorphy and have some evolutionary interest (Sigel *et al.*, 2011). But, the wholesome approach of character is giving a better clue of separation at both the generic and infrageneric level (Sen, 2014). The works at the molecular phylogeny also found some dispute for this group when the regional basis of works was performed (Gastony and Rollo, 1995, 1998; Zhang, 2007).

On the basis of phenetic study of the 5 genera, close relation or affinity with each other was noticed. The genus *Aleuritopteris* Fee and *Notholaena* R. Brown form a group and *Doryopteris* J. Sm and *Pellaea* Link. form another group; these two groups are allied with each other and form a broad group with *Cheilanthes* Sw. One important ambiguity which persisted so long regarding the generic segregation of *Aleuritopteris* from *Cheilanthes* gets some clear clue from this phenetic study in that they are quite apart from each other in phenogram but are linked. The characters as mentioned by Fraser-Jenkins and Dulawat (2009) to distinguish the genus *Cheilanthes* from *Aleuritopteris*, are narrow stipe scale and narrow leaf segment, which are very vague as is evident from our numerical data enlisted (Table 2).

To categorise the infrageneric taxa, phenetic study resolves some ambiguity. Placement of *C. subvillosa* Hook. under the genus *Cheilanthes* is also corroborated by our study as the taxon possesses similarity in generic characters with *Cheilanthes*. The placement of its synonym *Aleuritopteris subvillosa* (Hook.) Ching under the genus *Aleuritopteris* by Fraser-Jenkins and Dulawat (2009) is however not in conformity with the present study.

Table 1. List of taxa studied is mentioned here. For each taxon only one specimen is enlisted.

Sl.No.	Name of the taxa	Herbarium details
1.	*Aleuritopteris albomarginata* (Clarke) Ching	52761, 02.05.1975,R.D.Dixit,Takdah -Athmal Reserve, Darjeeling, West Bengal, 9125(CAL);
2.	*A. anceps* (Blanford) Panigrahi	59343,27.03.1985,B. Ghosh and S.R. Ghosh, K.T. Road,950 m. Manipur, CAL
3.	*A. argentea* (Gmel.)Fee	Zwa-Kabru, 6958 (CAL).
4.	*A. bicolor* (Roxb.) Fraser-Jenkins	KS – 183, 12.10.2012; Kakali Sen, Almora,
5.	*A. bullosa* (Kunze) Ching	1878, Zy. King, Nilgiri Hills,(CAL);
6.	*A. chrysophylla* (Hook.) Ching	08.08.1892, G.A. Gammie,Lachung, Sikkim, 7136 (CAL)
7.	*A. doniana* S.K.Wu	KS -146, 06.10.2010, Kakali Sen Dello Kalimpong(BURD)
8.	*A. formosana* (Hayata)Tagawa	KS -149,06.10.2010,Kakali Sen,Dello, Kalimpong(BURD)
9.	*A.grisea* (Blanf) Panigr.	15205, Feb. 1972, Panigrahi,Bilaspur, M.P. (CAL).
10.	*A. rufa* (Don) Ching	KS -204,09.10.2012, Kakali Sen, Samla Tal, Tanakpur, Uttarakhand(BURD)
11.	*A. subargentea* Ching ex Sk. Wu	July, 1904, J. Walton, Sangpo valley, 6952 (CAL)
12.	*A. subdimorpha* (C.B.Clarke and Baker) Fraser-Jenk.	KS -157,08.10.2010, Kakali Sen, Bhusuk, Gangtok(BURD)
13.	*Cheilanthes acrostica* (Balbis) Tod	02.01.1986, B.P.Uniyal Archi, Jammu and Kashmir, 80379, (CAL);
14.	*C. belangeri* (Bory) C.Chr.	21.10.1952, Rev. B. Godfrey, 7088, N.Lushai Hills, Assam (CAL);
15.	*C. keralansis* Nair and Ghosh.	49442, 29.07.1977, A.N. Henry, Kanyakumari, Keeriparai(CAL)
16.	*C. mysorensis* Wall. ex. Hook.	09.11.2001, P. Amrutalakshmi, Nellore, Andhrapradesh, 25119 (CAL);
17.	*C. subvillosa* Hook.	466, Tamilnadu, 9048 (CAL)
18.	*C. tenuifolia* (Burm.) Sw.	8614, 24.09.81, M.K. Mama and U.P. Samaddar, Netarhat, Palamau Dist., Bihar, 936 (CAL);
19.	*Doryopteris concolor* (Langsd. and Fisch.) Kuhn	04.11.1996, Sanchita Gangopadhyay, Kodaikanal (BURD);
20.	*Notholaena marantae* (L.)Desv.	3673, 18.09.1984, J.F.Duthie (BURD);
21.	*Pellaea bovinii* Hook.	7671, December, 1910, A. Meeblod, $6000^{ft.}$ Devicolani, S. India, 13487 (CAL).

Table 2. Characters taken for Generic segregation of cheilanthoid ferns by cluster analysis.

Characters taken	Character State	
	(0)	(1)
1. Indusium	absent	present
2. Farina	absent	present
3. Leaf texture	coriaceous	membranous or herbaceous
4. Pinna	sessile	stalked
5. Vascular commissure	absent	present
6. Indument on leaf surface	absent	present
7. Pinna dissection	unipinnate leaf present	always more than 1-pinnate
8. Pinna margin	dissected	entire
9. Non-perinate spores	present	absent

Table 3. Characters taken for cluster analysis of *Aleuritopteris spp.* (infrageneric level).

Characters taken	Character State	
	(0)	(1)
1. Stipe scale	base	throughout
2. Rachis scale	absent	present
3. Scales	present in costae & costule	absent
4. Rhizome scale	non-clathrate	clathrate
5. Rhizome scale	concolorous	bicolorous
6. Rhizome scale	non-glandular	glandular
7. Stipe scale	non-clathrate	clathrate
8. Stipe scale	concolorous	bicolorous
9. Farina	white	golden Yellow
10. Indusial	margin entire	with fimbriated projections
11. Stomatal type	polocytic	polo- and other type
12. Position of stomata	hypostomatic	amphistomatic
13. Pinnae	opposite	alternate
14. Pinnae	sessile	stalked
15. Lamina shape	lanceolate	boat shaped
16. Lamina	glabrous	indument present (except farina gland)
17. Petiole color	tan	black
18. Venation	open dichotomous	not
19. Vein ending	dilated	Not dilated
20. Dichotomization pattern	≤3	>3
21. Vein goes	upto the margin	not
22. Position of sorus	sorus at vein tip	some distance away from tip
23. Epidermal cell surface	convex	concave
24. Epidermal cellwall width	≥5µm	<5µm
25. Guard cell length	≥30µm	<30µm
26. Rhizome scale length	≥4mm	<4mm
27. Rhizome scale width	≥0.5	<0.5
28. Stipe scale length	≥4	<4
29. Stipe scale width	≥0.5	<0.5
30. Stipe/rachis length ratio	≥1	<1
31. Blade width	≥5cm	<5cm
32. Length/width ratio basalmost pinna of basal segments	≥4	<4
33. Length/width ratio median pinna of basal segments	≥3	<3
34. Length ratio of acroscopic/basiscopic segments of basal pinna	≥0.5	<0.5
35. Width ratio of acroscopic/basiscopic segments of basal pinna	≥0.6	<0.6
36. Spore Dia (P)	≥30µm	<30µm
37. Spore Dia (E)	≥50µm	<50µm
38. Exine thickness	≥2µm	<2µm
39. Laesural (L) longest arm	≥25µm	<25µm
40. Crassimarginate/tenuimarginate	Crassimarginate	tenuimarginate
41. Perine	not cristate	cristate
42. Tapetal depositions	absent	present
43. Perisporic strands	present	absent

Table 4. Characters taken for cluster analysis of *Cheilanthes spp.*(infrageneric level).

Characters taken	Character state	
	(0)	(1)
1. Stipe scale	base	throughout
2. Rachis scale	absent	present
3. Scales	present in costae & costule	absent
4. Rhizome scale	non-clathrate	clathrate
5. Rhizome scale	non-glandular	glandular
6. Stipe scale	non-clathrate	clathrate
7. Stipe scale	concolorous	bicolorous
8. Farina	white	golden Yellow
9. Indusial margin	entire	with fimbriated projections
10. Stomatal type	polocytic	polo and other type
11. Position of stomata	hypostomatic	amphistomatic
12. Pinnae	opposite	alternate
13. Pinnae	sessile	stalked
14. Lamina shape	lanceolate	boat shaped
15. Lamina	glabrous	indument present (except farina gland)
16. Petiole color	tan	black
17. Venation	open dichotomous	not
18. Vein ending	dilated	Not dilated
19. Dichotomization pattern	≤ 3	> 3
20. Vein goes	upto the margin	not
21. Position of sorus	sorus at vein tip	some distance away from tip
22. Epidermal cell surface	convex	concave
23. Epidermal cell wall width	$\geq 1 \mu m$	$< 1 \mu m$
24. Guard cell length	$\geq 40 \mu m$	$< 40 \mu m$
25. Rhizome scale length	$\geq 4mm$	$< 4mm$
26. Rhizome scale width	≥ 0.5	< 0.5
27. Stipe scale length	≥ 4	< 4
28. Stipe scale width	≥ 0.5	< 0.5
29. Stipe/rachis length ratio	≥ 1	< 1
30. Blade width	$\geq 5cm$	$< 5cm$
31. Length/width ratio basalmost pinna of basal segments	≥ 4	< 4
32. Length/width ratio median pinna of basal segments	≥ 3	< 3
33. Length ratio of acroscopic/basiscopic segments of basal pinna	≥ 0.5	< 0.5
34. Width ratio of acroscopic / basiscopic segments of basal pinna	≥ 0.6	< 0.6
35. Spore Dia(P)	$\geq 30 \mu m$	$< 30 \mu m$
36. Spore Dia(E)	$\geq 50 \mu m$	$< 50 \mu m$
37. Exine thickness	$\geq 2 \mu m$	$< 2 \mu m$
38. Laesural (L)longest arm	$\geq 25 \mu m$	$< 25 \mu m$
39. Crassimarginate/tenuimarginate	crassimarginate	tenuimarginate
40. Perine	absent	present
41. Tapetal depositions	absent	Present
42. Perisporic strands	present	absent

The placement of the genera *Hemionitis* L., *Parahemionitis* Panigrahi and *Pityrograma* Link. with cheilanthoid group of ferns (Fraser-Jenkins and Dulawat, 2009) must not be supported as their taxonomic positions were clarified earlier by Smith *et al.* (2006) on the basis of morphology as well as molecular taxonomy in other subfamilies Hemionitidae and Taenitidae respectively.

Phenetic study
Cluster analysis at generic level

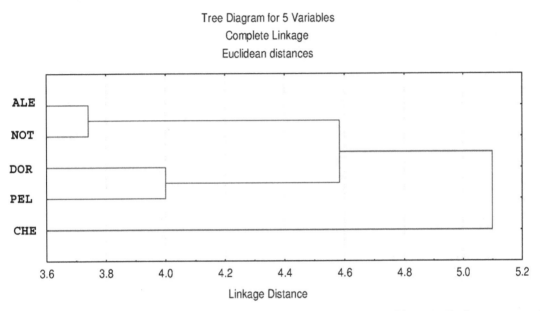

Fig. 1. Shows the relationship of cheilanthoid ferns at generic level (Abbrev.ALE-*Aleuritopteris*; CHE-*Cheilanthes*, DOR-*Doryopteris*, NOT-*Notholaena*, PEL-*Pellaea*).

Aleuritopteris (ALE) and *Notholaena* (NOT) have the nearest relation as is revealed from complete linkage at the Euclidean Distance of c.3.7. There is another closely related cluster formed by *Doryopteris* (DOR) and *Pellaea* (PEL) at the linkage distance of 4.0 which in its turn show a relationship with the first cluster at the distance of c.4.6 to form a larger cluster which shows a natural affinity with *Cheilanthes* (CHE) more or less at the Euclidean Distance of 5.1. All the OTUs under study, although individually distinct, are thus moderately related because of their overall similarity at the ED of 5.1

Cluster analysis at infrageneric level
Cluster analysis of *Aleuritopteris* spp.

A. argentea (AAR) and *A. subargentea* (ASR) have the nearest relation as is revealed from complete linkage at the Euclidean Distance of c.2.8. At the level of the linkage Distance of 3.5 as many as 8 clusters can be recognized of which 4 are with solitary OTUs (Operational taxonomic Unit), *viz. A. rufa* (ARU), *A. chrysophylla* (ACH), *A. formosana* (AFO), *A. subdimorpha* (ASD). However, at the Linkage Distance of 4.5 three large clusters are recognizable, *viz.* AAL, AAN, ARU; AAR, ASR, ACH, ABI, ADO, AFO; ABU, AGR, ASD. The relatedness of the last two clusters mentioned is greater than with the first cluster which is clearly expressed at the distance of 4.7. However all the OTUs under study, although individually distinct, are linked ultimately at the Ed of 4.8 because of their moderate overall similarity.

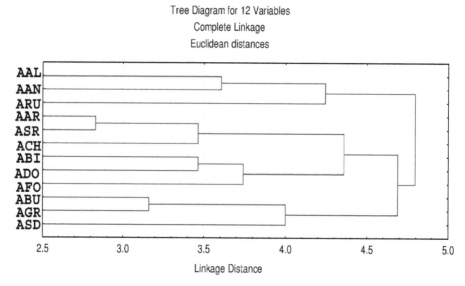

Fig. 2. Shows the infrageneric relationship of *Aleuritopteris* [Abbrev.: AAL- *Aleuritopteris albomarginata* (Clarke) Ching; AAN -*A. anceps* (Blanf.)Panigr. AAR- *A. argentea* (Gmel)Fee; ABI- *A. bicolor* (Roxb.) Fraser-Jenkins; ABU- *A. bullosa* (kze)Ching; ACH- *A. chrysophylla* (Hook) Ching; ADO- *A. doniana* S.K.Wu; AFO- *A. formosana* (Hay.) Tagawa; AGR- *A. grisea* (Blanf.)Panigr; ARU- *A. rufa* (D.Don)Ching; ASR- *A. subargentea* Ching ex Wu; ASD- *A. subdimorpha* (Clarke *et* Bak.)Fras.-Jenk.].

Cluster analysis of *Cheilanthes* spp.

Cheilanthes acrostica (Balbis)Tod (CAC) and *C. mysorensis* Wall ex.Hook. (CMY) have the nearest relation as is revealed from complete linkage at the Euclidean Distance of c.2.8. These OTUs in their turn show a relationship with *C. belangeri* (Bory) C.Chr. (CBE) at the distance of 4.0. At the same distance are linked *C.keralensis* Nair and Ghosh (CKE) and *C. subvillosa* Hook. (CSU). However, this cluster shows affinity with *C.tenuifolia* (CTE) more or less at the Euclidean Distance of 4.5. All the OTUs under study, although individually distinct, are related because of their overall similarity getting linked slightly above the Ed of 5.0.

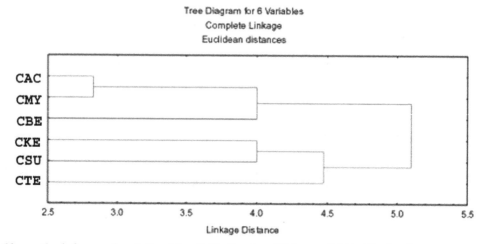

Fig. 3. Shows the infrageneric relationship of *Cheilanthes* (Abbrev.: CAC- *Cheilanthes acrostica*, CBE- *C. belangeri*, CKE- *C. keralensis*, CMY- *C. mysorensis*, CSU- *C. subvillosa*, CTE- *C. tenuifolia*).

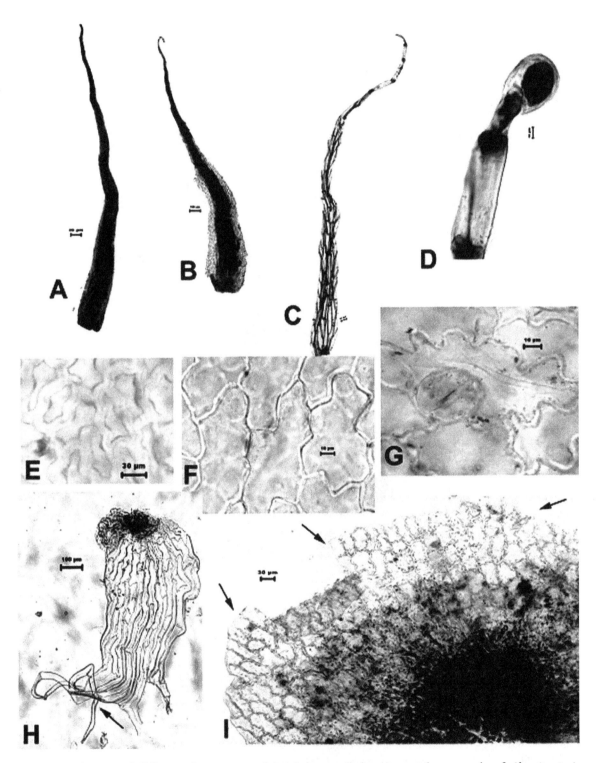

Fig. 4. LM images of different characters used in cluster analysis. A) concolorous scale of *Aleuritopteris subdimorpha* B) bicolorous scale of *A. formosana* C) clathrate non-glandular scale of *A. rufa* D) Glandular scale tip-*Cheilanthes mysorensis* E) epidermal cell-*A. chrysophylla* F) epidermal cell-*C. keralensis* G) polocytic stomata-*C. acrostica* H) fimbriated indusial margin-*A. rufa* (arrowhead shows the fimbriated margin) I) entire indusial margin-*A. chrysophylla* (arrowhead shows the entire margin).

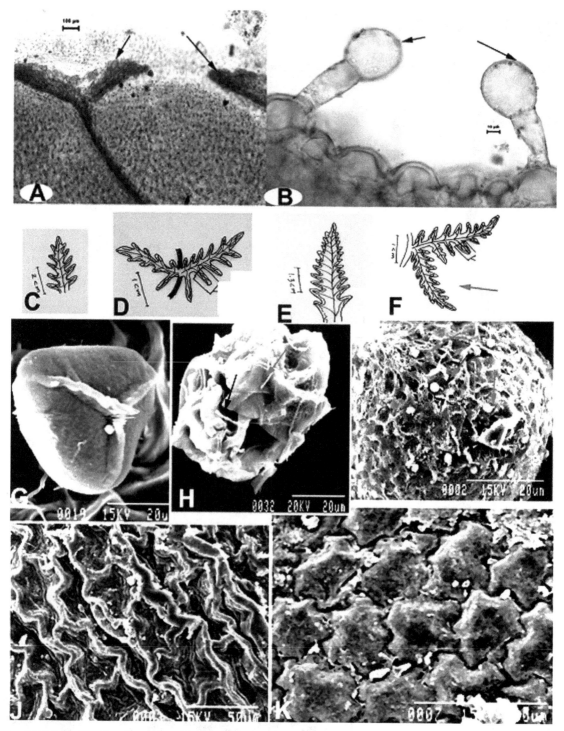

Fig. 5. LM (A-B), Free hand drawing (C-F) & SEM (G-K) images of characters used in cluster analysis. A) Commissural vein of *Doryopteris concolor* (arrowhead shows the marginal joining of veins) B) leaf gland of *Cheilanthes keralensis* C-D) ultimate & basal segment of leaf – *Aleuritopteris albomarginata* E-F) ultimate & basal segment of leaf- *A.bicolor* G) non-perinous spore-*C. tenuifolia* H) perisporic strands-*Pellaea falcata* I) tapetal deposits-*A.chrysophylla* (arrowhead shows the globular deposits) J) convex epidermal surface(adaxial)-*A. bullosa* K) concave epidermal surface(adaxial)-*A.formosana*.

The dendrogram based on phenetic study clearly revealed the interrelationships of five Indian cheilanthoid genera of arid region. Despite of their homoplasy of characters (Tryon and Tryon, 1973; Sen and Mukhopadhyay, 2014; Sen, 2014) they can be separated, though linked, clearly by using a multiple sets of characters. Present study is the first report describing the correlation between the cheilanthoid ferns of India at generic and infrageneric level and also establishes the generic segregation of *Aleuritopteris* and *Cheilanthes* by doing numerical taxonomic study.

Acknowledgements

Prof. Ambarish Mukherjee, Department of Botany, University of Burdwan, West Bengal is acknowledged for his kind care and help to perform this work. Also Mr. Kaushik Sarkar, Technical Assistant, & Dr. Srikanta Chakrabarty, USIC, University of Burdwan is acknowledged for taking the LM & SEM images. Two anonymous reviewers are also acknowledged here for their critical comment to improve the manuscript.

References

Dixit, R.D. 1984. A census of the Indian Pteridphytes. Flora India Series IV- Bot. Surv. India. pp.1-177.

Fraser-Jenkins, C.R. and Dulawat, C.S. 2009. A summary of Indian cheilanthoid ferns and the discovery of *Negripteris* (Pteridaceae), an afro-arabian fern genus new to India. Fern Gaz. **18**(5): 216-229.

Gastony, G.J. and Rollo, D.R.1995.Phylogeny and generic circumscriptions of cheilanthoid ferns (Pteridaceae:Cheilanthoideae) inferred from rbcL nucleotide sequences. Amer. Fern. J. **85**: 341-360.

Gastony, G.J. and Rollo, D.R. 1998. Cheilanthoid ferns (Pteridaceae: Cheilanthoideae) in the Southwestern United States and Adjacent Mexico – a molecular phylogenetic reassessment of generic lines. Aliso **17**:131-144.

Khullar, S.P. 1994.An illustrated fern flora of west Himalaya (Vol. I). International Book Distributors, Bishen Singh and Mahendra Pal singh, Dehra Dun, Indi, pp.1- 506.

Nayar, B.K. 1962. Ferns of India, no. VI, *Cheilanthes*. Nat. Bot. Gard. Lucknow, pp.1-35.

Pande, H.C. and Pande P.C., 2003. An illustrated fern flora of the Kumaon Himalaya. Vol.I, Bishen Singh Mahendra pal Singh, Dehra Dun, pp.1-372

Sen, K. and Mukhopadhyay R. 2011. LM and SEM Studies on Stomatal Morphotypes, Epidermal Characteristics and Spore Morphology of Some Indian Species of *Cheilanthes* Sw. Bioresearch Bulletin **5**: 304-310.

Sen, K. and Mukhopadhyay, R. 2014, New report of vessel elements in *Aleuritopteris* and *Cheilanthes*, Taiwania **59**(3): 231-239.

Sen, K. 2014, Ph.D. Thesis. "Studies in the morpho-anatomy & taxonomy of some Indian cheilanthoid ferns. Department of Botany, University of Burdwan.

Sigel, E. M., Windham, M. D., Huiet, L., Yatskievych, G.and Pryer, K. M. 2011. Species Relationships and Farina Evolution in the Cheilanthoid Fern Genus *Argyrochosma* (Pteridaceae). Syst. Bot. **36**(3): 554– 564.

Smith, A. R., Pryer, K.M., Schuettpelz, E., Korall, P., Schneider H. and Wolf, P.G. 2006. A classification of extant ferns. Taxon, **55**: 705-731.

Sneath, P.H.A. and Sokal, R.R.1973: *Numerical Taxonomy*. W.H. Freeman, San Francisco, pp.1-573.

Tryon, A.F., and Lugardon, B., 1990. Spores of the Pteridophyta: surface, wall structure, and diversity based on electron microscope studies. Springer–Verlag, New York, pp.1- 415.

Tryon, R.M. and Tryon, A.F. 1973. Geography, spores and evolutionary relations in the cheilanthoid ferns. *In:* Jermy,A.C.,Crabbe, J.A. and Thomas, B.A.(Eds.), "The Phylogeny and Classifications of the Ferns". Academic Press, London. pp. 145-153.

Tryon, R.M., and. Tryon. A.F 1982. Ferns and allied plants, with special reference to tropical America. Springer-Verlag, New York, pp.1-857.

Zhang, G., Zhang, X., Chen, Z., Liu, H., and Yang, W. 2007. First insights in the phylogeny of Asian cheilanthoid Ferns based on sequences of two chloroplast markers. Taxon **56**(2): 369-378.

NEW ANGIOSPERMIC TAXA FOR THE FLORA OF BANGLADESH

M. OLIUR RAHMAN[1] AND MD. ABUL HASSAN

Department of Botany, University of Dhaka, Dhaka 1000, Bangladesh

Keywords: New records; New species; Angiosperm; Flora of British India; Bengal Plants.

Abstract

This paper presents addition of 89 taxa under 64 genera distributed in 32 families for the flora of Bangladesh which are not included in the monumental works Flora of British India, Bengal Plants and Encyclopedia of Flora and Fauna of Bangladesh. Updated nomenclature, family name, references to the work and the precise localities have been furnished under each taxon.

Introduction

Bangladesh is a reservoir of plant resources comprising 3,611 angiosperm taxa occurring in the country (Ahmed *et al.*, 2007-2009). Despite several floristic studies were carried out over last four decades after the emergence of Bangladesh the botanical expedition throughout the country is yet to be completed. Hooker (1872-1897) and Prain (1903) made significant contribution on floristic studies in the Indian subcontinent and reported many species from the territory of present Bangladesh. Some regional flora of Bangladesh have also been produced by several workers (Heinig, 1925; Cowan, 1926; Raizada, 1941; Datta and Mitra, 1953; Sinclair, 1956).

In the recent past several authors paid attention to explore the flora of Bangladesh and recorded many taxa as new for the country. Mia and Khan (1995) published the first list of angiospermic taxa, not included in the Flora of British India (Hooker, 1872-1897) and Bengal Plants (Prain, 1903), adding a total of 325 species as recorded by several workers up to that time for Bangladesh. Rahman (2004a, b) added 138 angiosperm taxa to our knowledge on the flora of Bangladesh. Recently, Ahmed *et al.* (2007-2009) documented all species available in Bangladesh in the Encyclopedia of Flora and Fauna of Bangladesh adding several new angiosperm records for the country. After the momentous contribution made by Ahmed *et al.* (2007-2009) a large number of taxa have been added further, either as new records or as new taxa for the flora of Bangladesh. Therefore, the present study aimed at preparing a comprehensive checklist of all the new additions published either as new species or new records for the angiosperm flora of Bangladesh since Ahmed *et al.* (2007-2009).

Materials and Methods

This study was based on published works on new angiosperm records and new species for Bangladesh. All the relevant published papers and literature (Hooker, 1872-1897; Prain, 1903; Mia and Khan, 1995; Rahman, 2004a, b; Ahmed *et al.*, 2007-2009) have been consulted in order to update and finalize the newly reported taxa of Bangladesh. The new taxa are arranged in an alphabetical order and presented along with their updated nomenclature, family name, reference to the work and the precise locality.

Results

A total of 89 taxa under 64 genera belonging to 32 families have been added to the previous lists as new records or new species for Bangladesh. Despite some other new records or taxa

---

[1]Corresponding author. Email: prof.oliurrahman@gmail.com

identified for the flora of Bangladesh they are yet to be published and are not included in this paper. The taxa recorded here as new for the country are presented below.

1. **Alchornea mollis** Benth. *ex* Mull.-Arg. (Euphorbiaceae). Uddin *et al.* (2015d). p. 89. Moulvi Bazar: Srimangal, Harinchara.
2. **Allophylus samarensis** Merr. (Sapindaceae). Uddin *et al.* (2015c). p. 78. Moulvi Bazar: Madahbkundo Eco-park.
3. **Amorphophallus excentricus** Helt (Araceae). Ara and Hassan (2012). p. 18. Moulvi Bazar: Madhabkundo forest.
4. **Amorphophallus krausei** Engl. (Araceae). Ara and Hassan (2012). p. 18. Moulvi Bazar: Adampur beat, Kawargola forest; Lawachara reserve forest.
5. **Ancistrocladus tectorius** (Lour.) Merr. (Ancistrocladaceae). Uddin *et al.* (2015d). p. 90. Moulvi Bazar: Kamalganj, Lawachara National Park; Madhabkundo Eco-park.
6. **Argostemma sarmentosum** Wall. (Rubiaceae). Das and Rahman (2010). p. 216. Rangamati: Shubalong.
7. **Aristolochia coadunata** Back. (Aristolochiaceae). Uddin *et al.* (2015b). p. 69. Rangamati: Pharua reserve forest, Bilaichari.
8. **Aspidopteris tomentosa** (Bl.) Juss. (Malpighiaceae). Uddin and Hassan (2015). p. 35. Rangamati: Kaptai, Rampahar.
9. **Atalantia kwangtungensis** Merr. (Rutaceae). Uddin *et al.* (2015c). p. 79. Moulvi Bazar: Madahbkundo Eco-park.
10. **Begonia rubella** Buch.-Ham. *ex* D. Don (Begoniaceae). Uddin and Hassan (2015). p. 36. Rangamati: Kaptai, Sitapahar.
11. **Beilschmiedia sikkimensis** King *ex* Hook. f. (Lauraceae). Uddin and Hassan (2015). p. 38. Rangamati: Kaptai, Rampahar.
12. **Boehmeria aspera** Wedd. (Urticaceae). Uddin *et al.* (2015a). p. 2. Rangamati: Pharua reserve forest, Bilaichari.
13. **Boehmeria clidemioides** Miq. (Urticaceae). Uddin *et al.* (2015a). p. 3. East Bengal (CAL).
14. **Boehmeria hamiltoniana** Wedd. (Urticaceae). Uddin *et al.* (2015a). p. 5. East Bengal (CAL).
15. **Boehmeria manipurensis** Friis & Wilmot-Dear (Urticaceae). Uddin *et al.* (2015a). p. 6. Rangamati: Sapchari; Khagrachari: Gomoti, Panchari.
16. **Boeica filiformis** Clarke (Gesneriaceae). Uddin *et al.* (2015c). p. 81. Moulvi Bazar: Madahbkundo Eco-park.
17. **Brachycorythis obcordata** (Lindl.) Summerh (Orchidaceae). Hoque and Huda (2008). p. 193. Bandarban: Chimbuk.
18. **Catunaregam longispina** (Link) Tirveng. (Rubiaceae). Das *et al.* (2013). p. 258. Gazipur: Chandra forest; Habigonj: Chanbari, Rema-Kalenga Wildlife Sanctuary; Sherpur: Gajni forest.
19. **Chlorophytum nepalense** (Lindley) Baker (Liliaceae). Afroz *et al.* (2008). p. 193. Sherpur: Runctia sal forest.
20. **Colocasia virosa** Kunth (Araceae). Ara and Hassan (2012). p. 19. Moulvi Bazar: Muraichara beat, Ichachara forest.
21. **Colubrina javanica** Miq. (Rhamnaceae). Rahman *et al.* (2014). p. 199. Bagerhat: Sundarbans east forest division, Katka, near forest station.
22. **Cryptocarya calderi** M. Gangop. (Lauraceae). Uddin and Hassan (2015). p. 39. Cox's Bazar: Dulahazara safari park; Rangamati: Kaptai, Sitapahar.

23. **Cucumis hystrix** Chakravarty (Cucurbitaceae). Uddin *et al.* (2012). p. 205. Rangamati: Bilaichari, Pharua reserve forest.
24. **Cuphea carthagenensis** (Jacq.) J.F. Macbr. (Lythraceae). Hossain *et al.* (2015). p. 115. Sylhet: Lacctura; Shahjalal University of Science and Technology campus; Jaflong.
25. **Curcuma bakerii** Rahman & Yususf (Zingiberaceae). Rahman (2012). p. 121. Tangail: Madhupur sal forest.
26. **Curcuma hookerii** Rahman & Yusuf (Zingiberaceae). Rahman (2012). p. 123. Chittagong: Barabkundu.
27. **Curcuma roxburghii** Rahman *et* Yusuf (Zingiberaceae). Rahman and Yusuf (2012). p. 80. Rangamati: Rangapani.
28. **Curcuma wallichii** Rahman *et* Yusuf (Zingiberaceae). Rahman and Yusuf (2012). p. 82. Maulvi Bazar: Srimongal, Lawachara rain forest.
29. **Curcuma wilcockii** Rahman *et* Yusuf (Zingiberaceae). Rahman and Yusuf (2012). p. 83. Tangail: Madhupur sal forest, Rasulpur; Sylhet: Tamabil.
30. **Dianella ensifolia** (L.) DC. (Liliaceae). Uddin and Hassan (2009). p. 181. Rangamati: Kaptai, Rampahar.
31. **Diospyros albiflora** Alston (Ebenaceae). Sultana *et al.* (2010). p. 249. Patuakhali: Mirjagong.
32. **Egeria densa** Planchón (Hydrocharitaceae). Alfasane *et al.* (2010). p. 210. Bandarban: Bogakain lake.
33. **Elatostema dissectum** Wedd. (Urticaceae). Uddin *et al.* (2015a). p. 7. Chittagong Hill Tracts: Mynimukh (CAL).
34. **Elatostema ellipticum** Wedd. (Urticaceae). Uddin *et al.* (2015a). p. 8. East Bengal (K).
35. **Elatostema griffithii** Hook. f. (Urticaceae). Uddin *et al.* (2015a). p. 9. East Bengal (K).
36. **Elatostema obtusum** Wedd. (Urticaceae). Uddin *et al.* (2015a). p. 9. East Bengal (CAL).
37. **Elatostema procridioides** Wedd. (Urticaceae). Uddin *et al.* (2015a). p. 11. East Bengal (CAL).
38. **Elatostema subincisum** Wedd. (Urticaceae). Uddin *et al.* (2015a). p. 12. Chittagong (CAL).
39. **Embelia parviflora** Wall. ex A. DC. (Myrsinaceae). Uddin *et al.* (2015e). p. 96. Moulvi Bazar: Juri forest range, Lathitilla forest beat.
40. **Galium pusillosetosum** Hara (Rubiaceae). Das *et al.* (2013). p. 259. Chittagong: Baluchara.
41. **Guazuma ulmifolia** Lam. (Sterculiaceae). Mia *et al.* (2011). p. 154. Noakhali.
42. **Gynura nepalensis** DC. (Asteraceae). Afroz *et al.* (2014). p. 101. Dhaka: Dhaka University Botanical Garden; Netrakona: Kendua.
43. **Helicteres viscida** Bl. (Sterculiaceae). Mia *et al.* (2011). p. 154. Chittagong: Jaldi Range, Bilaichori.
44. **Hemiorchis rhodorrhachis** Schum. (Zingiberaceae). Srivastava and Ghoshal (2005). p. 59. Chittagong Hill Tracts: Barkal.
45. **Hydrocotyle verticillata** Thunb. (Apiaceae). Khatun *et al.* (2010). p. 105. Dhaka: Azimpur.
46. **Ilex glomerata** King (Aquifoliaceae). Uddin *et al.* (2015d). p. 91. Moulvi Bazar: Lawachara National Park.
47. **Illigera khasiana** C.B. Clarke (Hernandiaceae). Uddin *et al.* (2015b). p. 70. Rangamati: Pharua reserve forest, Bilaichari.
48. **Laportea bulbifera** (Siebold & Zuccarini) Wedd. (Urticaceae). Uddin *et al.* (2015a). p. 13. East Bengal (CAL).

49. **Lindera neesiana** (Wall. *ex* Nees) Kurz (Lauraceae). Ara and Khan (2015). p. 28. East Bengal (CAL).
50. **Litsea khasyana** Meissn. (Lauraceae). Ara and Khan (2015). p. 29. East Bengal (CAL, K).
51. **Litsea umbellata** (Lour.) Merr. (Lauraceae). Ara and Khan (2015). p. 29. East Bengal (CAL).
52. **Maytenus hookeri** Loes. (Celastraceae). Uddin and Hassan (2015). p. 40. Rangamati: Kaptai, Rampahar.
53. **Mitrephora grandiflora** Beddome (Annonaceae). Uddin and Hassan (2015). p. 42. Rangamati: Kaptai, Sitapahar.
54. **Mussaenda incana** Wall. *ex* Roxb. (Rubiaceae). Das and Rahman (2010). p. 217. Chittagong: Sitakunda, Eco-Park area; Moulvi Bazar: Srimongal, Bhanugach road.
55. **Mussaenda keenani** Hook. f. (Rubiaceae). Das *et al.* (2012). p. 22. Chittagong: Korerhat, Koila, Guichari; Rangamati: Kaptai, Rampahar, Madhabchari; Moulvi Bazar: Srimongal, Lawachora forest.
56. **Mycetia listeri** Deb. (Rubiaceae). Das *et al.* (2012). p. 23. Chittagong: Jamaichari; Rangamati: Kaptai, Rampahar.
57. **Mycetia malayana** (G. Don) Craib. (Rubiaceae). Uddin and Rahman (2015). p. 104. Chittagong: Dohazari, Lalutia; Moulvi Bazar: Madhabkundo Eco-park; Rangamati: Kaptai, Sitapahar west.
58. **Mycetia mukerjiana** Deb & Dutta (Rubiaceae). Das and Rahman (2010). p. 218. Rangamati: Kaptai, Sitapahar Wildlife Sanctuary.
59. **Mycetia sinensis** (Hemsley) Craib (Rubiaceae). Uddin and Rahman (2015). p. 107. Moulvi Bazar: Kamalganj, Lawachara National Park.
60. **Mycetia stipulata** (Hook. f.) O. Kuntze subsp. **macrostachya** (Hook. f.) Deb (Rubiaceae). Uddin and Rahman (2015). p. 108. East Bengal (K).
61. **Neodistemon indicum** (Wedd.) Babu & Henry (Urticaceae). Uddin *et al.* (2015a). p. 14. Moulvi Bazar: Madhabkundo Eco-park; Rangamati: Kaptai, Rampahar.
62. **Ophiorrhiza eriantha** Wight (Rubiaceae). Das *et al.* (2013). p. 260. Khagrachari: Shilchari, Alu tila.
63. **Ophiorrhiza fasciculata** D. Don (Rubiaceae). Das *et al.* (2012). p. 24. Chittagong: Dhopachari, Gondamara; Rangamati: Kutukchari, Chegaiya-chari.
64. **Oxyceros rugulosus** (Thwaites) Tirveng. (Rubiaceae). Das and Rahman (2010). p. 219. Chittagong: Hazarikhil; Cox's Bazar: Teknaf; Rangamati: Pablakhali; Sylhet: Jafflong.
65. **Pellionia heteroloba** Wedd. (Urticaceae). Uddin *et al.* (2015a). p. 16. East Bengal (CAL, K).
66. **Pellionia heyneana** Wedd. (Urticaceae). Uddin *et al.* (2015a). p. 17. Rangamati: Pharua reserve forest, Bilaichari.
67. **Pellionia repens** (Lour.) Merr. (Urticaceae). Uddin *et al.* (2015a). p. 18. Dhaka: Balda garden.
68. **Phenax mexicanus** Wedd. (Urticaceae). Uddin *et al.* (2015a). p. 19. Rangamati: Kaptai, Karnaphuli sadar beat; Pharua reserve forest, Bilaichari.
69. **Phyllanthus columnaris** Muell.-Arg. (Euphorbiaceae). Uddin and Hassan (2015). p. 43. Rangamati: Kaptai, Rampahar.
70. **Pilea anisophylla** Wedd. (Urticaceae). Uddin *et al.* (2015a). p. 20. East Bengal (K).
71. **Pilea bracteosa** Wedd. (Urticaceae). Uddin *et al.* (2015a). p. 22. East Bengal (K).

72. **Pilea insolens** Wedd. (Urticaceae). Uddin *et al.* (2015a). p. 23. East Bengal (K, CAL).
73. **Pollia thyrsiflora** (Bl.) Endley *ex* Hassk. (Commelinaceae). Uddin and Hassan (2015). p. 44. Rangamati: Kaptai, Sitapahar, Jamaichara.
74. **Psychotria stipulacea** Wall. (Rubiaceae). Das *et al.* (2012). p. 25. Rangamati: Kaptai, Sitapahar.
75. **Psydrax umbellata** (Wight) Bridson (Rubiaceae). Das *et al.* (2012). p. 26. Chittagong: Bomariaghona; Sylhet: Tamabil.
76. **Pulicaria vulgaris** Gaertn. (Asteraceae). Rahman *et al.* (2011). p. 205. Patuakhali: Galachipa.
77. **Pyrenaria diospyricarpa** Kurz (Theaceae). Uddin *et al.* (2015b). p. 71. Rangamati: Pharua reserve forest, Bilaichari.
78. **Sarcopyramis napalensis** Wall. (Melastomataceae). Uddin *et al.* (2015e). p. 98. Moulvi Bazar: Juri forest range, Lathitilla forest beat.
79. **Sida spinosa** L. (Malvaceae). Shetu *et al.* (2015). p. 111. Dhaka: Near Kafrul thana, Mirpur; Khulna: Khulna University campus.
80. **Spermacoce exilis** (Williams) Adams *ex* Burger *et* Taylor (Rubiaceae). Das *et al.* (2013). p. 261. Chittagong: Kumira, Dardorir chara; Cox's Bazar: Himchari, Bhangamura; Laksmipur: Ramgonj; Moulvi Bazar: Srimongal, Tea resort; Sylhet: Tibbi College campus.
81. **Staurogyne simonsii** (Anders.) O. Kuntze (Acanthaceae). Uddin *et al.* (2015e). p. 99. Moulvi Bazar: Juri forest range, Lathitilla forest beat.
82. **Sterculia urens** Roxb. (Sterculiaceae). Mia *et al.* (2011). p. 155. Chittagong.
83. **Steudnera gagei** Krause (Araceae). Ara and Hassan (2012). p. 20. Moulvi Bazar:Adampur beat, Gangpali.
84. **Tarenna helferi** (Kurz) N.P. Balakr. (Rubiaceae). Das and Rahman (2010). p. 220. Rangamati: Betbunia, Mahajan para; Moulvi Bazar: Srimongal, Lawachara forest.
85. **Tarenna stellulata** (Hook. f.) Ridl. (Rubiaceae). Das *et al.* (2013). p. 262. Chittagong: Sitakunda, Chandranath hill.
86. **Trigonostemon viridissimus** (Kurz) Airy Shaw (Euphorbiaceae). Uddin *et al.* (2015c). p. 82. Moulvi Bazar: Madahbkundo Eco-park.
87. **Vanilla havilandii** Rolfe (Orchidaceae). Uddin *et al.* (2015b). p. 73. Rangamati: Pharua reserve forest, Bilaichari, Monlovi chara.
88. **Xanthosoma undipes** (K. Koch) K. Kock (Araceae). Ara and Hassan (2012). p. 22. Gazipur: Kamesshor village.
89. **Zingiber salarkhanii** (Zingiberaceae). Rahman and Yusuf (2013). p. 240. Chittagong: Sitakundu, Chandranath hill; Khagrachari: Teen Tila, Marissa road, Moulvi Bazar: Srimongal, Lawachara reserve forest.

References

Afroz, S., Tutul, E., Uddin, M.Z. and Hassan, M.A. 2008. *Chlorophytum nepalense* (Lindley) Baker (Liliaceae) - A new angiospermic record for Bangladesh. Bangladesh J. Bot. **37**(2): 193–194.

Afroz, S., Uddin, M.Z. and Hassan, M.A. 2014. *Gynura nepalensis* DC. (Asteraceae) - A new angiosperm record for Bangladesh. Bangladesh J. Plant Taxon. **21**(1): 101–104.

Ahmed, Z.U., Begum, Z.N.T., Hassan, M.A., Khondker, M., Kabir, S.M.H., Ahmad, M., Ahmed, A.T.A., Rahman, A.K.A. and Haque, E.U. (Eds) 2007-2009. Encyclopedia of Flora and Fauna of Bangladesh, Vols. **6-12**. Asiatic Society of Bangladesh, Dhaka.

Alfasane, M.A., Khondker, M., Islam, M.S. and Bhuiyan, M.A.H. 2010. Egeria densa Planchón (Hydrocharitaceae) - A new angiospermic record for Bangladesh. Bangladesh J. Plant Taxon. **17**(2): 209–213.

Ara, H. and Hassan, M.A. 2012. Five new records of aroids for Bangladesh. Bangladesh J. Plant Taxon. **19**(1): 17–23.

Ara, H. and Khan, B. 2015. Three new records of Lauraceae from Bangladesh. Bull. Bangladesh National Herb. **4**: 27–32.

Cowan, J.M. 1926. The flora of Chakaria Sundarbans. Rec. Bot. Surv. Ind. **11**: 197–225.

Das, S.C. and Rahman, M.A. 2010. Notes on the Rubiaceae. 3. Five new records for Bangladesh. Bangladesh J. Bot. **39**(2): 215–222.

Das, S.C., Dev, P.K. and Rahman, M.A. 2012. Notes on the Rubiaceae - 4: Five new records for Bangladesh. Bangladesh J. Bot. **41**(1): 21–28.

Das, S.C., Dev, P.K. and Rahman, M.A. 2013. Notes on the Rubiaceae - 5: Five new records for Bangladesh. Bangladesh J. Bot. **42**(2): 257–264.

Datta, R.B. and Mitra, J.N. 1953. Common Plants in and around Dacca city. Bull. Bot. Soc. Beng. **7**(1&2): 1–110.

Heinig, R.L. 1925. List of the Plants of Chittagong Collectorate and Hill Tracts. Darjeeling.

Hooker, J.D. 1872-1897. The Flora of British India. Vols. **1–7**. L. Reeve & Co. Ltd., England.

Hoque, M.M. and Huda, M.K. 2008. *Brachycorythis obcordata* (Lindl.) Summerh. (Orchidaceae) - A new angiospermic record for Bangladesh. Bangladesh J. Bot. 37(2): 199–201.

Hossain, G.M., Khan, M.S.A., Rahman, M.S., Haque, A.K.M.K. and Rahim, M.A. 2015. *Cuphea carthagenensis* (Jacq.) J.F. Macbr. (Lythraceae) – A new angiosperm record for Bangladesh. Bull. Bangladesh National Herb. **4**: 115–117.

Khatun, B.M., Rahman, M.O. and Sultana, S.S. 2010. *Hydrocotyle verticillata* Thunb. (Apiaceae) - A new angiospermic record for Bangladesh. Bangladesh J. Plant Taxon. **17**(1): 105–108.

Mia, M.K. and Khan, B. 1995. First list of angiospermic taxa of Bangladesh not included in Hooker's Flora of British India and Prain's Bengal Plants. Bangladesh J. Plant Taxon. **2**(1&2): 25–45.

Mia, M.M.K., Rahman, M.O., Hassan, M.A. and Huq, A.M. 2011. Three new records of Sterculiaceae for Bangladesh. Bangladesh J. Plant Taxon. **18**(2): 153–157.

Prain, D. 1903. Bengal Plants. Vols. **1&2**. (Reprint edition 1963). Botanical Survey of India, Calcutta.

Rahman, M.A. 2012. Discovery of new species from Bangladesh. Plantae Discoverie **1**: 1–34.

Rahman, M.A. and Yusuf, M. 2012. Three new species of *Curcuma* L. (Zingiberaceae) from Bangladesh. Bangladesh J. Plant Taxon. **19**(1): 79–84.

Rahman, M.A. and Yusuf, M. 2013. *Zingiber salarkhanii* (Zingiberaceae) - A new species from Bangladesh. Bangladesh J. Plant Taxon. **20**(2): 239–242.

Rahman, M.O. 2004a. Second list of angiospermic taxa not included in Hooker's Flora of British India and Prain's Bengal Plants - Series I. Bangladesh J. Plant Taxon. **11**(1): 77–82.

Rahman, M.O. 2004b. Second list of angiospermic taxa not included in Hooker's Flora of British India and Prain's Bengal Plants - Series II. Bangladesh J. Plant Taxon. **11**(2): 49–56.

Rahman, M.O., Sultana, M., Begum, M. and Hassan, M.A. 2011. *Pulicaria vulgaris* Gaertn. (Asteraceae) -A new species record for Bangladesh. Bangladesh J. Plant Taxon. **18**(2): 205–208.

Rahman, M.S., Hossain, G.M., Khan, S.A. and Uddin, S.N. 2014. *Colubrina javanica* Miq. (Rhamnaceae) – A new angiosperm record for Bangladesh. Bangladesh J. Plant Taxon. **21**(2): 199–202.

Raizada, M.B. 1941. On the Flora of Chittagong. Indian Forester **67**(5): 245–254.

Shetu, S.S., Khan, M.S.A. and Uddin, S.N. 2015. *Sida spinosa* L. (Malvaceae) - A new angiosperm species record for Bangladesh. Bull. Bangladesh National Herb. **4**: 111–113.

Sinclair, J. 1956. Flora of Cox's Bazar, East Pakistan. Bull. Bot. Soc. Beng. **9**(2): 92–94.

Srivastava, S.C. and Ghoshal, P.P. 2005. *Hemiorchis rhodorrhachis* Schum. (Zingiberaceae) - A new record for Bangladesh. Bangladesh J. Plant Taxon. **12**(1): 59–61.

Sultana, M., Rahman, M.O., Begum, M. and Hassan, M.A. 2010. *Diospyros albiflora* Alston (Ebenaceae) - A new angiospermic record for Bangladesh. Bangladesh J. Bot. **39**(2): 249–251.

Uddin, S.N and Hassan, M.A. 2009. *Dianella ensifolia* (L.) DC. (Liliaceae) - A new angiospermic record for Bangladesh. Bangladesh J. Plant Taxon. **16**(2): 181–184.

Uddin, S.N., Khan, B. and Mirza, M.M. 2012. *Cucumis hystrix* Chakrav. (Cucurbitaceae) - A new angiospermic record for Bangladesh. Bangladesh J. Plant Taxon. **19**(2): 205–207.

Uddin, S.N. and Hassan, M.A. 2015. Discovery of eight angiosperm new records for Bangladesh from Rampahar and Sitapahar reserve forest under Rangamati district. Bull. Bangladesh National Herb. **4**: 33–49.

Uddin, S.N. and Rahman, N. 2015. Notes on occurrence of the genus *Mycetia* Reinwardt (Rubiaceae) in Bangladesh. Bull. Bangladesh National Herb. **4**: 103–110.

Uddin, S.N., Khan, B. and Hassan, M.A. 2015a. Nineteen new records of Urticaceae from Bangladesh. Bull. Bangladesh National Herb. **4**: 1–25.

Uddin, S.N., Khan, B. and Mirza, M.M. 2015b. Discovery of four angiospermic new records for Bangladesh from Pharua Reserve forest under Rangamati district. Bull. Bangladesh National Herb. **4**: 67–76.

Uddin, S.N., Khan, B. and Khokan, M.E.H. 2015c. Discovery of four angiosperm new records for Bangladesh from Madhabkundo Eco-park under Moulvi Bazar district. Bull. Bangladesh National Herb. **4**: 77–85.

Uddin, S.N., Khokan, M.E.H. and Khan, B. 2015d. Discovery of three angiosperm new records for Bangladesh from Lawachara National Park under Moulvi Bazar district. Bull. Bangladesh National Herb. **4**: 87–94.

Uddin, S.N., Khokan, M.E.H., Khan, B. and Islam, K.K. 2015e. Discovery of three new angiosperm records for Bangladesh from Juri forest range-1 under Moulvi Bazar district. Bull. Bangladesh National Herb. **4**: 95–102.

UPDATED NOMENCLATURE AND TAXONOMIC STATUS OF THE PLANTS OF BANGLADESH INCLUDED IN HOOK. f. *THE FLORA OF BRITISH INDIA*

M. Enamur Rashid and M. Atiqur Rahman

Department of Botany, University of Chittagong, Chittagong-4331, Bangladesh

Keywords: J.D. Hooker; *Flora of British India*; Bangladesh; Nomenclature; Taxonomic Status.

Abstract

One hundred seventy seven species belonging to 88 genera under 14 natural orders are determined to have been recorded in the third volume of J.D. Hooker's, the *Flora of British India* from the area now fall in Bangladesh. These taxa are enumerated with updated nomenclature and current taxonomic status following ICN and Cronquist's system of plant classification respectively resulting in 169 species under 93 genera and 14 families. Collection locality with collector's name of each species wherever available, as cited in protologue, is also included.

Introduction

The plants from the area of Bangladesh included in volumes I and II of J.D. Hooker's, the *Flora of British India* have already been determined and reported with their updated nomenclature and taxonomic status following ICN (Rashid and Rahman, 2011, 2012). The present report deals with the treatment of the taxa of the volume III of the *Flora of British India* (1880-1882). This volume includes three parts (VII-IX) published in 3 different dates consisting of a total of 22 natural orders, 354 genera and 2174 species. In this volume, J.D. Hooker was assisted by an eminent botanist, C.B. Clarke. Hooker alone described 4 natural orders while Clarke alone described 18 natural orders.

Among these taxa, described in this volume, 177 species belonging to 88 genera and 14 natural orders are determined to be recorded from the area now in Bangladesh. The objective of the study was to update the nomenclature and taxonomic status of the plants of Bangladesh which have been included in the *Flora of British India*.

Materials and Methods

In this study, volume III of the *Flora of British India* (Hook.f., 1880-1882) has been surveyed for determining the taxa recorded from the area now fall in Bangladesh following Rashid and Rahman (2011, 2012). Bangladesh Gazetteers (Ishaq, 1979) has been consulted to ascertain the collection localities fall within the area of Bangladesh. In case of Bengal and Jainta hills, mentioned as collection localities, relevant literature, such as, Roxburgh (1814, 1820, 1824, 1832), Wallich (1828-1849), Kurz (1877), Prain (1903), Brandis (1906), Heinig (1925), Cowan (1926), Kanjilal *et al.* (1939), Rhaizada (1941) and Sinclair (1956) have been consulted to confirm whether the taxon belongs to the area now fall in Bangladesh.

The current nomenclature of each species was determined by consulting ICN (Voss, 1983; McNeill *et al.*, 2012), Internet sources (I-III), Brummitt and Powell (1992). Taxonomic status of the taxa were determined by following Cronquist (1981) and to determine the synonyms of respective species relevant literature, *viz.,* Ali (1971), Hara and Williams (1979), Hara *et al.* (1982), Rahman and Wilcock (1991), Brummitt (1992), Wu and Raven (1994), Rahman and

Wilcock (1995), Mabberley (1997), Press *et al.* (2000), Wu *et al.* (2005), Ahmed *et al.* (2008a, b; 2009a, b, c) and updated Kew Plant list from internet sources (I-III) have been consulted.

Results and Discussion

The search on the third volume of the *Flora of the British India* revealed a total of 177 species in 88 genera under 14 natural orders from the area now in Bangladesh (Table 1).

Table 1, Natural Orders with contributors and distribution of taxa in the Volume III

Natural order as in Hook.f. (1880-1882)	Contributor	Total No. of genera/ species described	No. of genera/ species from the area of Bangladesh
1. Caprifoliaceae	C.B. Clarke	8/49	1/1
2. Rubiaceae	J.D. Hooker	91/640	31/81
3. Valerianeae	C.B. Clarke	4/17	0/0
4. Dipsaceae	C.B. Clarke	4/17	0/0
5. Compositae	J.D. Hooker	123/634	20/37
6. Stylidieae	C.B. Clarke	1/3	1/2
7. Goodenovieae	C.B. Clarke	1/2	0/0
8. Campanulaceae	C.B. Clarke	13/65	2/5
9. Vacciniaceae	C.B. Clarke	4/50	1/2
10. Ericaceae	C.B. Clarke	9/65	0/0
11. Monotropeae	C.B. Clarke	3/3	0/0
12. Epacrideae	C.B. Clarke	1/1	0/0
13. Diapensiaceae	C.B. Clarke	1/1	0/0
14. Plumbagineae	C.B. Clarke	6/9	1/1
15. Primulaceae	J.D. Hooker	9/80	2/2
16. Myrsineae	C.B. Clarke	11/93	4/6
17. Sapotaceae	C.B. Clarke	8/55	3/3
18. Ebenaceae	C.B. Clarke	2/75	1/7
19. Styraceae	C.B. Clarke	2/70	1/1
20. Oleaceae	C.B. Clarke	10/91	5/11
21. Salvadoraceae	C.B. Clarke	3/5	0/0
22. Apocynaceae	J.D. Hooker	40/149	15/18
Total: NO 22	Contributors 02	Genera/species 354/2174	Genera/species 88/177

After current nomenclatural treatment, the number of species reduced to 169, while the genera splited to 93. The species included in families Vaccinaceae and Styraceae have been transfered to Ericaceae and Symplocaceae respectively. It is determined, so far, that 29 generic names have been changed and 59 remain unchanged. On the other hand, 79 names of species have also been changed and 98 remain unchanged. Hence after updated nomenclatural treatment, 169 species and 93 genera under 14 families are recognized from the area of Bangladesh so far and presented in Table 2.

Table 2. List of taxa as in Hook.f, the *Flora of British India* Volume III from the area of Bangladesh with their current nomenclature and taxonomic status

Species, Natural Order, recorded area - collector's name/ Wall. Cat. no. as in Hook.f. (1880-1882)	Current nomenclature with *loc. cit.* and family as of Cronquist (1981)
1. *Sambucus javanica* Blume Natural Order: Caprifoliaceae East Bengal - Not mentioned	1. **Sambucus javanica** Reinw. *ex* Blume in Bijdr. Fl. Ned. Ind. 13: 657 (1825). **Family:** Caprifoliaceae
2. *Cephalanthus naucleoides* DC. Natural Order: Rubiaceae Silhet - Not mentioned	2. **Cephalanthus tetrandra** (Roxb.) Ridsdale & Bakh.f. in Blumea 23: 182 (1976). **Family:** Rubiaceae
3. *Adina sessilifolia* Hook.f. Natural Order: Rubiaceae Chittagong - Roxburgh & c.	3. **Neonauclea sessilifolia** (Roxb.) Merr. in J. Wash. Acad. Sci. 5: 542 (1915). **Family:** Rubiaceae
4. *A. polycephala* Benth. Natural Order: Rubiaceae Silhet - De Silva, Griffith and Chittagong - J.D.H. & T.T.	4. **Metadina trichotoma** (Zoll. & Mor.) Bakh.f., Taxon 19: 472 (1970). **Family:** Rubiaceae
5. *A. polycephala* Benth. var. *microphylla* Hook.f. Natural Order: Rubiaceae Silhet -Wallich	4. **Metadina trichotoma** (Zoll. & Mor.) Bakh.f., Taxon 19: 472 (1970). **Family**: Rubiaceae
6. *Stephegyne diversifolia* Hook.f. Natural Order: Rubiaceae Chittagong - Roxburgh & c.	5. **Mitragyna diversifolia** (Wall. *ex* G. Don) Havil. in J. Linn. Soc., Bot. 33: 71 (1897). **Family:** Rubiaceae
7. *Nauclea ovalifolia* Roxb. Natural Order: Rubiaceae Silhet - Not mentioned	3. **Neonauclea sessilifolia** (Roxb.) Merr. in J. Wash. Acad. Sci. 5: 542 (1915). **Family:** Rubiaceae
8. *Uncaria ovata* Br. Natural Order: Rubiaceae Silhet - Wall. Cat. 6112	6. **Uncaria canescens** Korth., Verh. Nat. Gesch. Ned. Bot.: 172 (1842). **Family:** Rubiaceae
9. *U. sessilifructus* Roxb. Natural Order: Rubiaceae Chittagong - Not mention	7. **Uncaria sessilifructus** Roxb., Fl. Ind. 2: 130 (1824). **Family:** Rubiaceae
10. *U. homomalla* Miq. Natural Order: Rubiaceae Eastern Bengal; Jyntea hills - Wall. Cat. 6108	8. **Uncaria homomalla** Miq. in Fl. Ned. Ind. 2: 343 (1857). **Family:** Rubiaceae
11. *U. pilosa* Roxb. Natural Order: Rubiaceae Chittagong - Roxburgh, J.D.H. & T.T.	9. **Uncaria scandens** (Smith) Hutch., Sarg. Pl. Wilson. 3: 406 (1916). **Family:** Rubiaceae
12. *Hymenodictyon excelsum* Wall. Natural Order: Rubiaceae Chittagong - Not mentioned	10. **Hymenodictyon orixense** (Roxb.) Mabb., Taxon 31: 66 (1982). **Family:** Rubiaceae

Species, Natural Order, recorded area - collector's name/ Wall. Cat. no. as in Hook.f. (1880-1882)	Current nomenclature with *loc. cit.* and family as of Cronquist (1981)
13. *Wendlandia tinctoria* DC. Natural Order: Rubiaceae Chittagong - Not mentioned	11. **Wendlandia tinctoria** (Roxb.) DC., Prodr. 4: 411 (1830). **Family:** Rubiaceae
14. *W. paniculata* DC. Natural Order: Rubiaceae Silhet - Not mentioned	12. **Wendlandia paniculata** (Roxb.) DC., Prodr. 4: 411 (1830). **Family:** Rubiaceae
15. *Dentella repens* Forst. Natural Order: Rubiaceae Throughout Bengal - Not mentioned	13. **Dentella repens** J. R. Forst. & G. Forst., Char. Gen. Pl. Ins. Mar. Austr.: 26, t. 13 (1776). **Family:** Rubiaceae
16. *Hedyotis scandens* Roxb. Natural Order: Rubiaceae Silhet and Chittagong- Not mentioned	14. **Hedyotis scandens** Roxb., Fl. Ind. 1: 369. (1820). **Family:** Rubiaceae
17. *H. uncinella* Hook. & Arn. Natural Order: Rubiaceae Jyntea hills. - Wall. Cat. 842	15. **Hedyotis uncinella** Hook. & Arn., Bot. Beechey Voy.: 192 (1833). **Family:** Rubiaceae
18. *H. auricularia* Linn. Natural Order: Rubiaceae Chittagong - Not mentioned	16. **Hedyotis auricularia** L., Sp. Pl.: 101 (1753). **Family:** Rubiaceae
19. *H. lineata* Roxb. Natural Order: Rubiaceae Silhet and Chittagong – Not mentioned	17. **Hedyotis lineata** Roxb., Fl. Ind. 1: 369 (1820). **Family:** Rubiaceae
20. *H. glabra* Br. Natural Order: Rubiaceae Silhet - de Silva	18. **Hedyotis insularis** (Spreng.) Deb & R.M. Dutta in Taxon 32(2): 285 (1983). **Family:** Rubiaceae
21. *H. hispida* Retz Natural Order: Rubiaceae Chittagong - Not mentioned	19. **Hedyotis verticillata** (L.) Lam., Tabl. Encycl. 1: 271 (1792). **Family:** Rubiaceae
22. *H. monocephala* Br. Natural Order: Rubiaceae Silhet – Wall. Cat. 846	20. **Hedyotis brunonis** Merr. in Philipp. J. Sci. 60: 35 (1936). **Family:** Rubiaceae
23. *H. thomsoni* Hook.f. Natural Order: Rubiaceae East Bengal - J.D.H. & T.T.	21. **Hedyotis thomsonii** Hook.f., Fl. Brit. India 3: 63 (1880). **Family:** Rubiaceae
24. *Oldenlandia diffusa* Roxb. var. *extensa* Hook.f. Natural Order: Rubiaceae Silhet - Wall. Cat. 869 & Griffith	22. **Hedyotis diffusa** var. **extensa** (Hook.f.) Dutta in Bot. Surv. India (2004). **Family:** Rubiaceae
25. *O. crystallina* Roxb. Natural Order: Rubiaceae East Bengal - Griffith, Chittagong - C.B. Clarke	23. **Hedyotis pumila** L.f., Suppl. Pl.: 119 (1781). **Family:** Rubiaceae

Species, Natural Order, recorded area - collector's name/ Wall. Cat. no. as in Hook.f. (1880-1882)	Current nomenclature with *loc. cit.* and family as of Cronquist (1981)
26. *O. trinervia* Retz. Natural Order: Rubiaceae Chittagong - J.D.H. & T.T.	24. **Hedyotis trinervia** (Retz.) Roem & Schult., Syst. Veg. 3: 197 (1818). **Family:** Rubiaceae
27. *O. paniculata* Linn. Natural Order: Rubiaceae Silhet - Not mentioned	25. **Hedyotis racemosa** Lam., Encycl. 3: 80 (1789). **Family:** Rubiaceae
28. *Anotis urophilla* Wall. Natural Order: Rubiaceae Jyntea - Gomez, Griffith, & c.	26. **Neanotis urophylla** (Wall. *ex* Wight & Arn.) W.H. Lewis, Ann. Missouri Bot. Gard. 53: 40 (1966). **Family:** Rubiaceae
29. *Ophiorrhiza harrisiana* Heyne Natural Order: Rubiaceae Silhet and Chittagong - Not mentioned	27. **Ophiorrhiza rugosa** Wall *ex* Roxb., Fl. Ind. 2: 547 (1824). **Family:** Rubiaceae
30. *O. harrisiana* Heyne var. *argentea* Wall. Natural Order: Rubiaceae Silhet and Chittagong - Not mentioned	28. **Ophiorrhiza rugosa** var. **argentea** (Wall. *ex* G. Don) Deb & Mondal in Bull. Bot. Surv. India 24 (1-4): 228 (1983). **Family:** Rubiaceae
31. *O. trichocarpa* Blume Natural Order: Rubiaceae Chittagong - Lister	29. **Ophiorrhiza trichocarpos** Blume, Bijdr.: 977 (1826). **Family:** Rubiaceae
32. *O. Wallichii* Hook.f. Natural Order: Rubiaceae Jyntea hills - Gomez	30. **Ophiorrhiza wallichii** Hook.f., Fl. Brit. India 3: 79 (1880); **Family:** Rubiaceae
33. *O. villosa* Roxb. Natural Order: Rubiaceae Chittagong hills - Roxburgh	31. **Ophiorrhiza villosa** Roxb., Fl. Ind. 2: 546 (1824). **Family:** Rubiaceae
34. *Silvianthus bracteatus* Hook.f. Natural Order: Rubiaceae Silhet - de Silva, Griffith & c.	32. **Silvianthus bracteatus** Hook.f., Icon. Pl. t. 1048: 36 (1868). **Family:** Caprifoliaceae
35. *Mussaenda Roxburghii* Hook.f. Natural Order: Rubiaceae Chittagong - J.D.H. & T.T.	33. **Mussaenda roxburghii** Hook.f., Fl. Brit. India 3: 87 (1880). **Family:** Rubiaceae
36. *M. glabra* Vahl Natural Order: Rubiaceae Chittagong - Griffith & Helfer	34. **Mussaenda glabra** Vahl, Symb. Bot. 3: 38 (1794). **Family:** Rubiaceae
37. *Adenosacme longifolia* Wall. Natural Order: Rubiaceae Chittagong - Not mentioned	35. **Mycetia longifolia** (Wall.) Kuntze, Revis. Gen. Pl. 1: 289 (1891). **Family:** Rubiaceae
38. *Myrioneuron nutans* Wall. Natural Order: Rubiaceae Chittagong hills - C.B. Clarke	36. **Myrioneuron nutans** Wall. *ex* Kurz, Fl. Brit. Burma 2: 55 (1874). **Family:** Rubiaceae

Species, Natural Order, recorded area - collector's name/ Wall. Cat. no. as in Hook.f. (1880-1882)	Current nomenclature with *loc. cit.* and family as of Cronquist (1981)
39. *M.Clarkei* Hook.f. Natural Order: Rubiaceae Chittagong - J.D.H. & T.T.	37. **Myrioneuron clarkei** Hook.f., Fl. Brit. India 3: 96 (1880). **Family:** Rubiaceae
40. *Webera odorata* Roxb. Natural Order: Rubiaceae Silhet - de Silva and Griffith	38. **Tarenna odorata** (Roxb.) B. L. Rob., Proc. Amer. Acad. Arts. 45: 405 (1910). **Family:** Rubiaceae
41. *W. disperma* Hook.f. Natural Order: Rubiaceae Silhet - Griffith and J.D.H. & T.T.	39. **Tarenna disperma** (Hook.f.) Pitard in Fl. Gen. Indo-China 3: 208 (1923). **Family:** Rubiaceae
42. *W. Campaniflora* Hook.f. Natural Order: Rubiaceae Chittagong-Bruce, Seetakoond - J.D.H. & T.T. and Burkul - C.B. Clarke	40. **Tarenna campaniflora** (Hook.f.) Balak., Bull. Bot. Surv. India 22 (1-4): 175 (1982). **Family:** Rubiaceae
43. *Randia tetrasperma* Roxb. Natural Order: Rubiaceae Silhet - Not mentioned	41. **Himalrandia tetrasperma** (Roxb.) Yamazaki in Jap. J. Bot. 45: 340 (1970). **Family:** Rubiaceae
44. *R. fasciculata* DC. Natural Order: Rubiaceae Silhet - Wallich & c.	42. **Benkara fasciculata** (Roxb.) Ridsdale, Reinwardtia 12: 298 (2008). **Family:** Rubiaceae
45. *R. dumetorum* Lamk. Natural Order: Rubiaceae Chittagong and Silhet - Not mentioned	43. **Catunaregam spinosa** (Thunb.) Tirveng. in Bull. Mus. Natl. Hist. Nat., Ser. 3, Bot. 35: 13 (1978). **Family:** Rubiaceae
46. *R. longiflora* Lamk. Natural Order: Rubiaceae Chittagong - Not mentioned	44. **Oxyceros longiflorus** (Lam.) T. Yamaz. in J. Jap. Bot. 45: 339 (1970). **Family:** Rubiaceae
47. *R. Wallichii* Hook.f. Natural Order: Rubiaceae Silhet-De Silva; Chittagong- J. D. H. & T.T.	45. **Tarennoidea wallichii** (Hook.f.) Tirveng. & Sastre in Mauritius Inst. Bull. 8(4): 90 (1979). **Family:** Rubiaceae
48. *Gardenia lucida* Roxb. Natural Order: Rubiaceae Chittagong - Roxburgh & c.	46. **Gardenia resinifera** Roth, Nov. Pl. Sp.: 150 (1821). **Family:** Rubiaceae
49. *G. coronaria* Ham. Natural Order: Rubiaceae Chittagong - Roxburgh & c.	47. **Gardenia coronaria** Buch.-Ham. in Embassy Ava ed. 2, 3: 307 (1809). **Family:** Rubiaceae
50. *G. turgida* Roxb. Natural Order: Rubiaceae Silhet - Not mentioned	48. **Ceriscoides turgida** (Roxb.) Tirveng., Bull. Mus. Natl. Hist. Nat., Ser. 3, Bot. 35: 15 (1978). **Family:** Rubiaceae
51. *G. campanulata* Roxb. Natural Order: Rubiaceae Chittagong and Silhet - Roxburgh, Griffith & c.	49. **Ceriscoides campanulata** (Roxb.) Tirveng, Bull. Mus. Natl. Hist. Nat., Ser. 3, Bot. 35: 16 (1978). **Family:** Rubiaceae

Species, Natural Order, recorded area - collector's name/ Wall. Cat. no. as in Hook.f. (1880-1882)	Current nomenclature with *loc. cit.* and family as of Cronquist (1981)
52. *Petunga Roxburghii* DC. Natural Order: Rubiaceae Chittagong and Silhet - Roxburgh & Wallich	50. **Hypobathrum racemosum** (Roxb.) Kurz, Prelim. Rep. Forest Pegu App. B: 59 (1875). **Family:** Rubiaceae
53. *Hyptianthera stricta* W. & A. Natural Order: Rubiaceae E. Bengal - Not mentioned	51. **Hyptianthera stricta** (Roxb. *ex* Schult.) Wight & Arn., Prodr. Fl. Ind. Orient. 399 (1834). **Family:** Rubiaceae
54. *Canthium didymum* Roxb. Natural Order: Rubiaceae Jyntea - de Silva	52. **Canthium dicoccum** (Gaertn.) Teijsm. & Binn., Cat. Herb. Bogor: 113 (1866). **Family:** Rubiaceae
55. *C. angustifolium* Roxb. Natural Order: Rubiaceae Sunderbunds, Chittagong and Silhet - Roxburgh & c.	53. **Canthium angustifolium** Roxb., Fl. Ind. 2: 169 (1824). **Family:** Rubiaceae
56. *C. parvifolium* Roxb. Natural Order: Rubiaceae Chittagong - Wall. Cat. 8257 and J.D.H. & T.T.	54. **Canthium parvifolium** Roxb., Fl. Ind. 2: 170 (1824). **Family:** Rubiaceae
57. *Ixora acuminata* Roxb. Natural Order: Rubiaceae Chittagong - C.B. Clarke	55. **Ixora acuminata** Roxb., Fl. Ind. 1: 383 (1820). **Family:** Rubiaceae
58. *I. parviflora* Vahl Natural Order: Rubiaceae Chittagong - C.B. Clarke	56. **Ixora pavetta** Andr., Bot. Repos. 2: t 78 (1799). **Family:** Rubiaceae
59. *I. villosa* Roxb. Natural Order: Rubiaceae Silhet - Wall. Cat. 6137	57. **Ixora balakrishnii** Deb & Rout in J. Bombay Nat. Hist. Soc. 89: 44 (1992). **Family:** Rubiaceae
60. *I. cuneifolia* Roxb. Natural Order: Rubiaceae Silhet - De Silva	58. **Ixora cuneifolia** Roxb., Fl. Ind. 1: 380 (1820). **Family:** Rubiaceae
61. *I. coccinea* Linn. Natural Order: Rubiaceae Chittagong - J.D.H. & T. T.	59. **Ixora coccinea** L., Sp. Pl.: 110 (1753). **Family:** Rubiaceae
62. *Pavetta subcapita* Hook.f. Natural Order: Rubiaceae Jyntea hills - Gomez	60. **Pavetta subcapita** Wall. *ex* Hook.f., Fl. Brit. India 3: 150 (1880). **Family:** Rubiaceae
63. *P. naucleiflora* Wall. Natural Order: Rubiaceae Silhet - Wall. Cat. 6171	61. **Pavetta naucleiflora** R. Br. *ex* G. Don, Gen. Hist. 3: 575 (1834). **Family:** Rubiaceae
64. *Coffea bengalensis* Roxb. Natural Order: Rubiaceae Silhet and Chittagong - Not mentioned	62. **Psilanthus bengalensis** (Roxb. *ex* Schult.) Leroy, Bull. Mus. Natl. Hist. Nat., B, Adansonia 3: 252 (1982). **Family:** Rubiaceae

Species, Natural Order, recorded area - collector's name/ Wall. Cat. no. as in Hook.f. (1880-1882)	Current nomenclature with *loc. cit.* and family as of Cronquist (1981)
65. *C. fragrans* Wall. Natural Order: Rubiaceae Silhet - Gomez	63. **Psilanthus fragrans** (Wall. *ex* Hook.f.) Leroy, Bull. Mus. Natl. Hist. Nat., B, Adansonia 3: 256 (1982). **Family:** Rubiaceae
66. *C. khasiana* Hook.f. Natural Order: Rubiaceae Jyntea hill - J.D.H. & T.T. & C.B. Clarke	64. **Nostolachma khasiana** (Korth.) Deb & Lahiri, Bull. Bot. Surv. India 17: 162 (1978). **Family:** Rubiaceae
67. *Morinda angustifolia* Roxb. Natural Order: Rubiaceae Chittagong - Not mentioned	65. **Morinda angustifolia** Roxb., Pl. Coromandel 3: 32 (1815). **Family:** Rubiaceae
68. *M. persicaefolia* Ham. Natural Order: Rubiaceae Chittagong - Hamilton	66. **Morinda persicaefolia** Buch.-Ham., Trans. Linn. Soc. London 13: 535 (1822). **Family:** Rubiaceae
69. *M. umbellata* Linn. Natural Order: Rubiaceae East Bengal - Not mentioned	67. **Morinda umbellata** L., Sp. Pl.: 176 (1753). **Family:** Rubiaceae
70. *Psychotria adenophylla* Wall. Natural Order: Rubiaceae Chittagong - J.D.H. & T.T.	68. **Psychotria adenophylla** Wall. in Roxb., Fl. Ind. 2: 166 (1824). **Family:** Rubiaceae
71. *P. calocarpa* Kurz Natural Order: Rubiaceae Chittagong - Gomez, & c.	69. **Psychotria calocarpa** Kurz in J. Asiat. Soc. Bengal, Pt. 2, Nat. Hist. 41(2): 315 (1872). **Family:** Rubiaceae
72. *P. silhetensis* Hook.f. Natural Order: Rubiaceae Silhet - de Silva & c.	70. **Psychotria silhetensis** Hook.f., Fl. Brit. India 3: 174 (1880). **Family:** Rubiaceae
73. *P. Montana* Blume Natural Order: Rubiaceae Silhet - De Silva	71. **Psychotria montana** Blume, Catalogus: 54 (1823). **Family: Rubiaceae**
74. *P. sphaerocarpa* Wall. Natural Order: Rubiaceae Silhet hills - Wallich	72. **Psychotria sphaerocarpa** Wall. in Roxb., Fl. Ind. 2: 161 (1820). **Family:** Rubiaceae
75. *Chasalia curviflora* Thw. var *ellipsoidea* Hook.f. Natural Order: Rubiaceae Jyntea - C.B. Clarke	73. **Chasalia curviflora** Thw. var. **ellipsoidea** Hook.f., Fl. Brit. India 3: 177 (1880). **Family:** Rubiaceae
76. *Geophila reniformis* Don Natural Order: Rubiaceae Silhet - Roxburgh and De Silva	74. **Geophila repens** (L.) Johnst., Sargentia 8: 281 (1949). **Family:** Rubiaceae
77. *Lasianthus cyanocarpus* Jack Natural Order: Rubiaceae Silhet and Chittagong - Wallich & c.	75. **Lasianthus cyanocarpus** Jack in Trans. Linn. Soc. London 14: 125 (1823). **Family:** Rubiaceae

Species, Natural Order, recorded area - collector's name/ Wall. Cat. no. as in Hook.f. (1880-1882)	Current nomenclature with *loc. cit.* and family as of Cronquist (1981)
78. *L. wallichii* Wight Natural Order: Rubiaceae Silhet - De Silva & c.	76. **Lasianthus attenuatus** Jack in Trans. Linn. Soc. London 14: 126 (1823). **Family:** Rubiaceae
79. *L. tentaculus* Hook.f. Natural Order: Rubiaceae Silhet -Wall. Cat. 8306	77. **Lasianthus rigidus** Miq., Fl. Ned. Ind. 2: 321 (1857). **Family:** Rubiaceae
80. *L. attenuates* Jack Natural Order: Rubiaceae Silhet - De Silva	76. **Lasianthus attenuatus** Jack in Trans. Linn. Soc. London 14: 126 (1823). **Family:** Rubiaceae
81. *L. tubiferus* Hook.f. Natural Order: Rubiaceae Jyntea hills - Griffith, & c.	78. **Lasianthus inodorus** Blume, Bijdr.: 998 (1826). **Family:** Rubiaceae
82. *L. inconspicuus* Hook.f. Natural Order: Rubiaceae Silhet - Wallich; Wall. Cat. 8313L	79. **Lasianthus lucidus** var. **inconspicuus** (Hook.f.) H. Zhu, Acta Bot. Yunnan. 20: 154 (1998). **Family:** Rubiaceae
83. *Ethulia conyzoides* Linn. Natural Order: Compositae Silhet- Not mentioned	80. **Ethulia conyzoides** L., Sp. Pl. ed. 2: 1171 (1762). **Family:** Asteraceae
84. *Vernonia Thomsoni* Hook.f. Natural Order: Compositae Chittagong, Seetakoond - J.D.H. & T. T.	81. **Vernonia thomsonii** Hook.f. Fl. Brit. India 3: 232 (1881). **Family:** Asteraceae
85. *V. saligna* DC. Natural Order: Compositae Chittagong - Not mentioned	82. **Vernonia saligna** DC., Prodr. 5: 33 (1836). **Family:** Asteraceae
86. *V. arborea* Ham. Natural Order: Compositae Silhet - Not mentioned	83. **Vernonia arborea** Buch.-Ham. *ex* Buch.-Ham., Trans. Linn. Soc. London 14: 218 (1825). **Family:** Asteraceae
87. *V. volkameriaefolia* DC. Natural Order: Compositae Jaintea hills - Griffith	84. **Vernonia volkameriaefolia** DC., Prodr. 5: 32 (1836). **Family:** Asteraceae
88. *V. scandens* DC. Natural Order: Compositae Silhet - Not mentioned	85. **Vernonia vagans** DC., Prodr. 5: 32 (1836). **Family:** Asteraceae
89. *Cyathocline lyrata* Cass. Natural Order: Compositae Chittagong - Not mentioned	86. **Cyathocline purpurea** (Buch.-Ham. *ex* D. Don) Kuntze, Rev. Gen. Pl.: 333 (1891). **Family:** Asteraceae
90. *Erigeron asteroides* Roxb. Natural Order: Compositae Bengal - Not mentioned	87. **Erigeron sublyratus** Roxb. *ex* DC. in Wight, Contr. Bot. Ind.: 9 (1834). **Family:** Asteraceae

Species, Natural Order, recorded area - collector's name/ Wall. Cat. no. as in Hook.f. (1880-1882)	Current nomenclature with *loc. cit.* and family as of Cronquist (1981)
91. *Conyza semipinnatifida* Wall. Natural Order: Compositae Soonderbunds at Burisal - Clarke	88. **Conyza semipinnatifida** Wall. *ex* DC., Prodr. 5: 382 (1836). **Family**: Asteraceae
92. *Thespis divaricata* DC. Natural Order: Compositae Silhet - Not mentioned	89. **Thespis divaricata** DC. in Guill. Arch. Bot. 2: 517 (1833). **Family**: Asteraceae
93. *Blumea amplectens* DC. Natural Order: Compositae Bengal - Not mentioned	90. **Blumea obliqua** (L.) Druce, Rep. Bot. Excu. Club Brit. Isles 4: 609 (1917). **Family**: Asteraceae
94. *B. amplectens* DC. var. *maritima* Hook.f. Natural Order: Compositae Soonderbunds - Not mentioned	90. **Blumea obliqua** (L.) Druce, Rep. Bot. Excu. Club Brit. Isles 4: 609 (1917). **Family**: Asteraceae
95. *Blumea bifoliata* DC. Natural Order: Compositae Bengal - Not mentioned	91. **Blumea bifoliata** (L.) DC. in Wight, Contr. Bot. Ind.: 14 (1834). **Family**: Asteraceae
96. *B. sericans* Hook.f. Natural Order: Compositae Chittagong hills - Clarke	92. **Blumea sericans** (Kurz) Hook.f., Fl. Brit. India 3: 262 (1881). **Family**: Asteraceae
97. *B. laciniata* DC. Natural Order: Compositae Bengal - Not mentioned	93. **Blumea laciniata** (Roxb.) DC., Prodr. 5: 436 (1836). **Family**: Asteraceae
98. *B. oxyodonta* DC. Natural Order: Compositae Bengal - Not mentioned	94. **Blumea oxyodonta** DC. in Wight, Contr. Bot. Ind.: 15 (1834). **Family**: Asteraceae
99. *B. myriocephala* DC. Natural Order: Compositae Chittagong - Clarke	95. **Blumea lanceolaria** (Roxb.) Druce, Bot. Soc. Exch. Club Br. Isles 4: 609 (1917). **Family**: Asteraceae
100. *B. balsamifera* DC. Natural Order: Compositae Chittagong - Not mentioned	96. **Blumea balsamifera** (L.) DC., Prodr. 5: 447 (1836). **Family**: Asteraceae
101. *Laggera flava* Benth. Natural Order: Compositae Chittagong - Not mentioned	97. **Blumeopsis falcata** (D. Don) Merr. in J. Arnold Arbor. Cambridge (1938). **Family**: Asteraceae
102. *L. aurita* Schultz-Bip. Natural Order: Compositae Chittagong - Not mentioned	98. **Blumea viscosa** (Mill.) V. M. Badillo, Revista Fac. Agron. (Maracay): 7 9 (1974). **Family**: Asteraceae
103. *Pluchea indica* Less. Natural Order: Compositae Sunderbunds - Not mentioned	99. **Pluchea indica** (L.) Less., Linnaea 6: 150 (1831). **Family**: Asteraceae

Species, Natural Order, recorded area - collector's name/ Wall. Cat. no. as in Hook.f. (1880-1882)	Current nomenclature with *loc. cit.* and family as of Cronquist (1981)
104. *Sphaeranthus africanus* Linn. Natural Order: Compositae Silhet - Not mentioned	100. **Sphaeranthus africanus** L., Sp. Pl. ed. 2: 1314 (1762). **Family:** Asteraceae
105. *S. indicus* Linn. Natural Order: Compositae Silhet - Not mentioned	101. **Sphaeranthus indicus** L., Sp. Pl.: 927 (1753). **Family:** Asteraceae
106. *Caesulia axillaris* Roxb. Natural Order: Compositae Chittagong - Not mentioned	102. **Caesulia axillaris** Roxb., Pl. Corom. 1: 64, t. 93 (1798). **Family:** Asteraceae
107. *Enhydra fluctuans* Lour. Natural Order: Compositae Silhet - Not mentioned	103. **Enhydra fluctuans** Lour., Fl. Cochinch. 511 (1790). **Family:** Asteraceae
108. *Wedelia calendulacea* Less. Natural Order: Compositae Silhet - Not mentioned	104. **Wedelia chinensis** (Osbeck) Merr., Philipp. J. Sci. 12: 111 (1917). **Family:** Asteraceae
109. *W. biflora* DC. Natural Order: Compositae Near the Sea from Bengal-Not mentioned	105. **Melanthera biflora** (L.) Wild, Kirkia: 54 (1965). **Family:** Asteraceae
110. *Cotula hemisphaerica* Wall. Natural Order: Compositae Dry rice field in Bengal - Not mentioned	106. **Cotula hemisphaerica** Wall. *ex* Benth. & Hook.f. Gen. Pl. 2: 429 (1873). **Family:** Asteraceae
111. *Artemisia caruifolia* Ham. Natural Order: Compositae Eastern Bengal - Not mentioned	107. **Artemisia carvifolia** Buch.-Ham. *ex* Roxb., Fl. Ind. 2: 422 (1820). **Family: Asteraceae**
112. *Senecio obtusatus* Wall. Natural Order: Compositae Jyntea hills - De Silva	108. **Senecio obtusatus** Wall. *ex* DC., Prodr. 6: 367 (1838). **Family:** Asteraceae
113. *S. ramosus* Wall. Natural Order: Compositae Silhet - Roxburgh	109. **Senecio ramosus** Wall. *ex* Hook.f., Fl. Brit. India 3: 342 (1881). **Family:** Asteraceae
114. *Cnicus arvensis* Hoffm. Natural Order: Compositae Soonderbunds - Not mentioned	110. **Cirsium arvense** (L.) Scop., Fl. Carn. ed. 2 (2): 126 (1772). **Family:** Asteraceae
115. *Saussurea affinis* Spreng Natural Order: Compositae Silhet - Not mentioned	111. **Hemistepta lyrata** (Bunge) Bunge in Dorp. Jahrb. Litt. 1: 222 (1833). **Family:** Asteraceae
116. *Crepis acaulis* Hook.f. Natural Order: Compositae Dinagepore - Not mentioned	112. **Launaea acaulis** (Roxb.) Kerr in Craib, Fl. Siam. Enum. 2: 299 (1936). **Family:** Asteraceae

Species, Natural Order, recorded area - collector's name/ Wall. Cat. no. as in Hook.f. (1880-1882)	Current nomenclature with *loc. cit.* and family as of Cronquist (1981)
117. *Lectuca polycephala* Benth. Natural Order: Compositae Bengal - Not mentioned	113. **Ixeris polycephala** Cass., Dict. Sci. Nat. 24: 50 (1822). **Family:** Asteraceae
118. *Launaea aspleniifolia* DC. Natural Order: Compositae Soonderbunds - Not mentioned	114. **Launaea aspleniifolia** (Willd.) DC., Prodr. 7: 181 (1838). **Family:** Asteraceae
119. *L. pinnatifida* Cass. Natural Order: Compositae Bengal - Not mentioned	115. **Launaea sarmentosa** (Willd.) Sch.-Bip. *ex* Kuntze, Rev. Gen. Pl. 1: 350 (1891). **Family:** Asteraceae
120. *Stylidium kunthii* Wall. Natural Order: Stylidieae Chittagong - H.f. & T.	116. **Stylidium kunthii** Wall. *ex* DC., Prodr., 7: 335 (1839). **Family:** Stylidiaceae
121. *S. tenellum* Swartz Natural Order: Stylidieae Dacca-Clarke and Chittagong - Kurz	117. **Stylidium tenellum** Sw., Mag. Ges. Naturf. Fr. Berlin 1: 51, pl. 2, t. 3, f. 3 (1807). **Family:** Stylidiaceae
122. *Lobelia trigona* Roxb. Natural Order: Campanulaceae Dacca - Clarke	118. **Lobelia alsinoides** Lam., Encycl. 3: 588 (1792). **Family:** Campanulaceae
123. *L. affinis* Wall. Natural Order: Campanulaceae Bengal - Not mentioned	119. **Lobelia zeylanica** L., Sp. Pl.: 932 (1753). **Family:** Campanulaceae
124. *L. terminalis* Clarke Natural Order: Campanulaceae Mymensingh - Clarke	120. **Lobelia terminalis** C.B. Clarke in Hook.f., Fl. Brit. India 3: 424 (1881). **Family:** Campanulaceae
125. *L. rosea* Wall. Natural Order: Campanulaceae North Bengal - Not mentioned	121. **Lobelia rosea** Wall. in Roxb., Fl. Ind. 2: 115 (1824). **Family:** Campanulaceae
126. *Campanomoea celebica* Blume Natural Order: Campanulaceae Chittagong - Not mentioned	122. **Cyclocodon celebicus** (Blume) D. Y. Hong, Acta Phytotax. Sin. 36(2): 109 (1998). **Family:** Campanulaceae
127. *Agapetes variegata* D. Don Natural Order: Vacciniaceae Chittagong hills - Not mentioned	123. **Agapetes variegata** (Roxb.) D. Don *ex* G. Don, Gen. Hist. 3: 862 (1834). **Family:** Ericaceae
128. *A. macrantha* Hook.f. Natural Order: Vacciniaceae Chittagong hills - Roxburgh	124. **Agapetes macrantha** (Hook.) Benth. & Hook.f., Gen. Pl. 2: 571 (1876). **Family:** Ericaceae
129. *Aegialitis rotundifolia* Roxb. Natural Order: Plumbagineae Bengal - Not mentioned	125. **Aegialitis rotundifolia** Roxb., Fl. Ind. 2: 111 (1832). **Family:** Plumbaginaceae

Species, Natural Order, recorded area - collector's name/ Wall. Cat. no. as in Hook.f. (1880-1882)	Current nomenclature with *loc. cit.* and family as of Cronquist (1981)
130. *Lysimachia javanica* Bl. Natural Order: Primulaceae Silhet – Not mentioned	126. **Lysimachia decurrens** G. Forst. in Fl. Ins. Austr. 12 12 (1786). **Family:** Primulaceae
131. *Anagallis arvensis* Linn. Natural Order: Primulaceae Bengal - Not mentioned	127. **Anagallis arvensis** L., Sp. Pl.: 148 (1753). **Family:** Primulaceae
132. *Maesa ramentacea* A. DC. Natural Order: Myrsineae Eastern Bengal – Not mentioned	128. **Maesa ramentacea** (Roxb.) A. DC., Trans. Linn. Soc. London 17: 133 (1834). **Family:** Myrsinaceae
133. *M. paniculata* A. DC. Natural Order: Myrsineae Silhet (Pundua) - H.f. & T.	129. **Maesa paniculata** A. DC., Trans. Linn. Soc. London 17: 133 (1834). **Family:** Myrsinaceae
134. *Embelia nutans* Wall. Natural Order: Myrsineae Silhet - Wall. Cat. 2303 & H.f. & T.	130. **Embelia nutans** Wall. in Roxb., Fl. Ind. 2: 290 (1824). **Family:** Myrsinaceae
135. *Ardisia paniculata* Roxb. Natural Order: Myrsineae Dacca -Clarke & Chittagong - Roxburgh	131. **Ardisia paniculata** Roxb., Fl. Ind. 2: 270 (1824). **Family:** Myrsinaceae
136. *A. icara* Ham. Natural Order: Myrsineae North-East Bengal; Mudhopoor - Hamilton	132. **Ardisia icara** Wall. *ex* DC., Trans. Linn. Soc. London 17: 125 (1834). **Family:** Myrsinaceae
137. *Amblyanthus glandulosus* A. DC. Natural Order: Myrsineae Silhet - Wall. Cat. 2265	133. **Amblyanthus glandulosus** (Roxb.) A. DC., Ann. Sci. Nat., Bot. II, 16: 83, t. 6 (1841). **Family:** Myrsinaceae
138. *Chrysophyllum Roxburghii* G. Don Natural Order: Sapotaceae Silhet - Wall. Cat. 4160	134. **Chrysophyllum roxburghii** G. Don, Gen. Hist. 4: 33 (1837). **Family:** Sapotaceae
139. *Sideroxylon grandifolium* Wall. Natural Order: Sapotaceae Silhet - Wall. Cat. 4155, 4156A	135. **Planchonella grandifolia** (Wall.) Pierre, Not. Bot.: 36 (1890). **Family:** Sapotaceae
140. *Dichopsis polyantha* Benth. Natural Order: Sapotaceae Silhet- Wall. Cat. 4166, 4156 & Chittagong - H.f. & T.	136. **Palaquium polyanthum** (Wall. *ex* G. Don) Baill in Traite Bot. Med. Phan. 1500 (1884). **Family:** Sapotaceae
141. *Diospyros Embryopteris* Pers. Natural Order: Ebenaceae Bengal - Not mentioned	137. **Diospyros peregrina** (Gaertn) Guerke in Nat. Pflanzenfam. 4(1): 164 (1891). **Family:** Ebenaceae
142. *D. Toposia* Ham. Natural Order: Ebenaceae Silhet & Chittagong - Roxburgh & Kurz	138. **Diospyros toposia** Buch.-Ham.in Trans. Linn. Soc. London 15: 115 (1827). **Family:** Ebenaceae

Species, Natural Order, recorded area - collector's name/ Wall. Cat. no. as in Hook.f. (1880-1882)	Current nomenclature with *loc. cit.* and family as of Cronquist (1981)
143. *D. nigricans* Wall. Natural Order: Ebenaceae Silhet - Wallich	139. **Diospyros nigricans** Wall. *ex* A. DC., Prodr. 8: 239 (1844). **Family:** Ebenaceae
144. *D. lanceaefolia* Roxb. Natural Order: Ebenaceae Silhet – Roxburgh	140. **Diospyros lanceifolia** Roxb., Fl. Ind. 2: 537 (1832). **Family:** Ebenaceae
145. *D. stricta* Roxb. Natural Order: Ebenaceae Silhet and Comilla - Roxburgh	141. **Diospyros stricta** Roxb., Fl. Ind. 2: 539 (1832). **Family:** Ebenaceae
146. *D. ramiflora* Roxb. Natural Order: Ebenaceae East Bengal - Roxburgh	142. **Diospyros ramiflora** Roxb., Fl. Ind. 2: 535 (1832). **Family:** Ebenaceae
147. *D. elegans* Clarke var. *Hookeri* Clarke Natural Order: Ebenaceae Chittagong; Seetakoond - H.f. & T.	143. **Diospyros elegans** C.B. Clarke in Hook.f., Fl. Brit. India 3: 571 (1882). **Family:** Ebenaceae
148. *Symplocos caudata* Wall. Natural Order: Styraceae Chittagong; Seetakoond - H.f. & T.	144. **Symplocos sumuntia** Buch.-Ham. *ex* D. Don, Prodr. Fl. Nepal. 145 (1825). **Family:** Symplocaceae
149. *Jasminum sambac* Ait. Natural Order: Oleaceae Bengal - Not mentioned	145. **Jasminum sambac** (L.) Sol., Hort. Kew 1: 8 (1789). **Family:** Oleaceae
150. *J. scandens* Vahl Natural Order: Oleaceae Chittagong - Not mentioned	146. **Jasminum scandens** (Retz.) Vahl, Symb. Bot. 3: 2 (1794). **Family:** Oleaceae
151. *J. anastomosans* Wall. Natural Order: Oleaceae Silhet; Chattuck - H.f. & T.	147. **Jasminum nervosum** Lour., Fl. Cochinch. 1: 20 (1790). **Family:** Oleaceae
152. *J. anastomosans* Wall. var. *silhetensis* Blume Natural Order: Oleaceae Silhet - Not mentioned	147. **Jasminum nervosum** Lour., Fl. Cochinch. 1: 20 (1790). **Family:** Oleaceae
153. *J. subtriplinerve* Blume Natural Order: Oleaceae Silhet - Wallich	147. **Jasminum nervosum** Lour., Fl. Cochinch. 1: 20 (1790). **Family:** Oleaceae
154. *J. auriculatum* Vahl Natural Order: Oleaceae Bengal - Not mentioned	148. **Jasminum auriculatum** Vahl, Symb. Bot. 3: 30 (1794). **Family:** Oleaceae
155. *J. lanceolaria* Roxb. Natural Order: Oleaceae Jaintea hills - Griffith, H.f. & T.	149. **Jasminum lanceolaria** Roxb., Fl. Ind. 1: 98 (1820). **Family:** Oleaceae

Species, Natural Order, recorded area - collector's name/ Wall. Cat. no. as in Hook.f. (1880-1882)	Current nomenclature with *loc. cit.* and family as of Cronquist (1981)
156. *Linociera macrophylla* Wall. Natural Order: Oleaceae Silhet - Wall. Cat. 2826	150. **Chionanthus ramiflorus** Roxb., Fl. Ind. 1: 106 (1820). **Family:** Oleaceae
157. *Olea dioica* Roxb. Natural Order: Oleaceae Chittagong hills - Roxburgh	151. **Olea dioica** Roxb., Fl. Ind. 1: 105 (1820). **Family:** Oleaceae
158. *Ligustrum robustum* Blume Natural Order: Oleaceae Silhet, Dacca, Chittagong & c. - Not mentioned	152. **Ligustrum robustum** (Roxb.) Blume, Mus. Bot. 1: 313 (1851). **Family:** Oleaceae
159. *Myxopyrum smilacifolium* Blume Natural Order: Oleaceae Silhet and Chittagong - Not mentioned	153. **Myxopyrum smilacifolium** (Wall.) Blume, Mus. Bot. 1: 320 (1851). **Family:** Oleaceae
160. *Willoughbeia edulis* Roxb. Natural Order: Apocynaceae Chittagong - Roxburgh, & c.	154. **Willoughbeia edulis** Roxb., Pl. Corom. 3: 77, t. 280 (1820). **Family:** Apocynaceae
161. *Melodinus monogynus* Roxb. Natural Order: Apocynaceae Silhet - Not mentioned	155. **Melodinus cochinchinensis** (Lour.) Merr., Trans. Amer. Philos. Soc., n.s. 24: 310 (1935). **Family:** Apocynaceae
162. *Tabernaemontana recurva* Roxb. Natural Order: Apocynaceae Chittagong - Roxburgh	156. **Tabernaemontana divaricata** (L.) R. Br. *ex* Roem. & Schult., Syst. Veg. 4: 427 (1819). **Family:** Apocynaceae
163. *Parsonia spiralis* Wall. Natural Order: Apocynaceae Silhet - Wall. Cat. 1631, 1632, 1633	157. **Parsonia alboflavescens** (Dennst.) Mabb., Taxon 26: 532 (1977). **Family:** Apocynaceae
164. *Vallaris Heynei* Spreng. Natural Order: Apocynaceae Silhet - Not mentioned	158. **Vallaris solanacea** (Roth) Kuntze, Rev. Gen. Pl. 2: 417 (1891). **Family:** Apocynaceae
165. *Pottsia cantonensis* Hook. & Arn. Natural Order: Apocynaceae Silhet - De Silva	159. **Pottsia laxiflora** (Blume) Kuntze, Rev. Gen. Pl. 2: 416 (1891). **Family:** Apocynaceae
166. *Wrightia coccinea* Sims Natural Order: Apocynaceae Silhet -Roxburgh, De Silva & Chittagong - Kurz	160. **Wrightia coccinea** (Roxb. *ex* Hornem.) Sims., Bot. Mag. 53: t. 2696 (1826). **Family:** Apocynaceae
167. *Strophanthus Wallichii* A. DC. Natural Order: Apocynaceae Chittagong- Seetakoond - H.f. & T.	161. **Strophanthus wallichii** A. DC., Prodr. 8: 418 (1844). **Family:** Apocynaceae
168. *Beaumontia grandiflora* Wall. Natural Order: Apocynaceae Silhet and Chittagong - Not mentioned	162. **Beaumontia grandiflora** Wall., Tent. Fl. Nepal 1: 15, t. 7 (1824). **Family:** Apocynaceae

Species, Natural Order, recorded area - collector's name/ Wall. Cat. no. as in Hook.f. (1880-1882)	Current nomenclature with *loc. cit.* and family as of Cronquist (1981)
169. *Ecdysanthera micrantha* A. DC. Natural Order: Apocynaceae Silhet - Kurz	163. **Urceola micrantha** (Wall. *ex* G. Don) Middlton, Novon 4: 51 (1994). **Family**: Apocynaceae
170. *Aganosma marginata* G. Don Natural Order: Apocynaceae Silhet and Chittagong - Not mentioned	164. **Amphineurion marginatum** (Roxb.) Middleton, Taxon 55: 502 (2006). **Family**: Apocynaceae
171. *A. cymosa* G. Don Natural Order: Apocynaceae Silhet - Roxburgh & c.	165. **Aganosma cymosa** (Roxb.) G. Don, Gen. Hist. 4: 77 (1837). **Family**: Apocynaceae
172. *A. cymosa* G. Don var. *cymosa* Hook.f. Natural Order: Apocynaceae Silhet - Not mentioned	165. **Aganosma cymosa** (Roxb.) G. Don, Gen. Hist. 4: 77 (1837). **Family**: Apocynaceae
173. *Epigynum laevigatum* Hook.f. Natural Order: Apocynaceae Silhet at Pundua - Wall. Cat. 1669	166. **Anodendron affine** (Hook. & Arn.) Druce, Rep. Bot. Soc. Exch. Club Brit. Isles 1916: 605 (1917). **Family**: Apocynaceae
174. *Rhynchodia Wallichii* Benth. Natural Order: Apocynaceae Silhet - De Silva	167. **Chonemorpha verrucosa** (Blume) Middlton, Novon 3: 455 (1993). **Family**: Apocynaceae
175. *Anodendron paniculatum* A. DC. Natural Order: Apocynaceae Silhet - Not mentioned	168. **Anodendron paniculatum** (Roxb.) A. DC., Prodr. 8: 444 (1844). **Family**: Apocynaceae
176. *Ichnocarpus frutescens* Br. Natural Order: Apocynaceae Silhet and Chittagong - Not mentioned	169. **Ichnocarpus frutescens** (L.) R. Br., Mem. Wern. Nat. Hist. Soc. 1: 62 (1809). **Family**: Apocynaceae
177. *I. ovatifolius* A. DC. Natural Order: Apocynaceae Silhet - Not mentioned	169. **Ichnocarpus frutescens** (L.) R. Br., Mem. Wern. Nat. Hist. Soc. 1: 62 (1809). **Family**: Apocynaceae

Acknowledgement

We are thankful to Professor Dr. Md. Abul Hassan, Department of Botany, University of Dhaka for his support to conduct this research.

References

Ahmed, Z.U., Begum, Z.N.T., Hassan, M.A., Khondker, M., Kabir, S.M.H., Ahmad, M., Ahmed, A.T.A., Rahman, A.K.A. and Haque, E.U. (Eds.). 2008a. Encyclopedia of Flora and Fauna of Bangladesh **6**: 1-408. Asiatic Society of Bangladesh, Dhaka.

Ahmed, Z.U., Hassan, M.A., Begum, Z.N.T., Khondker, M., Kabir, S.M.H., Ahmad, M., Ahmed, A.T.A., Rahman, A.K.A. and Haque, E.U. (Eds.). 2008b. Encyclopedia of Flora and Fauna of Bangladesh **7**: 1-546. Asiatic Society of Bangladesh, Dhaka.

Ahmed, Z.U., Hassan, M.A., Begum, Z.N.T., Khondker, M., Kabir, S.M.H., Ahmad, M., Ahmed, A.T.A., Rahman, A.K.A. and Haque, E.U. (Eds.). 2009a. Encyclopedia of Flora and Fauna of Bangladesh **8**: 1-478. Angiosperms: Dicotyledons (Fabaceae-Lythraceae). Asiatic Society of Bangladesh, Dhaka.

Ahmed, Z.U., Hassan, M.A., Begum, Z.N.T., Khondker, M., Kabir, S.M.H., Ahmad, M., Ahmed, A.T.A. (Eds.). 2009b. Encyclopedia of Flora and Fauna of Bangladesh **9**: 1-488. Asiatic Society of Bangladesh, Dhaka.

Ahmed, Z.U., Hassan, M.A., Begum, Z.N.T., Khondker, M., Kabir, S.M.H., Ahmad, M., Ahmed, A.T.A. (Eds.). 2009c. Encyclopedia of Flora and Fauna of Bangladesh **10**: 1-580. Asiatic Society of Bangladesh, Dhaka.

Ali, Z. 1971. Ericaceae. *In:* Nasir, E. and Ali, S.I. (Eds.), Flora of Pakistan. Fasc. **5**: 1- 72. Stewart Herbarium, Gordon College, Rawalpindi, Pakistan.

Brandis, D. 1906. Indian Trees. Bishen Singh Mahendra Pal Singh, Dehra Dun, India- 767pp.

Brumitt, R.K. 1992. Vascular Plants Families and Genera- Royal Botanic Gardens. England, 804 pp.

Brummitt, R.K. and Powell, C.E. 1992. Authors of Plant names. Royal Botanic Gardens, Kew, England, 732 pp.

Cowan, J.M. 1926. The Flora of Chakaria Sundarbans. Rec. Bot. Surv. Ind. **11**(2): 197-225.

Cronquist, A. 1981. An Integrated System of Classification of Flowering Plants. Columbia University Press- 1262 pp.

Hara, H. and Williams, L.H.J. 1979. An Enumeration of the Flowering Palnts of Nepal **2**: 1 - 220. Trustee of British Museum (Natural history), London.

Hara, H., Chater, O.A. and Williams, L.H.J. 1982. An Enumeration of the Flowering Plants of Nepal. **3**: 1-226. Trustee of British Museum (Natural history), London.

Heinig, R.L. 1925. List of Plants of Chittagong Collectorate and the Hill Tracts - The Bengal Government Branch Press, Darjeeling, India, 84 pp.

Hooker, J.D. 1880-1882. The Flora of British India **3**: 1-712. L. Reeve & Co. Ltd., Kent, England.

Internet sources: I. The Plant List, KEW II. IPNI for plant name query III. Google search: plant names and others

Ishaq, M. 1979. Bangladesh Gazetteers, Ministry of Cabinet Affairs, Establishment Division, Bangladesh Government press, Dhaka, Bangladesh.

Kanjilal, U.N., Das, A., Kanjilal, P.C. and De, R.N. (Eds.) 1939. Flora of Assam. **3**: 1-578. Government of Assam, Shillong, India.

Kurz, W.S. 1877. Forest Flora of British Burma, **2**: 1-550. Reprt.,1974. Bishen Sing Mahendra Pal Singh, Dhera Dun - India.

Mabberley, D.J. 1997. The Plant-Book, a portable dictionary of the vascular plants- (2nd edition). Cambridge University Press, Cambridge, UK- 858 pp.

McNeill, J., Barrie, F.R., Buck, W.R., Demoulin, V., Greuter, W., Hawksworth, D.L., Herendeen, P.S., Knapp, S., Marhold, K., Prado, J., Prudhomme Van Reine, W.F., Smith, G.F., Wiersema, J.H. and Turland, N.J. (Eds.). 2012. International Code of Nomenclature for algae, fungi, and plants (Melbourne Code), Adopted by the Eighteenth International Botanical Congress Melbourne, Australia, July 2011 (electronic ed.). Bratislava: *International Association for Plant Taxonomy.* Regnum Vegetabile, 154. Koeltz Scientific Books, Melbourne, Australia.

Prain, D. 1903. Bengal Plants, **I**: 1-490; **II**: 491-1013. Botanical Survey of India.

Press, J.R., Shrestha, K.K. and Sutton, D.A. 2000. Annotated Checklist of the Flowering Plants of Nepal. The Natural History Museum, London- 430 pp.

Rahman, M.A. and Wilcock, C.C. 1991. Periplocaceae *In:* Khan, M.S. and Rahman, M.M. (Eds.) Flora of Bangladesh. Fasc. **47**: 1-16. Bangladesh National Herbarium, Dhaka.

Rahman, M.A. and Wilcock, C.C. 1995. Asclepiadaceae *In:* Khan, M.S. and Rahman, M.M. (Eds.) Flora of Bangladesh. Fasc. **48**: 1-71. Bangladesh National Herbarium, Ministry of Environment and Forest, Dhaka.

Raizada, M.B. 1941. On the flora of Chittagong. The Indian Forester **67**(5): 245-254.

Rashid, M.E. and Rahman, M.A. 2011. Updated nomenclature and taxonomic status of the plants of Bangladesh included in Hook.f., the Flora of British India: Volume-I. Bangladesh J. Plant Taxon. **18**(2): 177-197.

Rashid, M.E. and Rahman, M.A. 2012. Updated nomenclature and taxonomic status of the plants of Bangladesh included in Hook.f., the Flora of British India: Volume-II. Bangladesh J. Plant Taxon. **19**(2): 173-190.

Roxburgh, W. 1814. Hortus Bengalensis. Boerhaave Press, Leiden (Holland). pp.1-105.

Roxburgh, W. 1820. Flora Indica. Carey, W. and Wallich, N. (Eds.) Vol. **1**. Mission Press, Serampore, Calcutta, India.

Roxburgh, W. 1824. Flora Indica. Carey, W. and Wallich, N. (Eds.) Vol. **2**. Mission Press, Serampore, Calcutta, India.

Roxburgh, W. 1832. Flora Indica. Carey, W. (Ed.) Vol. **3**. Mission Press, Serampore, Calcutta, India.

Sinclair, J. 1956. The Flora of Cox's Bazar, East Pakistan. Bull. Bot. Soc.Beng. **9**(2): 1-116.

Voss, E.G. (Ed.) 1983. International Code of Botanical Nomenclature (ICBN), Regnum Vegetabile, Vol. **97**. Bohn, Scheltema & Holkema, Utecht.

Wallich, N. 1828-1849. A Numerical list of dried specimens of plants in the East Indian Company's Museum, *Ined.*

Wu, Z.Y. and Raven, P.H. (Eds.). 1994. Flora of China- Vol. **17** (Verbenaceae through Solanaceae). Science Press, Beijing, and Missouri Botanical Garden Press, St. Louis.

Wu, Z.Y., Raven, P. H and Hong, D.Y. (Eds.). 2005. Flora of China- Vol. **14** (Apiaceae through Ericaceae). Science Press, Beijing, and Missouri Botanical Garden Press, St. Louis.

AN ANNOTATED CHECKLIST OF THE ANGIOSPERMIC FLORA OF RAJKANDI RESERVE FOREST OF MOULVIBAZAR, BANGLADESH

A.K.M. Kamrul Haque[1], Saleh Ahammad Khan, Sarder Nasir Uddin[2]
AND Shayla Sharmin Shetu

Department of Botany, Jahangirnagar University, Savar, Dhaka 1342, Bangladesh

Keywords: Checklist; Angiosperms; Rajkandi Reserve Forest; Moulvibazar.

Abstract

This study was carried out to provide the baseline data on the composition and distribution of the angiosperms and to assess their current status in Rajkandi Reserve Forest of Moulvibazar, Bangladesh. The study reports a total of 549 angiosperm species belonging to 123 families, 98 (79.67%) of which consisting of 418 species under 316 genera belong to Magnoliopsida (dicotyledons), and the remaining 25 (20.33%) comprising 132 species of 96 genera to Liliopsida (monocotyledons). Rubiaceae with 30 species is recognized as the largest family in Magnoliopsida followed by Euphorbiaceae with 24 and Fabaceae with 22 species; whereas, in Lilliopsida Poaceae with 32 species is found to be the largest family followed by Cyperaceae and Araceae with 17 and 15 species, respectively. *Ficus* is found to be the largest genus with 12 species followed by *Ipomoea*, *Cyperus* and *Dioscorea* with five species each. Rajkandi Reserve Forest is dominated by the herbs (284 species) followed by trees (130 species), shrubs (125 species), and lianas (10 species). Woodlands are found to be the most common habitat of angiosperms. A total of 387 species growing in this area are found to be economically useful. 25 species listed in Red Data Book of Bangladesh under different threatened categories are found under Lower Risk (LR) category in this study area.

Introduction

Rajkandi Reserve Forest (RRF) is located in Kamalganj upazilla under Moulvibazar district of Bangladesh. This forest area consists of ca. 2,450 hectares' land of Rajkandi forest range that lies between the 24°12′-24°17′N and 91°51′-91°55′E, and comprises diverse habitats and ecosystems. This tropical semi-evergreen forest falls within the Indo-Burma hot-spot of biodiversity (Myers *et al.*, 2000).

The extensive floristic exploration throughout British India conducted by J.D. Hooker (1872-1897) included the Sylhet region of the present political boundary of Bangladesh. Later, David Prain (1903) covered different regions of Bangladesh including Sylhet under his floristic exploration. Kanjilal *et al.* (1934, 1938-1940) included some areas of Sylhet region too. In those studies, any specific or detail information on local distribution and voucher specimens of the taxa described are missing. Later on, various sporadic inventories have been completed in different areas of greater Sylhet region, such as Das (1968), Alam (1988), Arefin *et al.* (2011), Uddin and Hassan (2004) and Sobuj and Rahman (2011). However, the flora and plant diversity of Moulvibazar district, have not yet been explored, except the plant diversity (Uddin and Hassan, 2010) or a plant group (Haque *et al.*, 2016) of a particular area. Taxonomic data on the current floristic composition of RRF collected through field inventories and examination of representative plant specimens are still lacking, though such data are important for the sustainable use and

---

[1]Corresponding author. Email: kamrulhaque1234@gmail.com
[2]Bangladesh National Herbarium, Zoo Road, Mirpur-1, Dhaka 1216, Bangladesh.

conservation of plant resources and resource-based development of the area. Therefore, an inventory on the floristic composition of RRF was conducted with the objectives to produce an annotated checklist of the angiospermic species of the area; to determine the current status of threatened species of angiosperms from Bangladesh in this forest; and to collect and preserve representative plant specimens for future reference.

Materials and Methods

Taxonomic inventories were conducted during 2010 to 2015 through 25 field trips in different seasons throughout the study area (Fig. 1). Necessary field data and representative plant specimens were collected and preserved following standard herbarium techniques (Bridson and Forman, 1989; Singh and Subramaniam, 2008). All plant specimens were preliminarily identified through consulting the experts and matching with relevant voucher specimens preserved at Jahangirnagar University Herbarium (JUH), and Bangladesh National Herbarium (DACB). Some critical specimens were identified at Central National Herbarium, Howrah, India (CAL) during the visit of one of the authors.

The identification of the plant specimens were verfied by matching with the images of pertinent type specimens available in the websites of international herbaria and consulting taxonomic descriptions and keys available in the relevant literatures (Hooker, 1872-1897; Prain, 1903; Wu and Raven, 1994-2001; Wu *et al.*, 1999-2013).

Nomenclature of each taxon was verified following Flora of China (Wu and Raven, 1994-2001; Wu *et al.*, 1999-2013) and the nomenclatural databases of The Plant List (2013) and TROPICOS (2017). The common names have been cited based on Huq (1986), Pasha and Uddin (2013) and interview with the local people. The families have been arranged following Cronquist (1981), and the genera and species under each family have been arranged alphabetically (Table 1). The economic uses of the species were recorded through interviews with the local people during the field surveys, and consulting the relevant literatures (e.g., Ghani, 1998; van Valkenburg and Bunyapraphatsara, 2002). Status of threatened plant species listed in Red Data Book of Bangladesh (Khan *et al.*, 2001; Ara *et al.*, 2013) was asseesed in context to RRF through field observation on natural distribution and regeneration of each species throughout the area, and IUCN threatened category was estimated consulting IUCN (2001). The voucher specimens have been preserved at JUH and DACB.

Results and Discussion

A total of 549 species of angiosperms under 412 genera and 123 families have been recorded from RRF with their natural distribution (Table 1). Among these families, 98 (79.67%) representing 316 genera and 418 species are identified as dicotyledons (Magnoliopsida), whereas, only 25 (20.33%) families consisting of 96 genera and 132 species as monocotyledons (Liliopsida). Among these families, 46 are represented by single species each and only 10 families by more than 10 (10-33) species. This reserve forest is dominated by the herbs comprising 284 species that are followed by trees of 130 species, shrubs of 125 species, and lianas of 10 species. These data indicate that Rajkandi Reserve Forest is floristically rich.

In Magnoliopsida, Rubiaceae with 30 species of 21 genera is recognized as the largest family in RRF followed by Euphorbiaceae with 24 species of 19 genera and Fabaceae with 22 species belonging to 17 genera. Poaceae consisting of 32 species of 25 genera is found to be the largest family in Lilliopsida, followed by Cyperaceae with 17 species of nine genera and Araceae with 15 species of 13 genera. *Ficus* with 12 species is found as the largest genus in the area, which is followed by *Ipomoea, Cyperus* and *Dioscorea* with five species each and *Maesa, Piper, Senna, Terminalia, Phyllanthus, Mussaenda, Ixora* and *Bambusa* with four species each.

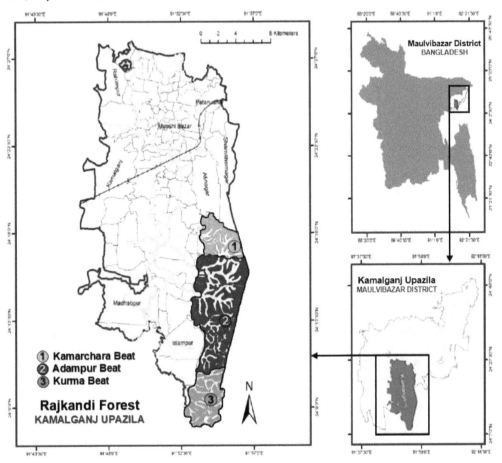

Fig. 1. Rajkandi Forest Range, Kamalganj, Moulvibazar, Bangladesh. (*Source*: Modified from Haque *et al.*, 2016).

The composition and distribution of species in all of the three forest beats of RRF, namely Adampur, Kurma and Kamarchara, were found to be variable remarkebly. A total of 231, 26 and five species were found to occur exclusively in Adampur-, Kamarchara- and Kurma beats, respectively. The occurance of total 538 species in Adampur beat, with 231 species exclusive and additional 307 species overlapping in other two beats (240 in Kurma and Kamarchara, 46 in Kurma and 21 in Kamarchara), indicates that this forest beat could be considered as a hotspot of biodiversity.

In this forest, woodlands were found to be the most common habitats harbouring the highest number of species (157 species), and this might be due to accumulation of nutrient components and humus-rich soil. In contrast, the finding of hill top to harbour relatively lower number of species (16 species) might be due to their poor humus and nutrient components.

The total number of angiosperm species (549) found in Rajkanndi Reserve Forest during this study is 15.20% of the total 3,611 species, and that of angiosperm families (123) is 59.42% of the total 207 families reported for Bangladesh (Ahmed *et al.*, 2008-2009).

Table 1. List of angiosperm species of Ranjkandi Reserve Forest under Maulvibazar district of Bangladesh.

Scientific name	Bangla name	Habit	Habitat	Distrib.	Uses	RSE
MAGNOLIOPSIDA Brongn.						
MAGNOLIACEAE Juss.						
Magnolia champaca (L.) Baill. *ex* Pierre	Champa	Tree, m	wd	All beats	T	Kamrul 594 (JUH)
ANNONACEAE Juss.						
Annona squamosa L.	Ata	Tree, s	ml (cu)	Ad	Fr	Kamrul 2140 (JUH)
A. reticulata L.	Nona	Tree, s	ml (cu)	All beats	Fr	Kamrul 2085 (JUH)
Alphonsea lutea (Roxb.) Hook.f. & Thomson	Fonseti	Tree, s	wd	Ad	Fr	S.N.Uddin N4625 (DACB)
Dasymaschalon longiflorum (Roxb.) Finet & Gagnep. *	Kulla	Shrub	wd	All beats	-	Kamrul 707 (JUH)
Fissistigma bicolor (Roxb.) Merr.	Hed-bheduli	Herb, cl	wd, fe	All beats	-	Kamrul 123 (JUH)
Miliusa velutina (Dunal) Hook.f. & Thomson	Gandhi gajari	Tree, m	wd	Ad	Fr	Kamrul 1523 (JUH)
Polyalthia longifolia (Sonn.) Thwaites	Debdaru	Tree, l	ml	Ka	O,T, M	Kamrul 2127 (JUH)
MYRISTICACEAE R.Br.						
Knema cinerea Warb.	Mota pasuti	Tree, m	wd	Ad	-	Kamrul 1574 (JUH)
LAURACEAE Juss						
Actinodaphne gullavara (Buch.-Ham. *ex* Nees) M.R. Almeida.	Modon mosta	Tree, m	hs	All beats	-	Kamrul 647 (JUH)
Cinnamomum tamala (Buch.-Ham.) T. Nees & Nees	Tejpata	Tree, m	sj	All beats	Sp, M	Kamrul 1285 (JUH)
Dehaasia kurzii Kingex Hook.f. *	Modon-mosto	Tree, s	hs	Ad	-	Kamrul 1022 (JUH)
Litsea monopetala (Roxb.) Pers.	Baro kukurchita	Tree, s	wd	All beats	M, Fr, T	Kamrul 1445 (JUH)
Ocotea lancifolia (Schott) Mez.	Dulia	Tree, s	wd	Ad	T	Kamrul 298 (JUH)
CHLORANTHACEAE R.Br. *ex* Sims						
Chloranthus erectus (Buch.-Ham.) Verdc.	Rantas	Shrub	fv	Ad	O, M	Kamrul 807 (JUH)
PIPERACEAE Giseke						
Peperomia pellucida (L.) Kunth	Pithapata	Herb, e	ml	Ad	M	Kamrul 316 (JUH)
Piper betle L.	Pan	Herb, cl	ml	Ad	M	Kamrul 11 (JUH)
P. longum L.	Pipul	Herb, cl	wd	Ad	M	Kamrul 2058 (JUH)
P. nigrum L.	Gol morich	Herb, cl	wd	Ad	M	Kamrul 723 (JUH)
P. sylvaticum Roxb.	Ban pan	Herb, cr	sj, fe	Ad, Ku	M	Kamrul 83 (JUH)
ARISTOLOCHIACEAE Juss.						
Aristolochia tagala Cham.	Iswararmul	Herb, cl	wd, hs	Ad	-	Kamrul 2139 (JUH)
SCHISANDRACEAE Blume						
Kadsura heteroclita (Roxb.) Craib	Kadsuta	Shrub	wd	Ad	M	Kamrul 1432 (JUH)
NYMPHAEACEAE Salisb.						
Nymphaea nouchali Burm. f.	Nilshapla	Herb, aq	ml	Ad	M	Kamrul 2233 (JUH)
N. rubra Roxb. *ex* Andrews	Lal shapla	Herb, aq	ml	Ad	V	Kamrul 2236 (JUH)
CERATOPHYLLACEAE Gray						
Ceratophyllum demersum L.	Sheola	Herb, aq	ml	Ad	M	Kamrul 2237 (JUH)
RANUNCULACEAE Juss.						
Nigella sativa L.	Kala-jeera	Herb, e	ml (cu)	Ka	M	Kamrul 2234 (JUH)
MENISPERMACEAE Juss.						
Cissampelos pareira L.	Akanadi	Herb, vi	wd	All beats	-	Kamrul 512 (JUH)
Cyclea barbata Miers	Patalpur	Herb, cl	wd	All beats	M	Kamrul 299 (JUH)

Table 1 Contd.

Scientific name	Bangla name	Habit	Habitat	Distrib.	Uses	RSE
Diploclisia glaucescens (Blume) Diels	Sonatola	Herb, vi	wd	Ad	M	Kamrul 1295 (JUH)
Pericampylus glaucus (Lam.) Merr.	Goria lata	Herb, cl	fe	Ad, Ku	Du	Kamrul 1209 (JUH)
Pycnarrhena planiflora Miers *ex* Hook. f. & Thomson *	Henalora	Shrub	ml	Ad, Ku	-	Kamrul 505 (JUH)
Stephania japonica (Thunb.) Miers	Nimukha	Herb, cl	Sj, ml	All beats	M	Kamrul 317 (JUH)
Tinospora sinensis (Lour.) Merr.	Padma gulancha	Shrub, cl	rb, sj	All beats	-	Kamrul 82 (JUH)
SABIACEAE Blume						
Meliosma pinnata (Roxb.) Maxim.	Bativa	Tree	wd	Ad	T, Fr	S.N.Uddin N5025 (DACB)
M. simplicifolia (Roxb.) Walp.	Dibru	Tree	wd	Ad	T	Kamrul 1433 (JUH)
Sabia lanceolata Colebr.	Sajba lat	Herb, cl	wd	All beats	T	Kamrul 1157 (JUH)
S. limoniacea Wall. *ex* Hook.f. & Thomson	Limo soobja	Herb, cl	wd	All beats	-	Kamrul 134 (JUH)
ULMACEAE Mirb.						
Trema orientalis (L.) Blume	Banjiga	Tree, m	rb	Ad, Ku	Fw, Fd	Kamrul 312 (JUH)
MORACEAE Gaudich.						
Artocarpus chama Buch.-Ham.	Chapalish	Tree, l	wd	All beats	T, Fr	Kamrul 2174 (JUH)
A. heterophyllus Lam.	Kanthal	Tree, l	ml	Ad, Ka	T, Fr	Kamrul 2037 (JUH)
A. lacucha Buch.-Ham.	Deua	Tree, m	wd	Ad	Fr, T, M	Kamrul 2070 (JUH)
Ficus benghalensis L.	Bot	Tree, l	ml	Ad	O, M	Kamrul 2235 (JUH)
F. benjamina L.	Pakur	Tree, l	fe	Ad, Ku	O, Fw	Kamrul 841 (JUH)
F. elastic Roxb. *ex* Hornem.	Rubber gach	Tree, s	ml	Ad	O	Kamrul 1458 (JUH)
F. heterophylla L.f.	Bhui dumur	Shrub	rb	All beats	M	Kamrul 440 (JUH)
F. hirta Vahl	Dadhuri	Shrub	fe	Ad	Fd	Kamrul 138 (JUH)
F. hispida L.f.	Kakdumur	Shrub	ml	Ad	Fr, M	Kamrul 303 (JUH)
F. pumila L.	Lata dumur	Herb, c	wd	Ad	O	Kamrul 2179 (JUH)
F. racemosa L.	Jagyadumur	Tree, s	ml	Ad	M	Kamrul 49 (JUH)
F. religiosa L.	Ashwath	Tree, l	ml	Ad	O, Fr, M	Kamrul 861 (JUH)
F. sagittata Vahl	Karat-bot	Herb, e	wd	Ad, Ku	-	Kamrul 1368 (JUH)
F. semicordata Buch.-Ham. *ex* Sm.	Sadimadi dumur	Tree, s	wd	Ad, Ku	-	Kamrul 806 (JUH)
F. variegata Blume	Bichitrabat	Tree, s	fv	Ad	Fw	Kamrul 843 (JUH)
Streblus asper Lour.	Sheora	Tree, s	rb	All beats	-	Kamrul 2180 (JUH)
CECROPIACEAE C.C. Berg.						
Poikilospermum suaveolens (Blume) Merr.	Dolia sat	Tree, m	wd	Ad	-	Kamrul 157 (JUH)
URTICACEAE Juss.						
Boehmeria glomerulifera Miq.	Borthurthuri	Shrub	wd	All beats	-	Kamrul 187 (JUH)
B. macrophylla Hornem.	Ulichara	Shrub	wd	Ad	-	Kamrul 285 (JUH)
Dendrocnide sinuata (Blume) Chew	Chutra	Shrub	wd	Ad	Fb, Fw, M	Kamrul 172 (JUH)
Elatostema clarkei Hook. f.	Clarkejhara	Herb, e	wd	Ad, Ka	-	Kamrul 820 (JUH)
Laportea interrupta (L.) Chew	Lal bichuti	Herb, e	ml	All beats	M	Kamrul 2201 (JUH)
Oreocnide integrifolia (Gaudich.) Miq.	Horhutta	Tree, m	wd	All beats	Fb	Kamrul 183 (JUH)
Pouzolzia zeylanica (L.) Benn.	Kullaruki	Herb, pr	ml, gl	Ad	M	Kamrul 837 (JUH)
Pilea glaberrima (Blume) Blume	Glabrum	Shrub	wd	All beats	-	Kamrul 404 (JUH)
Sarcochlamys pulcherrima Gaudich.	Marich	Tree, m	wd	Ad	Fw	Kamrul 1356 (JUH)

Table 1 Contd.

Scientific name	Bangla name	Habit	Habitat	Distrib.	Uses	RSE
JUGLANDACEAE DC. *ex* Perleb						
Engelhardia spicata Lesch. *ex* Blume	Jhumka bhadi	Tree, l	hs, fv	All beats	Sw	Kamrul 1457 (JUH)
FAGACEAE Dumort.						
Quercus obtusata Bonpl.	Batna	Tree, m	hs	Ad	-	Kamrul 1228 (JUH)
NYCTAGINACEAE Juss.						
Boerhavia defusa L.	Punarnava	Herb, pr	ml	All beats	M	Kamrul 2183 (JUH)
Bougainvillea spectabilis Willd.	Bagan bilash	Shrub, c	ml	Ad, Ka	O	Kamrul 1147 (JUH)
CHENOPODIACEAE Vent.						
Chenopodium album L.	Bathua shak	Herb, cr	ml	Ad	M	Kamrul 1241 (JUH)
AMARANTHACEAE Juss.						
Achyranthes aspera L.	Apang	Herb, e	fe	All beats	M	Kamrul 2128 (JUH)
Alternanthera paronychioides A.St.-Hil.	Jhuli khata	Herb, a	ml	Ad	M	Kamrul 2100 (JUH)
A. philoxeroides (Mart.) Griseb.	Henchi	Herb, a	ml	Ad	V	Kamrul 2137 (JUH)
A. sessilis (L.) R.Br. *ex* DC.	Malancha	Herb, a	ml	Ad	V	Kamrul 2143 (JUH)
Amaranthus spinosus L.	Kantanotey	Herb, e	ml	All beats	V, M	Kamrul 282 (JUH)
A. viridis L.	Notey shak	Herb, e	ml	Ad	V	Kamrul 1639 (JUH)
Cyathula prostrata (L.) Blume	Shyontula	Herb, pr	fv	All beats	M	Kamrul 886 (JUH)
POLYGONACEAE Juss.						
Persicaria hydropiper (L.) Delarbre	Biskatali	Herb, e	wd	All beats	-	Kamrul 38 (JUH)
Polygonum effusum Meisn.	Raniphul	Herb, e	ml, gl	Ad, Ku	M	Kamrul 1716 (JUH)
P. lapathifolium L.	Panibishkatali	Herb, e	wd	Ad, Ku	M	Kamrul 903 (JUH)
Rumex maritimus L.	Bon-palang	Herb, e	wd, rb	Ad	M	Kamrul 153 (JUH)
DILLENIACEAE Salisb.						
Dillenia indica L.	Chalta	Tree, s	ml	All beats	Fr, M	Kamrul 669 (JUH)
D. pentagyna Roxb.	Ban chalta	Tree, s	hs	Ad	T, M	Kamrul 1453 (JUH)
Tetracera sarmentosa (L.) Vahl	Lata chalta	Herb, cr	wd	All beats	-	Kamrul 804 (JUH)
DIPTEROCARPACEAE Blume						
Dipterocarpus turbinatus Gaertn.	Kali garjan	Tree, l	hs	Ad	T	Kamrul 2156 (JUH)
Hopea odorata Roxb.	Telsur	Tree, l	wd	Ad	T	Kamrul 2158 (JUH)
Shorea robusta Gaertn.	Sal	Tree, l	ht	Ad, Ka	T	Kamrul 1175 (JUH)
THEACEAE Mirb.						
Eurya acuminata DC.	Sagoler bori	Shrub	wd	Ad	Fw	S.N.Uddin N4813 (DACB)
Schima wallichii (DC.) Korth.	Bonak	Tree, s	ht, hs	All beats	T	Kamrul 551 (JUH)
ACTINIDIACEAE Gilg & Werderm.						
Saurauia roxburghii Wall.	Dalup	Tree, m	wd	Ad	Fr, Co	Kamrul 534 (JUH)
CLUSIACEAE Lindl.						
Garcinia cowa Roxb. *ex* Choisy	Cowa	Tree, s	wd	Ad	Fr	Kamrul 1075 (JUH)
Mesua ferrea L.	Nagessawar	Tree, s	wd	Ad	M	Kamrul 2151 (JUH)
ELAEOCARPACEAE Juss.						
Elaeocarpus petiolatus (Jack) Wall. *	Petipai	Tree, m	wd	All beats	M	Kamrul 1246 (JUH)
TILIACEAE Juss.						
Corchorus aestuans L.	Titapat	Herb, e	ml	Ka	M	Kamrul 2238 (JUH)
Grewia nervosa (Lour.) Panigrahi	Pichandi	Tree, s	sj	All beats	M	Kamrul 1519 (JUH)
G. serrrulata DC.	Pichandi	Shrub	sj	Ad	M	Kamrul 560 (JUH)
G. tiliifolia Vahl	Jonli pholsa	Tree, s	sj	Ad		Kamrul 784 (JUH)
Triumfetta rhomboidea Jacq.	Bon okua	Herb, e	gl, ml	All beats	M	Kamrul 98 (JUH)
T. pilosa Roth	Plofetta	Herb, e	sj	All beats		Kamrul 1873 (JUH)

Table 1 Contd.

Scientific name	Bangla name	Habit	Habitat	Distrib.	Uses	RSE
STERCULIACEAE Vent.						
Abroma augusta (L.) L.f.	Ulatkambal	Shrub	sj	Ad	M, Fb	Kamrul 2199 (JUH)
Byttneria aspera Colebr. *ex* Wall.	-	Liana	e	Ad	-	Kamrul 743 (JUH)
B. pilosa Roxb.	Harjora lata	Liana	e	All beats	M	Kamrul 128 (JUH)
Melochia corchorifolia L.	Tiki-okua	Herb, e	fv	All beats	M	Kamrul 382 (JUH)
Pterosprmum acerifolium (L.) Willd.	Kanackchampa	Tree, m	wd	All beats	O, M, I	Kamrul 1896 (JUH)
Sterculia villosa Roxb.	Udal	Tree, m	hs	All beats	Fr	Kamrul 1278 (JUH)
BOMBACACEAE Kunth						
Bombax ceiba L.	Simul	Tree, l	wd	Ad	Sw, Fb	Kamrul 190 (JUH)
MALVACEAE Juss.						
Abelmoschus moschatus Medik.	Mushak-dana	Herb, e	ml	All beats	M, V, O	Kamrul 410 (JUH)
Abutilon indicum (L.) Sweet	Petari	Herb, e	ml, gl	All beats	M, Fb	Kamrul 1710 (JUH)
Hibiscus rosa-sinensis L.	Jaba	Shrub, e	ml	Ad	O	Kamrul 985 (JUH)
H. macrophyllus Roxb. *ex* Hornem.	Udal	Tree, m	wd	All beats		Kamrul 1771 (JUH)
H. surattensis L.	Ram bhindi	Shrub	hs	Ad	M	Kamrul 1601 (JUH)
Malvaviscus arboreus Cav.	Marich jaba	Shrub	ml	Ad	O	Kamrul 2171 (JUH)
Sida acuta Burm. f.	Kureta	Herb, e	rs, ml	All beats	M	Kamrul 1009 (JUH)
S. cordata (Burm. f.) Borss.	Jumka	Herb, e	ml	All beats	-	Kamrul 335 (JUH)
S. rhombifolia L.	Lal berela	Herb, e	ml, rs	Ad, Ka	M	Kamrul 388 (JUH)
Urena lobata L.	Banghagra	Shrub	rs, ml	All beats	M	Kamrul 685 (JUH)
LECYTHIDACEAE A. Rich.						
Barringtonia acutangula (L.) Gaertn.	Hijol	Tree, m	ml	Ad	M	Kamrul 2133 (JUH)
FLACOURTIACEAE Rich. *ex* DC.						
Flacourtia indica (Burm. f.) Merr.	Bauchi	Shrub	sj	Ad, Ka	M, Fr, T	Kamrul 1624 (JUH)
F. jangomas (Lour.) Raeusch.	Lukluki	Tree, s	wd	All beats	Fr	Kamrul 1332 (JUH)
PASSILORACEAE Juss. *ex* Roussel						
Adenia trilobata (Roxb.) Engl.	Akandaphal	Herb, cl	wd, sj	Ad, Ku	M	Kamrul 1324 (JUH)
CARICACEAE Dumort.						
Carica papaya L.	Pape	Tree, s	ml	Ad	Fr, V	Kamrul 2259 (JUH)
CUCURBITACEAE Juss.						
Coccinia grandis (L.) Voigt	Telakucha	Herb, cl	gl, rs	Ad, Ku	M	Kamrul 2055 (JUH)
Gynostemma pentaphyllum (Thunb.) Makino	Gymnopada	Herb, cr	fv	Ad	M	Kamrul 466 (JUH)
Hodgsonia macrocarpa (Blume) Cogn.*	Makal maco	Liana	sb	Ad	M	Kamrul 2153 (JUH)
Luffa cylindrica (L.) M. Roem.	Dhundal	Herb, cl	ht	Ku	V, M	Kamrul 1569 (JUH)
Momordica charantia L.	Karola	Herb, cl	ml (cl)	Ad	V	Kamrul 2053 (JUH)
M. dioica Roxb. *ex* Willd.	Ghee korolla	Herb, cl	ml (cl)	Ad	V, M	Kamrul 2054 (JUH)
Mukia maderaspatana (L.) M. Roem.	Bilari	Herb, cl	gl, wd	Ad	M	Kamrul 1425 (JUH)
Thladiantha cordifolia (Blume) Cogn.	-	Herb, cl	sj	Ad	-	Kamrul 1423 (JUH)
Trichosanthes tricuspidata Lour.	Makal	Herb, cl	wd	All beats	-	Kamrul 1438 (JUH)
BEGONIACEAE C. Agardh						
Begonia annulata K. Koch.	Gonibata	Herb, e	sj	Ad	O, V	Kamrul 1908 (JUH)
B. roxburghii (Miq.) A.DC.	Gonorakto	Herb, e	sj	All beats	O, V	Kamrul 825 (JUH)
CAPPARACEAE Juss.						
Stixis suaveolens (Roxb.) Pierre	Madhumaloti	Herb, cl	wd, rb	All beats	Fr	Kamrul 05 (JUH)

Table 1 Contd.

Scientific name	Bangla name	Habit	Habitat	Distrib.	Uses	RSE
BRASSICACEAE Burnett						
Rorippa indica (L.) Hiern	Bansarisha	Herb, e	wd	Ka	-	Kamrul 2240 (JUH)
SAPOTACEAE Juss.						
Madhuca longifolia (J. Koenig *ex* L.) J.F. Macbr.	Mohua	Tree, m	ml, (pl)	Ad	M	Kamrul 2190 (JUH)
Mimusops elengi L.	Bokul	Tree, m	ml, (pl)	Ad	M, T	Kamrul 2192 (JUH)
EBENACEAE Gürke						
Diospyros malabarica (Desr.) Kostel.	Deshi gab	Tree, l	ml	Ad	M	Kamrul 2241 (JUH)
STYRACACEAE DC. & Spreng.						
Styrax serrulatus Roxb.	Kumjomeva	Tree, m	wd	All beats	-	Kamrul 486 (JUH)
SYMPLOCACEAE Desf.						
Symplocos macrophylla Wall. *ex* A.DC.*	Barabahuri	Tree, m	hs	Ad	-	Kamrul 536 (JUH)
PRIMULACEAE Batsch *ex* Borkh.						
Ardisia sanguinolenta Blume	-	Shrub	wd	Ad, Ka	M	Kamrul 1121 (JUH)
Embelia ribes Burm. f.	Bakul lata	Shrub	fv	Ad	M, Fr	Kamrul 1252 (JUH)
Hymenandra wallichii A. DC.	Bhau jawa	Shrub	wd	Ad, Ku	-	Kamrul 301 (JUH)
Maesa bengalensis Mez.	Banglauni	Tree, s	wd	Ad	-	Kamrul 1411 (JUH)
M. chisia Buch.-Ham. *ex* D. Don	Gangu lata	Shrub	wd	All beats	-	Kamrul 1660 (JUH)
M. indica (Roxb.) A. DC.	Sesu	Shrub	wd	All beats	M, V	Kamrul 966 (JUH)
M. ramentacea (Roxb.) A. DC.	Noa-maricha	Shrub	wd	All beats	M	Kamrul 544 (JUH)
CRASSULACEAE J.St.-Hil.						
Bryophyllum pinnatum (Lam.) Oken	Pathorkuchi	Herb, cr	ml	Ad	O, M	Kamrul 2242 (JUH)
ROSACEAE Juss.						
Prunus ceylanica Miq.	Ceylon cherry	Tree, s	sj	Ad	O, Sw	Kamrul 1587 (JUH)
Rubus hexagynus Roxb.	Hirachura	Shrub	hs	Ad	-	Kamrul 1328 (JUH)
Rosa chinensis Jacq	Kata golap	Shrub	ml	Ad	O	Kamrul 1106 (JUH)
MIMOSACEAE R.Br.						
Mimosa pudica L.	Lajjaboti	Shrub	ml, sj	All beats	M	Kamrul 433 (JUH)
Acacia auriculiformis A.Cunn. *ex* Benth.	Akashmoni	Tree, l	ml (pl)	All beats	T	Kamrul 1450 (JUH)
A. mangium Willd.	Mangium	Tree, l	ml (pl)	All beats	T	Kamrul 2112 (JUH)
Entada phaseoloides (L.) Merr.	Gila	Liana	wd	All beats	M, Co	Kamrul 117 (JUH)
CAESALPINIACEAE R.Br.						
Cassia fistula L.	Bandar lathi	Tree, m	wd	All beats	M, St, T	Kamrul 1212 (JUH)
Caesalpinia bonduc (L.) Roxb.	Nata	Shrub, sc	sj	Ka	M	Kamrul 1223 (JUH)
C. enneaphyllum Roxb.	Nataine	Herb, e	sj	Ka	O	Kamrul 1014 (JUH)
Delonix regia (Bojer *ex* Hook.) Raf.	Kuishnachura	Tree, m	ml (pl)	Ka	O, Sw	Kamrul 2150 (JUH)
Senna alata (L.) Roxb.	Dadmardan	Shrub	ml, rs	Ka	M	Kamrul 734 (JUH)
S. siamea (Lam.) H.S. Irwin & Barneby	Minjiri	Tree, s	rs	All beats	O	Kamrul 561 (JUH)
S. sophera (L.) Roxb.	Kalkashunda	Shrub	ml, sj	Ka	M	Kamrul 1807 (JUH)
S. tora (L.) Roxb.	Terasena	Herb, e	ml, rs	All beats	M	Kamrul 375 (JUH)
Tamarindus indica L.	Tetul	Tree, l	ml (pl)	Ka	Fr, M, T	Kamrul 580 (JUH)
FABACEAE Lindl.						
Cajanus scarabaeoides (L.) Thouars	Orhor	Shrub	hs	Ka	M	Kamrul 2134 (JUH)
Dalbergia stipulacea Roxb.	Dadbari	Herb, e	wd	All beats	M	Kamrul 1111 (JUH)
Dalhousiea bracteata (Roxb.) Graham *ex* Benth.*	Gupuri	Shrub	rb, wd	Ad	-	Kamrul 1586 (JUH)
Derris robusta (Roxb. *ex* DC.) Benth.	Korai	Tree, l	hs	All beats	-	Kamrul 1286 (JUH)

Table 1 Contd.

Scientific name	Bangla name	Habit	Habitat	Distrib.	Uses	RSE
Desmodium gangeticum (L.) DC.	Chalani	Shrub	rs, ml	All beats	M	Kamrul 500 (JUH)
D. heterophyllum (Willd.) DC.	Bon motorsuti	Herb, pr	rs	All beats	M	Kamrul 1006 (JUH)
D. heterocarpon (L.) DC.	Karpo modi	Shrub	fe	All beats	-	Kamrul 752 (JUH)
D. laxiflorum DC.	Laximodi	Shrub	wd	All beats	M	Kamrul 839 (JUH)
Tadehagi triquetrum (L.) H. Ohashi	Kanimanda	Shrub	hs	Ad	-	Kamrul 1153 (JUH)
Erythrina variegata L.	Bahari mander	Tree, m	ml	Ad	O, T, M	Kamrul 1461 (JUH)
Flemingia involucrata Benth.	Vuluk phan	Shrub	wd	All beats	-	Kamrul 2115 (JUH)
F. macrophylla (Willd.) Kuntze *ex* Merr.	Baro salpan	Shrub	wd	All beats	Dy	Kamrul 147 (JUH)
F. strobilifera (L.) W.T. Aiton	Chingri pata	Shrub	ml	All beats	M	Kamrul 925 (JUH)
Indigofera zollingeriana Miq.	Gerina nil	Shrub	fe	All beats	St	Kamrul 56 (JUH)
Lathyrus sativus L.	Khesari	Herb, e	ml (cl)	Ad	Fd, Pu	Kamrul 1708 (JUH)
Millettia pachycarpa Benth.	Bish lata	Liana	hs	Ad	I	Kamrul 1388 (JUH)
Mucuna pruriens (L.) DC.	Alkushi	Herb, c	sj	Ad	M	Kamrul 392 (JUH)
Pueraria phaseoloides (Roxb.) Benth.	Mugi kunch	Herb, e	sj	All beats	Fd, Gm	Kamrul 1091 (JUH)
Spatholobus parviflorus (DC.) Kuntze	Pan lata	Herb, c	sj	Ad	M, Du	Kamrul 1389 (JUH)
Tephrosia candida (Roxb.) DC.	Bilakshani	Herb, e	rs	Ad	M, Gm	Kamrul 1034 (JUH)
Uraria crinita (L.) Desv. *ex* DC.	Diangleja	Shrub	sj	Ad	M	Kamrul 424 (JUH)
Vigna mungo (L.) Hepper	Maskalay	Herb, e	ml (cl)	Ad	Pu	Kamrul 1087 (JUH)
SONNERATIACEAE Engl.						
Duabanga grandiflora (Roxb. *ex* DC.) Walp.	Bandorhola	Tree, l	fv, rb	Ad	T	Kamrul 1481 (JUH)
LYTHRACEAE J.St.-Hil.						
Ammannia multiflora Roxb.	Acidpata	Herb, e	fv	Ad	-	Kamrul 1858 (JUH)
Lagerstroemia speciosa (L.) Pers.	Jarul	Tree, m	ml	All beats	O, T, M	Kamrul 289 (JUH)
Lawsonia inermis L.	Mendi	Shrub	ml	Ka	Dy	Kamrul 1290 (JUH)
Rotala indica (Willd.) Koehne	Deshi ghurni	Herb, cr	ml	All beats	-	Kamrul 1140 (JUH)
R. rotundifolia (Buch.-Ham. *ex* Roxb.) Koehne	Dim ghurni	Herb, cr	ml	All beats	-	Kamrul 6 (JUH)
THYMELAEACEAE Juss.						
Aquilaria agallocha Roxb. *	Agar	Tree, m	ml	Ad, Ka	Pe	Kamrul 1144 (JUH)
MYRTACEAE Juss.						
Corymbia citriodora (Hook.) K.D. Hill & L.A.S. Johnson	Eucalyptus	Tree, l	ml (pl)	Ka	T	Kamrul 2249 (JUH)
Psidium guajava L.	Peyara	Tree, s	ml	Ad, Ka	Fr, M	Kamrul 2095 (JUH)
Syzygium cumini (L.) Skeels	Kalojam	Tree, l	rs, ml	All beats	Fr, T	Kamrul 1136 (JUH)
S. fruticosum DC.	Ban Jam	Tree, l	wd	Ad, Ku	Fr, T	Kamrul 2096 (JUH)
S. grande (Wight) Walp.	Dhaki Jam	Tree, l	fe	Ad, Ku	Fr, T	Kamrul 1020 (JUH)
ONAGRACEAE Juss.						
Ludwigia octovalvis (Jacq.) P.H.Raven	Bhuikura	Herb, e	ml	Ad	-	S.N.Uddin N4779 (DACB)
L. perennis L.	Amorkura	Herb, e	ml	All beats	-	Kamrul 144 (JUH)
MELASTOMATACEAE Juss.						
Melastoma malabathricum L.	Ban tejpata	Shrub	Wd, fe, ml	All beats	M	Kamrul 10 (JUH)
Osbeckia nepalensis Hook.f.	Nepaligachi	Shrub	Wd, fe	Ad	-	Kamrul 1828 (JUH)

Table 1 Contd.

Scientific name	Bangla name	Habit	Habitat	Distrib.	Uses	RSE
COMBRETACEAE R. Br.						
Combretum acuminatum Roxb.	Patyuni	Shrub	wd	All beats	-	Kamrul 2248 (JUH)
C. roxburghii Spreng.	Kaligaichi	Herb, e	wd	Ad	-	Kamrul 1033 (JUH)
C. wallichii DC.	Yunanlata	Herb, e	wd	Ad	-	Kamrul 178 (JUH)
Terminalia arjuna (Roxb. *ex* DC.) Wight & Arn.	Arjun	Tree, l	wd	Ka	M	Kamrul 566 (JUH)
T. bellirica (Gaertn.) Roxb.	Bohera	Tree, l	rs	Ka	M	Kamrul 582 (JUH)
T. catappa L.	Kathbadam	Tree, l	rs	Ka	M	Kamrul 2152 (JUH)
T. chebula Retz.	Horitoki	Tree, l	wd	Ka	M	Kamrul 583 (JUH)
RHIZOPHORACEAE Pers.						
Carallia brachiata (Lour.) Merr.	Rascow	Tree, m	wd	Ad	T, M	Kamrul 860 (JUH)
OLACACEAE Juss. *ex* R. Br.						
Olax acuminata Wall. *ex* Benth.	Capsul gach	Shrub	sj	All beats	-	Kamrul 349 (JUH)
LORANTHACEAE Juss.						
Helixanthera parasitica Lour.	Xanthric	Shrub	hs	Ad	-	Kamrul 1299 (JUH)
Macrosolen cochinchinensis (Lour.) Tiegh.	Chota banda	Shrub	wd	Ad	-	Kamrul 1104 (JUH)
Scurrula parasitica L.	Pargacha	Shrub	hs	Ad	-	Kamrul 1088 (JUH)
CELASTRACEAE R.Br.						
Bhesa robusta (Roxb.) Ding Hou*	Madhu-phal	Tree, m	wd	Ad	T	Kamrul 1097 (JUH)
HIPPOCRATEACEAE Juss.						
Salacia chinensis L.	Vesa	Shrub, sc	wd	Ad	M, Fr	Kamrul 1237 (JUH)
AQUIFOLIACEAE Bercht. & J. Presl						
Ilex godajam (Colebr. ex Wall.) Wall. ex Hook. f.	Raktim	Shrub	hs	Ad	Fw	Kamrul 1777 (JUH)
EUPHORBIACEAE Juss.						
Acalypha indica L.	Muktajhuri	Herb, e	ml	All beats	M	Kamrul 2159 (JUH)
Actephila excelsa (Dalzell) Müll.-Arg.	Lalsa	Shrub	hs	Ku	V, M	Kamrul 1577 (JUH)
Alchornea tiliifolia (Benth.) Müll.-Arg.	Alkotil	Shrub	sj, fv	Ad	Fw	Kamrul 2162 (JUH)
Antidesma acidum Retz.	Titij am	Shrub	wd	All beats	Fr	Kamrul 649 (JUH)
A. ghaesembilla Gaertn.	Khudi jam	Tree, s	wd	Ad	M	Kamrul 921 (JUH)
A. montanum Blume	Shial buka	Shrub	wd	All beats	Fr	Kamrul 366 (JUH)
Aporosa wallichii Hook. f.	Kokua	Tree, s	wd	All beats	-	Kamrul 154 (JUH)
Bischofia javanica Blume	Kainjal	Tree, s	fe	Ad	T, M	Kamrul 1396 (JUH)
Bridelia tomentosa Blume	Khoi	Shrub	sj	All beats	M	Kamrul 1103 (JUH)
Baccaurea ramiflora Lour.	Latkan	Tree, m	wd	Ad	Fr, M	Kamrul 1476 (JUH)
Chaetocarpus castanocarpus (Roxb.) Thwaites	Dhala kakua	Tree, m	wd	Ad	-	Kamrul 143 (JUH)
Cnesmone javanica Blume	Chutra	Shrub	Sj, ht	All beats	M	Kamrul 383 (JUH)
Croton bonplandianus Baill.	Banmarich	Herb, e	ml, rs	All beats	M	Kamrul 481 (JUH)
Baliospermum solanifolium (Burm.) Suresh	Chuka	Shrub	rb	Ad	M	Kamrul 125 (JUH)
Euphorbia hirta L.	Bara dudhia	Herb, pr	ml, rs	All beats	M	Kamrul 2164 (JUH)
Glochidion multiloculare (Rottler *ex* Willd.) Voigt	Aniatori	Shrub	hs	All beats	T	Kamrul 32 (JUH)
Macaranga indica Wight.	Gulle	Tree, s	wd	All beats	M	Kamrul 1231 (JUH)
Phyllanthus emblica L.	Amloki	Tree, m	sj	Ka	Fr, M	Kamrul 554 (JUH)
P. niruri L.	Bhuiamla	Herb, e	ml	Ad	M	Kamrul 760 (JUH)
P. reticulatus Poir.	Pankushi	Shrub	sj, rb	All beats	M	Kamrul 1499 (JUH)

Table 1 Contd.

Scientific name	Bangla name	Habit	Habitat	Distrib.	Uses	RSE
P. urinaria L.	Hazarmani	Herb, e	sj	Ad	M	Kamrul 79 (JUH)
Suregada multiflora (A.Juss.) Baill.	Ban naranga	Tree, s	wd	Ad	T	Kamrul 226 (JUH)
Ricinus communis L.	Bherenda	Shrub	wd	Ad	M	Kamrul 2167 (JUH)
Sauropus androgynus (L.) Merr.	Mithapotro	Shrub	wd	Ad	M	Kamrul 1397 (JUH)
RHAMNACEAE Juss.						
Gouania tiliifolia Lam.	Herjengota	Shrub	sj	Ad, Ku	V	Kamrul 1901 (JUH)
Ziziphus mauritiana Lam.	Bol boroi	Tree, s	ml	All beats	M	Kamrul 571 (JUH)
Z. oenopolia (L.) Mill.	Ban boroi	Shrub	sj	All beats	M	Kamrul 482 (JUH)
LEEACEAE Dumort.						
Leea guineensis G. Don.	-	Shrub	hs	All beats	-	Kamrul 604 (JUH)
L. indica (Burm.f.) Merr.	Kurkurjhibba	Shrub	fv	All beats	M	Kamrul 241 (JUH)
VITACEAE Juss.						
Ampelocissus latifolia (Roxb.) Planch.	Gowalia lata	Herb, c	rb, sj	Ad	-	Kamrul 2204 (JUH)
Cissus javana DC.	Dukhu lata	Liana	wd	All beats	O	Kamrul 702 (JUH)
C. adnata Roxb.	Bhatia lata	Liana	wd	All beats	M	Kamrul 639 (JUH);
Cayratia japonica (Thunb.) Gagnep.	Japani goali lata	Herb, cl	rb	Ad	M	Kamrul 508 (JUH)
C. trifolia (L.) Domin	Amal lata	Herb, cl	rb	All beats	M	Kamrul 1118 (JUH)
Tetrastigma lanceolarium (Roxb.) Planch.	-	Herb, e	ht	Ad	-	Kamrul 1589 (JUH)
T. leucostaphylum (Dennst.) Alston	Horina lata	Shrub	wd	Ad, Ku	-	Kamrul 739 (JUH)
POLYGALACEAE Hoffeanns. & Link						
Polygala chinensis L.	Meradu	Herb, pr	hs, sj	Ka	M	Kamrul 1428 (JUH)
P. erioptera DC.	Teradudhi	Herb, pr	ml	All beats	-	Kamrul 2188 (JUH)
STAPHYLEACEAE Martinov						
Turpinia pomifera (Roxb.) DC.	Bhola	Tree	wd	Ad	Fd, Fw	Kamrul 1459 (JUH)
SAPINDACEAE Juss.						
Allophylus cobbe (L.) Raeusch.	Chita	Shrub	wd	All beats	-	Kamrul 719 (JUH)
Lepisanthes senegalensis (Poir.) Leenh.	Chita	Shrub	wd	Ad, Ku	-	Kamrul 1029 (JUH)
Litchi chinensis Sonn.	Lichu	Tree, l	ml (pl)	Ad	Fr	Kamrul 2189 (JUH)
BURSERACEAE Kunth						
Garuga pinnata Roxb.	Paharijiga	Tree, m	wd	All beats	T, M	Kamrul 780 (JUH)
ANACARDIACEAE R. Br.						
Anacardium occidentale L.	Kaju.	Tree, m	rs	Ka	Fr	Kamrul 1201 (JUH)
Lannea coromandelica (Houtt.) Merr.	Jiga	Tree, m	rs	Ad	M	Kamrul 1451 (JUH)
Mangifera indica L.	Aam	Tree, l	ml	All beats	Fr, Fw	Kamrul 1195 (JUH)
M. sylvatica. Roxb.*	Jangli aam	Tree, m	wd	Ad	Fr	Kamrul 1444 (JUH)
Holigarna caustica (Dennst.) Oken.*	Jaowa	Tree, m	hs	Ad	T	Kamrul 220 (JUH)
Rhus succedanea L	Kakuasingh	Tree, l	wd	Ad	Fr	Kamrul 1377 (JUH)
Pegia nitida Colebr.	Tapir	Shrub	hs, wd	Ad	Fr, M	Kamrul 168 (JUH)
MELIACEAE Juss.						
Aphanamixis polystachya (Wall.) R.Parker	Pitraj	Tree, l	wd	All beats	T, M	Kamrul 850 (JUH)
Azadirachta indica A.Juss.	Neem	Tree, l	ml, rs	Ad	M, T	Kamrul 1240 (JUH)
Swietenia mahagoni (L.) Jacq.	Mahogini	Tree, l	ml, rs	Ad, Ka	T	Kamrul 1283 (JUH)
S. macrophylla King	Bara mahogoni	Tree, l	ml	Ad, Ka	T	Kamrul 1089 (JUH)
Dysoxylum gotadhora (Buch.-Ham.) Mabb.	Rata	Tree, l	ht	Ad, Ku	T	Kamrul 1415 (JUH)

Table 1 Contd.

Scientific name	Bangla name	Habit	Habitat	Distrib.	Uses	RSE
Toona ciliata M. Roem.	Toon	Tree, m	wd	Ad	T, M	Kamrul 1078 (JUH)
Walsura robusta Roxb.	Bonlichu	Tree, m	rb	Ad	T	Kamrul 1358 (JUH)
RUTACEAE Juss.						
Aegle marmelos (L.) Corrêa	Bel	Tree, m	ml	Ad	M, Fr	Kamrul 2078 (JUH)
Acronychia pedunculata (L.) Miq.	Bon jamir	Shrub	hs	All beats	Fr, Fw	Kamrul 541 (JUH)
Clausena anisata (Willd.) Hook. f. *ex* Benth.	Kalo-maricha	Shrub	wd	All beats	-	Kamrul 112 (JUH)
Citrus medica L.	Pani lebu	Shrub	wd	Ad, Ku	Pe, M	Kamrul 1171 (JUH)
Glycosmis pentaphylla (Retz.) DC.	Datmajani	Shrub	wd	Ad, Ku	M	Kamrul 815 (JUH)
Micromelum minutum Wight & Arn.	Dulia	Tree, s	wd	All beats	M	Kamrul 244 (JUH)
Murraya koenigii (L.) Spreng.	Curry pata	Shrub	wd	Ku	Fr, Sp, M	Kamrul 1576 (JUH)
OXALIDACEAE R. Br.						
Oxalis corniculata L.	Amrul	Herb, pr	op, rs	Ad, Ka	V, M	Kamrul 1155 (JUH)
Averrhoa bilimbi L.	Bilimbi	Tree, s	ml (pl)	Ad	V	Kamrul 2184 (JUH)
A. carambola L.	Kamranga	Tree, s	ml (pl)	Ad	Fr, M	Kamrul 2186 (JUH)
ARALIACEAE Juss.						
Brassaiopsis glomerulata (Blume) Regel	Kurila	Tree, s	wd, hs	All beats	M, O	Kamrul 1384 (JUH)
Trevesia palmata (Roxb. *ex* Lind.) Vis.	Argoja	Tree, m	hs	Ad	M, O	Kamrul 1095 (JUH)
APIACEAE Lindl.						
Centella asiatica (L.) Urb.	Thankuni	Herb, cr	ml	All beats	M	Kamrul 972 (JUH)
Eryngium foetidum L.	Bilati dhania	Shrub	ml (cl)	All beats	Sp, M	Kamrul 1134 (JUH)
Hydrocotyle sibthorpioides Lam.	Gimashak	Herb, cr	wv	Ad	-	Kamrul 1581 (JUH)
BUDDLEJACEAE Wilh.						
Buddleja asiatica Lour.	Neemda	Shrub	wd	All beats	M, Pe	Kamrul 775 (JUH)
GENTIANACEAE Juss.						
Canscora andrographioides Griff. *ex* C.B.Clarke *	Andakuni	Herb, cr	fv, ml	Ad	-	Kamrul 164 (JUH)
C. alata (Roth *ex* Roem. & Schult.) Wall.	Dhankuni	Herb, e	ml, rs	All beats	M	Kamrul 2068 (JUH)
C. diffusa (Vahl) R.Br. *ex* Roem. & Schult.	Fusakoni	Herb, cr	ml	Ad, Ku	-	Kamrul 1116 (JUH)
APOCYNACEAE Juss.						
Alstonia scholaris (L.) R. Br.	Chhatim	Tree, l	wd	Ka	M, Sw	Kamrul 2138 (JUH)
Holarrhena pubescens Wall. *ex* G. Don	Kurchi	Shrub	sj	Ad	M	Kamrul 1823 (JUH)
Ichnocarpus frutescens (L.) W.T. Aiton	Parallia lata	Herb, c	fe	All beats	M	Kamrul 431 (JUH)
Rauvolfia serpentina (L.) Benth. *ex* Kurz *	Sarpagandha	Herb, e	ht	Ad	M	Kamrul 737 (JUH)
Tabernaemontana divaricata (L.) R.Br. *ex* Roem. & Schult.	Tagar	Shrub	sj	All beats	M	Kamrul 21 (JUH)
ASCLEPIADACEAE Borkh.						
Gymnema acuminatum Wall.	Nimakumina	Liana	fe	Ad		Kamrul 981 (JUH)
Hoya parasitica Wall. *ex* Wight	Pargacha	Herb, ps	wd	All beats	M	Kamrul 1276 (JUH)
SOLANACEAE Juss.						
Datura metel L.	Dhutra	Shrub	sj	Ad	M	Kamrul 2079 (JUH)
Nicotiana plumbaginifolia Viv.	Ban tamak	Herb, e		All beats	-	Kamrul 2198 (JUH)
Solanum torvum Sw.	Gota begun	Herb, e	sj	All beats	-	Kamrul 2196 (JUH)

Table 1 Contd.

Scientific name	Bangla name	Habit	Habitat	Distrib.	Uses	RSE
S. americanum Mill.	Tit-begun	Herb, e	sj	All beats	M	Kamrul 1348 (JUH)
Physalis angulata L.	Futka	Herb, e	rs	Ad	-	Kamrul 2200 (JUH)
CONVOLVULACEAE Juss.						
Argyreia argentea (Roxb.) Sweet	Boro rupatala	Herb, c	sj	All beats	-	Kamrul 394 (JUH)
A. capitiformis (Poir.) Ooststr.	Bijtarak	Herb, c	fe	All beats	M	Kamrul 124 (JUH)
Cuscuta reflexa Roxb.	Sarnalata	Herb, ps	ml	All beats	M	Kamrul 1463 (JUH)
Evolvulus nummularius (L.) L.	Bhuiokua	Herb, Cr	sj	Ad	M	Kamrul 2146 (JUH)
Ipomoea aquatica Forssk.	Kolmishak.	Herb, a	ml	Ad	V	Kamrul 2131 (JUH)
I. alba L.	Dudh kolmi	Herb, cr	rb	Ad	M	Kamrul 2130 (JUH)
I. batatas (L.) Lam.	Shakalu	Herb, cr	rb	Ad	M	Kamrul 2132 (JUH)
I. cairica (L.) Sweet	Rail lata	Herb, cr	sj	Ad	-	Kamrul 1049 (JUH)
I. carnea Jacq.	Dhol kolmi	Shrub	ml	Ad	-	Kamrul 980 (JUH)
Merremia umbellata (L.) Hallier f.	Sada kalmi	Herb, cr	rs, fv	All beats	M	Kamrul 17 (JUH)
MENYANTHACEAE Dumort.						
Nymphoides hydrophylla (Lour.) Kuntze	Chadmala	Herb, a	ml	Ad, Ku	Fd	Kamrul 728 (JUH)
HYDROPHYLLACEAE R. Br.						
Hydrolea zeylanica (L.) Vahl	Kasschara	Herb, a	ml	All beats	M	Kamrul 901 (JUH)
BORAGINACEAE Juss.						
Cordia dichotoma G. Forst.	Boula	Tree, m	rb	All beats	M	Kamrul 2144 (JUH)
Heliotropium indicum L.	Hatisur	Herb, e	sj	All beats	M	Kamrul 2147 (JUH)
VERBENACEAE J.St.-Hil.						
Lantana camara L.	Kutuskanta	Shrub	wd, fe, ml	All beats	M	Kamrul 16 (JUH)
LAMIACEAE Martinov						
Callicarpa arborea Roxb.	Barmala	Tree	ht	All beats	M	Kamrul 499 (JUH)
C. longifolia Lam.	Lamarck	Shrub	wd	Ad, Ku	-	Kamrul 1354 (JUH)
Clerodendrum indicum (L.) Kuntze	Bamunhatti	Shrub	wd, fe	Ad	M	Kamrul 1059 (JUH)
C. infortunatum L.	Bhat	Shrub	wd, fe	All beats	-	Kamrul 13 (JUH)
C. laevifolium Blume	Mali bong	Shrub	wd, fe	Ad	O	Kamrul 59 (JUH)
Gmelina arborea Roxb. *ex* Sm.	Gamari	Tree, m	hs	All beats	T, M	Kamrul 1219 (JUH)
Gomphostema salarkhaniana Khanam & Hassan *	Kanimala	Herb, e	wd	Ad	-	Kamrul 84 (JUH)
Hyptis brevipes Poit.	Gol-tokma	Herb, e	sj, wd	All beats	M	Kamrul 43 (JUH)
H. suaveolens (L.) Poit.	Tokma	Herb, e	sj	Ad, Ka	M	Kamrul 994 (JUH)
Leucas zeylanica (L.) W. T. Aiton	Dondokalosh	Herb, e	ml	All beats	M	Kamrul 256 (JUH
Mosla dianthera (Buch.-Ham. *ex* Roxb.) Maxim.	Moshla	Herb, e	sj	Ka	M	Kamrul 1064 (JUH)
Ocimum gratissimum L.	Ram tulsi	Shrub	ml	All beats	M	Kamrul 1131 (JUH)
O. tenuiflorum L.	Kalo tulsi	Herb, e	sj	Ad	M	Kamrul 2061 (JUH)
Pogostemon auricularius (L.) Hassk.	Aripachuli	Herb, e	wt	Ad, Ku	M	Kamrul 37 (JUH)
Rotheca serrata (L.) Steane & Mabb.	Bamanhati	Shrub	wd	Ad	M	Kamrul 120 (JUH)
OLEACEAE Hoffeanns. & Link						
Jasminum sambac (L.) Aiton	Beli	Shrub	fe	All beats	O	Kamrul 1004 (JUH)
J. scandens (Retz.) Vahl	Jua	Shrub	hs	All beats	-	Kamrul 142 (JUH)
Myxopyrum smilacifolium (Wall.) Blume	Chiknabizi	Herb, e	wd	All beats	-	Kamrul 607 (JUH)
Nyctanthes arbor-tristis L.	Sheuly	Tree, m	wd	Ad	T, M	Kamrul 1149 (JUH)
Premna esculenta Roxb.	Lalong	Shrub	ht, rs	All beats	V, O	Kamrul 323 (JUH)
Tectona grandis L.f.	Shegun	Tree, l	wd	All beats	T	Kamrul 489 (JUH)
Vitex negundo L.	Nishinda	Tree, s	rb	All beats	M	Kamrul 588 (JUH)

Table 1 Contd.

Scientific name	Bangla name	Habit	Habitat	Distrib.	Uses	RSE
V. peduncularis Wall. *ex* Schauer	Awal	Tree, m	ht, hs	All beats	T	Kamrul 1257 (JUH)
V. pinnata L.	Seliawal	Tree, m	hs, fe	Ad	-	Kamrul 1413 (JUH)
PLANTAGINACEAE Juss.						
Limnophila rugosa (Roth) Merr.	Bandha keshori	Herb, e	ml	All beats	M	Kamrul 1128 (JUH)
Mecardonia procumbens (Mill.) Small	Ada birni	Herb, e	ml	Ad	-	Kamrul 135 (JUH)
Scoparia dulcis L.	Bondhone	Herb, e	ml, rs	All beats	M	Kamrul 796 (JUH)
LINDERNIACEAE Borsch, Kai Müll. & Eb. Fisch.						
Lindernia ciliata (Colsm.) Pennell	Bhui papri	Herb, pr	rb, fe, hs	All beats	-	Kamrul 755 (JUH)
L. antipoda (L.) Alston		Herb, pr	sj, gl	All beats	-	Kamrul 373 (JUH)
Torenia fournieri Linden *ex* E.Fourn.	Neritoren	Herb, pr	sj	All beats	O	Kamrul 19 (JUH)
OROBANCHACEAE Vent.						
Aeginetia indica L.	Agienata	Herb, ps	hs	Ad	O	Kamrul 1380 (JUH)
ACANTHACEAE Juss.						
Acanthus leucostachyus Wall. *ex* Nees *	Kastacha	Shrub	wd	Ad	M	Kamrul 1217 (JUH)
Andrographis paniculata (Burm.f.) Wall. *ex* Nees *	Kalomegh	Herb, e	sj	Ad, Ka	M	Kamrul 2081 (JUH)
Blepharis integrifolia (L.f.) E.Mey. & Drège *ex* Schinz.	Deshi blephar	Herb, e	sj	Ad	-	Kamrul 1143 (JUH)
Eranthemum strictum Colebr. *ex* Roxb.	Khara murali	Shrub	wd	All beats	-	Kamrul 1161 (JUH)
Hygrophila polysperma (Roxb.) T. Anderson	Murmura	Herb, e	ml	Ad	-	Kamrul 1661 (JUH)
Justicia adhatoda L.	Bashak	Shrub	fe	Ad	M	Kamrul 2136 (JUH)
J. diffusa Willd.	Pitapapra	Herb, e	rs, sj	Ad		Kamrul 2135 (JUH)
Lepidagathis incurva Buch.-Ham. *ex* D.Don	Linagathis	Herb, e	sj	All beats	-	Kamrul 905 (JUH)
Nelsonia canescens (Lam.) Spreng.	Paramul	Herb, e	ml	All beats	-	Kamrul 182 (JUH)
Phlogacanthus curviflorus (Wall.) Nees	Agnilora	Shrub	fe	Ad	O	Kamrul 1163 (JUH)
P. thyrsiformis (Roxb. *ex* Hardw.) Mabb.	Rambasak	Shrub	fe	Ad	M	Kamrul 108 (JUH)
P. tubiflorus Nees	Agnibasak	Shrub	fe	Ad	-	Kamrul 121 (JUH)
Phaulopsis imbricata (Forssk.) Sweet Hort.	Kantasi	Herb, e	sj	All beats	-	Kamrul 2067 (JUH)
Rungia pectinata (L.) Nees	Pindi	Herb, pr	ml	All beats	M	Kamrul 1975 (JUH)
Staurogyne argentea Wall.	-	Herb, e	hf	All beats	-	Kamrul 811 (JUH)
S. polybotrya Kuntze	Polygyne	Herb, e	ml	Ad	-	Kamrul 1124 (JUH)
S. zeylanica Kuntze	Cylongyne	Herb, e	ht	Ad	-	Kamrul 1579 (JUH)
Strobilanthes scaber Nees	Khaskhasabila	Herb, e	sj	All beats	-	Kamrul 132 (JUH)
Thunbergia grandiflora (Roxb. *ex* Rottl.) Roxb.	Neel lata	Herb, cl	sp	All beats	O	Kamrul 638 (JUH)
BIGNONIACEAE Juss.						
Stereospermum tetragonum DC.	Awal	Tree, l	wd	Ad	-	Kamrul 1804 (JUH)
CAMPALUNACEAE Juss.						
Lobelia zeylanica L.	Cylon lobel	Herb, pr	sj	All beats	-	Kamrul 883 (JUH)
RUBIACEAE Juss.						
Catunaregam spinosa (Thunb.) Tirveng.	Mankanta	Shrub	rb	Ad	M	Kamrul 1322 (JUH)
Dentella repens (L.) J.R. Forst. & G.Forst.	Bhuipat	Herb, pr	sj	Ad, Ku	-	Kamrul 45 (JUH)
Gardenia coronaria Buch.-Ham.	Sitgach	Tree, m	hs	Ad	T, Fr	Kamrul 1254 (JUH)
Scleromitrion scabrum (Wall. *ex* Kurz) Neupane & N. Wikstrom	-	Herb, e	hs	Ad, Ku	-	Kamrul 46 (JUH)

Table 1 Contd.

Scientific name	Bangla name	Habit	Habitat	Distrib.	Uses	RSE
Hedyotis scandens Roxb.	Bish lata	Herb, pr	wd	All beats	M	Kamrul 1486 (JUH)
H. verticillata (L.) Lamk.	-	Herb, pr	wd	All beats	-	Kamrul 1559 (JUH)
Ixora acuminata Roxb.	Nata rangan	Shrub	wd	All beats	O	Kamrul 1164 (JUH)
I. coccinea L.	Rangan	Shrub	wd	Ad, Ku	O	Kamrul 1288 (JUH)
I. pavetta Andr.	Ganghalrangan	Shrub	wd	All beats	-	Kamrul 434 (JUH)
I. spectabilis Wall. *ex* G.Don	Shum rangan	Shrub	wd	Ad, Ku	-	Kamrul 31 (JUH)
Knoxia sumatrensis (Retz.) DC	Sumatranoxi	Herb, cr	wd	Ad	-	S.N.Uddin N4744 (DACB)
Lasianthus chrysoneurus (Korth.) Miq.	Sony lasi	Shrub	wd	Ad	-	Kamrul 1167 (JUH)
Morinda angustifolia Roxb.	Jangli basok	Tree, s	sj	All beats	M	Kamrul 171 (JUH)
Mussaenda roxburghii Hook. f.	Sil daura	Shrub	hs	All beats	M	Kamrul 1505 (JUH)
M. frondosa L.	Nagabali	Shrub	wd	All beats	M	Kamrul 1259 (JUH)
M. macrophylla Wall.	Baropata muchenda	Shrub	wd	All beats	-	Kamrul 25 (JUH)
Mussaenda sp.	-	Shrub, cl	wd	Ad	-	Kamrul 630 (JUH)
Mycetia longifolia (Wall.) Kuntze	Mycetelon	Shrub	sj	Ad, Ku	-	Kamrul 401 (JUH)
Myrioneuron nutans R. Br. *ex* Kurz	Natanuran	Shrub	wd	Ad	-	Kamrul 1518 (JUH)
Mitragyna parviflora (Roxb.) Korth.	Dakrom	Tree, m	ht	Ad	-	Kamrul 2243 (JUH)
Neolamarckia cadamba (Roxb.) Bosser	Kadam	Tree, l	rs	Ad, Ku	O	Kamrul 297 (JUH)
Ophiorrhiza mungos L.	Gandahanakuli.	Herb, e	sj	All beats	M	Kamrul 23 (JUH)
Pavetta indica L.	Kathchapa	Shrub	wd	Ad	M	Kamrul 1309 (JUH)
P. polyantha (Hook.f.) R. Br. *ex* Bremek.	Polinakli	Shrub	wd	Ad	-	Kamrul 1609 (JUH)
Psychotria adenophylla Wall.	Baro sudma	Shrub	wd	All beats	M	Kamrul 131 (JUH)
P. calocarpa Kurz	Ranga bhutta	Shrub	wd	Ad	M	Kamrul 917 (JUH)
Coffea benghalensis B.Heyne *ex* Schult.	Bonnya kofee	Shrub	hs	Ad	-	Kamrul 1617 (JUH)
Richardia scabra L.	Nakli ipecac	Herb, e	sj	Ad	-	Kamrul 880 (JUH)
Oxyceros kunstleri (King & Gamble) Tirveng.	Ichuri	Shrub	wd	Ad	M	Kamrul 874 (JUH)
Wendlandia grandis (Hook. f.) Cowan	Tulaload	Tree, l	wd	Ad	-	Kamrul 149 (JUH)
ASTERACEAE Bercht. & J. Presl						
Acmella caulirhiza Delile	Mahatitinga	Herb, pr	fe	All beats	-	Kamrul 194 (JUH)
Adenostemma lavenia (L.) Kuntze	Baro-kesuti	Herb, e	fe	All beats	-	Kamrul 963 (JUH)
Ageratum conyzoides L.	Fulkuri	Herb, e	rs, sj, ml	All beats	M	Kamrul 08 (JUH)
Blumea lacera (Burm.f.) DC.	Barokukshim	Herb, pr	rs, sj	All beats	M	Kamrul 1186 (JUH)
Crassocephalum crepidioides (Benth.) S. Moore	Duubbecrepi	Herb, e	fe	All beats	M, V	Kamrul 417 (JUH)
Chromolaena odorata (L.) R.M.King & H. Rob.	Rail lata	Shrub	sj, fe	All beats	M	Kamrul 923 (JUH)
Cosmos sulphureus Cav.	Tara gada	Herb, e	ml (pl)	Ka	O	Kamrul 2251 (JUH)
Cyanthillium cinereum (L.) H. Rob.	Shialmutra	Herb, e	ml	All beats	M	Kamrul 2142 (JUH)
Eclipta prostrata (L.) L.	Kalokeshi	Herb, pr	ml, fe	All beats	M	Kamrul 357 (JUH)
Elephantopus scaber L.	Hastipadi	Herb, e	ml, fe, sj	All beats	M	Kamrul 567 (JUH)
Enhydra fluctuans Lour.	Helencha	Herb, aq	ml	All beats		Kamrul 2145 (JUH)
Grangea maderaspatana (L.) Poir.	Namuti	Herb, e	rb	All beats	M	Kamrul 1192 (JUH)
Launaea asplenifolia (Willd.) Hook.f.	Tikadana	Herb, e	hs	Ad		Kamrul 1604 (JUH)
Mikania micrantha Kunth	Assam lata	Herb, c	ml	All beats	M	Kamrul 74 (JUH)
Pseudognaphalium luteoalbum (L.) Hilliard & B.L. Burtt	Bara kamra	Herb, e	sj	Ad	M	Kamrul 155 (JUH)
Synedrella nodiflora (L.) Gaertn.	Nakphul	Herb, e	rs, fe	All beats	-	Kamrul 2141 (JUH)
Xanthium strumarium L.	Ghagra	Herb, e	fe	Ad	-	Kamrul 1753 (JUH)

Table 1 Contd.

Scientific name	Bangla name	Habit	Habitat	Distrib.	Uses	RSE
LILIOPSIDA Batsch						
HYDROCHARITACEAE Juss.						
Blyxa japonica (Miquel) Maxim. *ex* Asch. & Gürke	Japani blixa	Herb, aq	ml	Ka	-	Kamrul 1503 (JUH)
Ottelia alismoides (L.) Pers.	Panikola	Herb, aq	ml	Ku	Fr, M	Kamrul 1504 (JUH)
ARECACEAE Bercht. & J.Presl						
Areca catechu L.	Supari	Tree, m	ml (pl)	Ad	M, T	Kamrul 2108 (JUH)
Borassus flabellifer L.	Tal	Tree, l	ml (pl)	Ad	Fr, Du, Ju	Kamrul 2110 (JUH)
Caryota urens L.	Chau-gota	Tree, m	ht	Ad	-	Kamrul 625 (JUH)
Calamus tenuis Roxb.	Jali bet	Herb, cl	wd	All beats	Du	Kamrul 1558 (JUH)
C. erectus Roxb. *	Sungota	Herb, e	wd	Ad, Ku	Du	Kamrul 574 (JUH)
C. longisetus Griff. *	Bet	Shrub	wd	Ka	Du	S.N.Uddin N4485 (DACB)
Cocos nucifera L.	Narikel	Tree, l	ml	Ad	Fr, Ol, Du	Kamrul 2111 (JUH)
Daemonorops jenkinsiana (Griff.) Mart.	Golla	Shrub	hs	Ad	Du	Kamrul 1809 (JUH)
Phoenix sylvestris (L.) Roxb.	Deshi khejur	Tree, m	wd	Ad	Du, Ju	Kamrul 2212 (JUH)
PANDANACEAE R.Br.						
Pandanus foetidus Roxb.	Keya kanta	Shrub	fv	Ad	-	Kamrul 1042 (JUH)
ARACEAE Juss.						
Aglaonema hookerianum Schott *	Nimahook	Herb, e	hs, rb	Ad	M	Kamrul 1597 (JUH)
Alocasia cucullata (Lour.) G. Don	Bishkachu	Herb, e	hs, rb	All beats	M	Kamrul 225 (JUH)
A. macrorrhizos (L.) G. Don	Mankachu	Herb, e	rb, ml	Ad	M	Kamrul 2073 (JUH)
Amorphophallus bulbifer (Roxb.) Blume	Owl	Herb, e	hs, ml	Ad, Ku	V	Kamrul 250 (JUH)
Colocasia esculenta (L.) Schott	Jangli kachu	Herb, e	rb, rf	All beats	V, M	Kamrul 1373 (JUH)
C. gigantea (Blume) Hook.f.	Salad kachu	Herb, e	ml	Ad	V	Kamrul 2031 (JUH)
Homalomena aromatica (Spreng.) Schott	Gandhabi kochu	Herb, e	hs	Ad	V	Kamrul 325 (JUH)
Lasia spinosa (L.) Thwaites	Kanta kachu	Herb, e	wd	Ad, Ku	V, M	Kamrul 624 (JUH)
Pistia stratiotes L.	Topapana	Herb, aq	ml	Ku	M	Kamrul 2002 (JUH)
Pothos scandens L.	Hatilata	Liana, ep	sj	All beats	M	Kamrul 9 (JUH)
Epipremnum aureum (Linden & André) G.S. Bunting	Money plant	Liana	ml	Ad	O	Kamrul 2004 (JUH)
Rhaphidophora glauca (Wall.) Schott	Fidoka	Herb, cl	wd	Ad	-	Kamrul 1386 (JUH)
Steudnera colocasioides Hook. f. *	Biskachu	Herb, e	ml	Ad	V	Kamrul 2027 (JUH)
Typhonium flagelliforme (Lodd.) Blume	Ghechu	Herb, e	sj, wt	Ad		Kamrul 2205 (JUH)
T. trilobatum (L.) Schott	Ghetkul	Herb, e	sj	Ad	V	Kamrul 2207 (JUH)
LEMNACEAE Martinov						
Lemna perpusilla Torr.	Khudipana	Herb, aq	wt	Ad	Fd, Gm	Kamrul 2244 (JUH)
COMMELINACEAE Mirb.						
Amischotolype mollissima (Blume) Hassk.	Molosima	Herb, pr	hs, rb	Ad, Ku	-	Kamrul 399 (JUH)
Commelina diffusa Burm.f.	Monayna kanshira	Herb, e	hs, rb	All beats	M	Kamrul 764 (JUH)
C. erecta L.	Jata kanchira	Herb, e	sj	Ad	V	Kamrul 283 (JUH)
Floscopa scandens Lour.	-	Herb, p	fv	All beats	M	Kamrul 1024 (JUH)
Murdannia nudiflora (L.) Brenan	Kureli	Herb, e	fe, rs	All beats	-	Kamrul 91 (JUH)
Pollia secundiflora (Blume) Bakh.f.	Kandopoli	Herb, e	wd	Ad	-	Kamrul 809 (JUH)

Table 1 Contd.

Scientific name	Bangla name	Habit	Habitat	Distrib.	Uses	RSE
ERIOCAULACEAE Martinov						
Eriocaulon quinquangulare L.	Guri	Herb, e	ml	All beats	-	Kamrul 762 (JUH)
JUNCACEAE Juss.						
Juncus prismatocarpus R.Br.	Atoshi junca	Herb, e	rb	Ad	-	Kamrul 2247 (JUH)
CYPERACEAE Juss.						
Bulbostylis barbata (Rottb.) C.B. Clarke.	Balbobata	Herb, e	wd	All beats	-	Kamrul 1607 (JUH)
Cyperus compactus Retz.	Bandorghasi	Herb, e	sj, gl	All beats	Sb	Kamrul 2109 (JUH)
C. iria L.	Barachucha	Herb, e	gl	Ad	Sb	Kamrul 1125 (JUH)
C. laxus Lam.	Alga ghasi	Herb, e	ht	All beats	Sb	Kamrul 63 (JUH)
C. pilosus Vahl	Pasham kathai	Herb, e	fe	All beats	Sb	Kamrul 193 (JUH)
C. tenuispica Steud.	Paikamutha	Herb, e	sj, rs	All beats	Sb	Kamrul 452 (JUH)
Eleocharis geniculata (L.) Roem. & Schult.	Joraghasi	Herb, e	fe, gl	Ad	-	Kamrul 447 (JUH)
Fimbristylis schoenoides (Retz.) Vahl	Kesari malanga	Herb, e	gl	All beats	-	Kamrul 1456 (JUH)
Hypolytrum nemorum (Vahl) Spreng.	Trumram ghasi	Herb, e	fe, gl	All beats	-	Kamrul 1269 (JUH)
Kyllinga nemoralis (J.R.Forst. & G.Forst.) Dandy *ex* Hutch. & Dalziel	Subasinirbisa	Herb, e	sj, fe, gl	Ad	Fd	Kamrul 2215 (JUH)
K. brevifolia Rottb.	Shabujnirbisa	Herb, e	rs, gl	All beats	Fd	Kamrul 64 (JUH)
K. bulbosa P.Beauv.	Golanirbisa	Herb, e	rs, gl,	Ad	Fd	Kamrul 2217 (JUH)
Pycreus polystachyos (Rottb.) P. Beauv.	Paikpoli ghasi	Herb, e	fv, sj	Ad	Sb	Kamrul 446 (JUH)
Scleria levis Retz.	Rialevi ghasi	Herb, e	gl	All beats	-	Kamrul 546 (JUH)
S. biflora Roxb.	Riaflora ghasi	Herb, e	ml, fe	Ad, Ka	-	Kamrul 1512 (JUH)
S. terrestris (L.) Fassett	Dharal ghasi	Herb, e	ml, gl	All beats	-	Kamrul 731 (JUH)
Rhynchospora corymbosa (L.) Britton	Shonathuti ghasi	Herb, e	ml, gl	Ad	-	Kamrul 1404 (JUH)
POACEAE Barnhart.						
Bambusa balcooa Roxb.	Borak bans	Herb, e	ml	Ad	Du	Kamrul 2218 (JUH)
B. bambos (L.) Voss	Ban bans	Herb, e	ml	All beats	Du	Kamrul 2219 (JUH)
B. polymorpha Munro	Parua	Herb, e	hs, ht	Ad, Ku	Du	Kamrul 2221 (JUH)
B. tulda Roxb.	Mirtinga	Herb, e	ml	All beats	Du	Kamrul 457 (JUH)
Brachiaria kurzii (Hook. f.) A. Camus	Kurokti ghas	Herb, e	sj	Ad, Ku	-	Kamrul 1550 (JUH)
Centotheca lappacea (L.) Desv.	Centughas	Herb, e	sp	All beats	Fd	Kamrul 941 (JUH)
Chrysopogon aciculatus (Retz.) Trin.	Premkata	Herb, e	gl, rs	Ad, Ku	Sb	Kamrul 210 (JUH)
C. zizanioides (L.) Roberty	Khaskhas	Herb, e	rs, gl	All beats	Fd, Sb	Kamrul 1513 (JUH)
Cynodon dactylon (L.) Pers.	Durba ghas	Herb, cr	ml, rs,	All beats	O, Fd, Sb	Kamrul 2222 (JUH)
Cyrtococcum oxyphyllum (Steud.) Stapf	Oxycocca ghas	Herb, e	ml, rs,	Ad, Ka	-	Kamrul 209 (JUH)
C. patens (L.) A. Camus	Patcocca ghas	Herb, cr	gl	All beats	-	Kamrul 557 (JUH)
Dendrocalamus longispathus (Kurz) Kurz	Rupai	Herb, e	ml	All beats	-	Kamrul 1818 (JUH)
Dactyloctenium aegyptium (L.) Willd.	Kakpaya	Herb, e	ml, rs	Ad	-	Kamrul 2057 (JUH)
Digitaria ciliaris (Retz.) Koeler	Kokjachira	Herb, e	fe	Ad	-	Kamrul 2572 (JUH)
Eleusine indica (L.) Gaertn.	Malankuri	Herb, e	sj	All beats	Fd	Kamrul 1129 (JUH)
Eragrostis ciliaris (L.) R. Br.	Lomkoni	Herb, e	ml	Ad	-	Kamrul 1563 (JUH)
E. unioloides (Retz.) Nees *ex* Steud.	Chirakoni	Herb, e	sj	All beats	Fd, Gm	Kamrul 777 (JUH)
Imperata cylindrica (L.) Raeusch.	Chhan	Herb, e	sj	Ad	Fd, M	Kamrul 211 (JUH)
Hildaea pallens (Sw.) C. Silva & R.P. Oliveira	-	Herb, e	fv	Ad	-	Kamrul 832 (JUH)
Lophatherum gracile Brongn.	Lolphali ghas	Herb, e	hs, rs	All beats	Fd	Kamrul 635 (JUH)
Leersia hexandra Sw.	Fulka ghas	Herb, pr	ml	Ad	Fd	Kamrul 1130 (JUH)
Melocanna baccifera (Roxb.) Kurz	Muli	Herb, e	hs	All beats	V, Du	Kamrul 622 (JUH)

Table 1 Contd.

Scientific name	Bangla name	Habit	Habitat	Distrib.	Uses	RSE
Oryza rufipogon Griff.	Bunodhan	Herb, e	ml	Ad, Ka	Fd	Kamrul 461 (JUH)
Oplismenus compositus (L.) P. Beauv.	Gohur durba	Herb, e	fe, hs	All beats	-	Kamrul 2101 (JUH)
Panicum notatum Retz.	Panita ghas	Herb, e	fe	All beats	-	Kamrul 776 (JUH)
P. brevifolium L.	Panibrevi ghas	Herb, e	fe	All beats	-	Kamrul 2039 (JUH)
Paspalum conjugatum P.J.Bergius.	Moisshya ghas	Herb, e	sj, fe	Ad, Ka	-	Kamrul 455 (JUH)
Phragmites karka (Retz.) Trin. *ex* Steud.	Khakra ghas	Herb, er	sj, rb	Ad, Ku	Du	Kamrul 999 (JUH)
Saccharum spontaneum L.	Kash	Herb, e	sj	Ad, Ka	-	Kamrul 2060 (JUH)
Setaria palmifolia (J. Koenig) Stapf.	Urodhan	Herb, e	fe	Ad, Ku	Fd, M	Kamrul 858 (JUH)
Thysanolaena latifolia (Roxb. *ex* Hornem.) Honda	Jharu phul	Herb, e	fe, sj	All beats	Du	Kamrul 141 (JUH)
Echinochloa colona (L.) Link.	Shama ghas	Herb, e	iv	Ad	Fd	Kamrul 1409 (JUH)
BROMELIACEAE Juss.						
Ananas comosus (L.) Merr.	Anarash	Herb, e	ml	Ad	Fr	Kamrul 2087 (JUH)
MUSACEAE Juss.						
Musa paradisiaca L.	Kola	Herb, e	ml	Ad, Ka	Fr, V, M	Kamrul 2080 (JUH)
M. acuminata Colla	Pahari kola	Herb, e	hs	Ad	Fr	Kamrul 160 (JUH)
M. ornata Roxb.	Jangli kola	Herb, e	hs	Ad, Ku	Fr, V	Kamrul 799 (JUH)
ZINGIBERACEAE Martinov						
Amomum aromaticum Roxb. *	Alachi	Herb, e	hs, sj	All beats	Fr, M	Kamrul 51 (JUH)
Alpinia malaccensis (Burm. f.) Roscoe	Deotara.	Herb, e	hs	All beats	M	Kamrul 769 (JUH)
Curcuma aromatica Salisb.	Ban haldi	Herb, e	hs	Ad	M	Kamrul 2226 (JUH)
C. caesia Roxb.	Kala holdi	Herb, e	hs	Ad	M	Kamrul 1234 (JUH)
C. phaeocaulis Valeton	Shoti	Herb, e	hs	All beats	M	Kamrul 2075 (JUH)
Globba bracteolata Wall. *ex* Baker	Dhaki globba	Herb, e	hs	Ad	M	Kamrul 22 (JUH)
G. marantina L.	Maran globba	Herb, e	hs	Ad, Ku	-	Kamrul 479 (JUH)
G. multiflora Wall. *ex* Baker *	Shukh globba	Herb, e	hs	All beats	-	Kamrul 247 (JUH)
Hedychium coronarium J. Koen.	Dolon chapa	Herb, e	ml	Ad	O	Kamrul 2227 (JUH)
H. thyrsiforme Sm. *	Pala ada	Herb, e	ml	All beats	-	Kamrul 611 (JUH)
Zingiber officinale Roscoe	Ada	Herb, e	ml	Ad, Ka	Sp, M	Kamrul 2232 (JUH)
Z. zerumbet (L.) Roscoe *ex* Sm.	Bon ada	Herb, e	ml	All beats	M	Kamrul 2229 (JUH)
COSTACEAE Nakai.						
Hellenia speciosa (J. Koenig) Govaerts	Bandugi	Herb, e	sj	Ad	M	Kamrul 395 (JUH)
CANNACEAE Juss.						
Canna indica L.	Kolabati	Herb, e	ml	Ad	M, O	Kamrul 2213 (JUH)
MARANTACEAE R.Br.						
Phrynium pubinerve Blume	Pashompitali	Herb, e	fe	All beats	Du	Kamrul 176 (JUH)
P. placentarium (Lour.) Merr.	-	Herb, e	fe	Ad, Ku	Du	Kamrul 618 (JUH)
Schumannianthus dichotomus (Roxb.) Gagnep.	Murta	Shrub	fe	Ad	Du	Kamrul 2276 (JUH)
PONTEDERIACEAE Kunth.						
Eichhornia crassipes (Mart.) Solms.	Kachuripana	Herb, aq	ml	Ad	Fd	Kamrul 1904 (JUH)
Monochoria hastata (L.) Solms	Baranukha	Herb, aq	ml	All beats	Fd	Kamrul 226 (JUH)
M. vaginalis (Burm.f.) C. Presl	Bara nukha	Herb, aq	ml	All beats	-	Kamrul 445 (JUH)
HAEMODORACEAE R. Br.						
Peliosanthes teta Andrews	Napigach	Herb, e	wd	Ad	-	Kamrul 2253 (JUH)

Table 1 Contd.

Scientific name	Bangla name	Habit	Habitat	Distrib.	Uses	RSE
LILIACEAE Juss.						
Asparagus racemosus Wild.	Shatamuli	Shrub	sj	Ad	M	Kamrul 2255 (JUH)
Crinum amoenum Ker Gawl. *ex* Roxb.	Gang kachu	Herb, e	hs	All beats	O, M	Kamrul 1910 (JUH)
Molineria latifolia (Dryand. *ex* W.T. Aiton) Herb. *ex* Kurz	Molinpasna	Herb, e	sj	Ad	-	Kamrul 2258 (JUH)
Gloriosa superba L.	Ulatchandal	Herb, c	ml	Ad	M	Kamrul 2257 (JUH)
Molineria capitulata (Lour.) Herb.	Satipata	Herb, e	fe	Ad	-	Kamrul 327 (JUH)
AGAVACEAE Dumort.						
Dracaena spicata Roxb.	Kado drakan	Shrub	hs	Ad, Ku	-	Kamrul 309 (JUH)
Sansevieria trifasciata Prain	Sutahara	Herb, e	ml	Ad	M, O	Kamrul 1797 (JUH)
TACCACEAE Dumort.						
Tacca integrifolia Ker-Gawl.	Mati munda	Herb, e	Fe, sj	All beats	O	Kamrul 158 (JUH)
STEMONACEAE Caruel						
Stichoneuron membranaceum Hook. f. *	Koniron	Herb, e	ht, fv	Ad, Ku	-	Kamrul 662 (JUH)
Stemona tuberosa Lour.	Lalguraniya alu	Herb, e	fe	Ad	-	Kamrul 640 (JUH)
SMILACACEAE Vent.						
Smilax ovalifolia Roxb. *ex* D. Don	Kumari lata	Herb, cl	wd	All beats	M	Kamrul 33 (JUH)
S. perfoliata Lour.	Kumarika	Herb, cl	hs	Ad, Ku	-	Kamrul 1114 (JUH)
DIOSCOREACEAE R.Br.						
Dioscorea alata L.	Chupri alu	Herb, cl	wd	All beats	V	Kamrul 2028 (JUH)
D. bulbifera L.	Sora alu	Herb, cl	fe, sj	Ad	V, M	Kamrul 2046 (JUH)
D. glabra Roxb.	Sora alu	Herb, cl	wd	All beats	M	Kamrul 1044 (JUH)
D. hamiltonii Hook. f.	Thakan budo	Herb, cl	sj	Ad	V	Kamrul 2066 (JUH)
D. pentaphylla L.	Jum alu	Herb, cl	sj, rs	All beats	M	Kamrul 319 (JUH)
ORCHIDACEAE Juss.						
Aerides odorata Lour.	Churi	Herb, ep	wd	Ad	O	Kamrul 1327 (JUH)
Cymbidium aloifolium (L.) Sw. *	Churi	Herb, ep	wd	Ad	O	Kamrul 180 (JUH)
Dendrobium lindleyi Steud.	Linrium	Herb, ep	wd	All beats	-	Kamrul 1190 (JUH)
Peristylus sp.	-	Herb, e	wd	Ka	-	Kamrul 1585 (JUH)
Vanda tessellata (Roxb.) Hook. *ex* G.Don.	Rasna	Herb, ep	wd	Ad	M	Kamrul 584 (JUH)

LEGEND: **Habit**. cl = climbing, cr = creeping, de = decumbent, ep = epiphytic, e = erect, er = erect reed, l = large, m = medium, pr = prostrate, ps = parasitic, aq = aquatic, s = small, sc = scandent, vi = vine. HABITAT. cu = cultivated, fe = forest margin, fv = forest valley, gl = grassland, ht = hill top, hs = hill slope, ml = marginal land, pl = planted, rb = river bank, rs = roadsides, sj = scrub jungle, wd = in forest. **Distrib.** = Distribution. Ad = Adampur beat, Ka = Kamarchara beat, Ku = Kurma beat. **Use**. Co = cosmetics, Dy = dye, Du = domestic uses, Fb = Fibre, Fd = fodder, Fr = Fruit, Fw = fuel wood, Gm = green manure, I = insecticide, Ju = juice, M = medicine, O = ornamental, Ol = oil, Pe = perfume, Pu = pulse, St = shade tree, Sb = soil binder, Sp = spice, Sw = soft wood, T = timber, and V = vegetable. **RSE** = Representative Specimens Examined. * = Species listed in Red Data Book of Vascular Plants of Bangladesh.

In respect to land area, the number of angiosperm species found in RRF during this study seems higher than that reported by few studies on the protected areas of Bangladesh, e.g., Teknaf Game Reserve (Khan *et al.*, 1994), and Chunuti Wildlife Sanctuary (Khan and Huq, 2001). In contrast, the number of angiosperm species found in RRF is lower than that of some other protected areas of the country, e.g., Rema-Kalenga Wildlife Sanctuary (Uddin and Hassan, 2004), and Lawachara National Park (Uddin and Hassan, 2010). These data indicate that this reserve forest houses an important portion of the flora of Bangladesh.

The study recognizes a total of 387 angiosperm species of RRF as economically useful and among these species 82 are useful in two and 17 in three categories. The major categories of these economically useful species are medicinal (231 species), timber (50 species), fruit (49 species), ornamental (47 species), vegetable (34 species), fodder (23 species), domestic uses (19 species), fuel wood (11 species), and soil binder (9 species).

The RRF area houses 25 species included as threatened in the Red Data Book of Vascular Plants of Bangladesh (Table 1; Khan *et al.*, 2001; Ara *et al.*, 2013). This emumeration of threatened species is higher in respect to that reported for few forest areas, e.g., Lawachara National Park (Uddin and Hassan, 2010) and Sundarban Mangrove Forest (Rahman *et al.*, 2016), and on the other hand, lower in respect to that recorded from other forest areas of Bangladesh (*e.g.*, Rema-Kalenga forest; Uddin and Hassan, 2004). All of these 25 species were found in many localities of RRF with normal natural regeneration and any threat or stress exclusive for these species could not be recognized there, and therefore, they were categorized under the Lower Risk (LR) category for Rajkandi Reserve Forest.

This checklist provides basic information on all angiosperm species currently occurring in the Rajkandi Reserve Forest, which can be considered as an important database as well as baseline to track the trend of changes in the floristic composition of this reserve forest in course of time and different biogeographical processes. This study also informs the current status of 25 threatened species in RRF. These data might be useful in planning, management, conservation and sustainable development of this valuable forest resource of Bangladesh.

Acknowledgements

The authors are grateful to the authorities of the Bangladesh Forest Department, Bangladesh National Herbarium (DACB) and Dhaka University Salar Khan Herbarium (DUSH) for their co-operation during conducting this study. The authors are thankful to the Chief Editor and the Reviewers of the Journal for their critical review of the manuscript.

References

Ahmed, Z.U., Begum, Z.N.T., Hassan, M.A., Khondker, M., Kabir, S.M.H., Ahmed, M., Ahmed, A.T.A., Rahman, A.K.T. and Haque, E.U. (Eds) 2008-2009. Encyclopedia of Flora and Fauna of Bangladesh, Vols. **6-10, 11, 12**. Asiatic Society of Bangladesh, Dhaka.

Alam, M.K. 1988. Annotated checklist of woody flora of Sylhet forests. Bulletinn 5, Plant Taxonomy Series. Bangladesh Forest Research Institute, Chittagong, pp. 1–153.

Ara, H., Khan, B. and Uddin, S.N. (Eds) 2013. Red Data Book of Vascular Plants of Bangladesh. Bangladesh National Herbarium, Dhaka, pp. 1–280.

Arefin, M.K., Rahman, M.M., Uddin, M.Z. and Hassan, M.A. 2011. Angiosperm flora of Satchari National Park, Habiganj, Bangladesh. Bangladesh J. Plant Taxon. **18**(2): 117–140.

Bridson, D.M. and Forman, F. 1989. *In*: Bridson, D.M. and Forman, F. (Eds), The Herbarium Handbook. Royal Botanic Gardens, Kew, 214 pp.

Cronquist, A. 1981. An integrated system of classification of flowering plants. Columbia University Press, New York, pp. 1-1262.

Das, D.K. 1968. The vegetation of Sylhet forests. Pak. J. Forest. **18**(3): 307–316.

Ghani, A. 1998. Medicinal Plants of Bangladesh with Chemical Constituents and Uses. Asiatic Society of Bangladesh, pp. 1–467.

Huq, A.M. 1986. Plant Names of Bangladesh. Bangladesh National Herbarium, BARC, Dhaka, Bangladesh, pp. 1–289.

Haque, A.K.M.K., Khan, S.A., Uddin, S.N. and Rahim, M.A. 2016. Taxonomic checklist of the pteridophytes of Rajkandi Reserve Forest, Moulvibazar, Bangladesh. Jahangirnagar University J. Biol. Sci. **5**(2): 27–40.

Hooker, J.D. 1872-1897. The Flora of British India. Vols. **1**–**7**. L. Reeve & Co., Ashford, Kent, UK.

IUCN. 2001. IUCN Red List Categories and Criteria. Version **3.1**. IUCN, Gland, Switzerland and Cambridge, U.K.

Kanjilal, U.N., Kanjilal, P.C. and Das, A. 1934. Flora of Assam. Vol. **1**. (Reprint 1982). A Von Book Company, Delhi.

Kanjilal, U.N., Kanjilal, P.C. and Das, A. 1938-1940. Flora of Assam. Vol. **2-4**. (Reprint 1982). A Von Book Company, Delhi.

Khan, M.S., Rahman, M.M. and Ali, M.A. (Eds) 2001. Red Data Book of Vascular Plants of Bangladesh, Vol. **1**. Bangladesh National Herbarium, Dhaka, Bangladesh, pp. 1–179.

Khan, M.S. and Huq, A.M. 2001. The vascular flora of Chunati wildlife sanctuary in south Chittagong. Bangladesh J. Plant Taxon. **8**(1): 47–64.

Khan, M.S., Rahman, M.M., Huq. A.M., Mia, M.M.K. and Hassan, M.A. 1994. Assessment of Biodiversity of Teknaf Game Reserve in Bangladesh focussing on economically and ecologically important plant species. Bangladesh J. Plant Taxon. **1**(1): 21–33.

Myers, N., Mittermeier, R.A., Mittermeier, C.G., da Fonseca, G.A.B. and Kent, J. 2000. Biodiversity hotspots for conservation priorities. Nature **403**: 853–858.

Pasha, M.K and Uddin, S.B. 2013. Dictionary of Plant Names of Bangladesh (Vascular Plants). Janokalyan Prokashani, Chittagong, pp. 1–320.

Prain, D. 1903. Bengal Plants. Vols. **1 & 2**. (Reprint 1963). Botanical Survey of India, Calcutta.

Rahman, M.S., Hossain, M.G., Khan, S.A. and Uddin, S.N. 2016. An annotated checklist of the vascular plants of Sundarban mangrove forest of Bangladesh. Bangladesh J. Plant Taxon. **22**(1): 17–41.

Singh, H.B. and Subramaniam, B. 2008. Field Manual on Herbarium Techniques. National Institute of Science Communication and Information Resources, pp. 1–297.

Sobuj, N.A, and Rahman, M. 2011. Assessment of plant diversity in Khadimnagar National Park of Bangladesh. International J. Environ. Sci. **2**(1): 79–91.

The Plant List, 2013. The Plant List, a working list of all plant species. Version 1.1 < http://www.the-plantlist.org/>. Accessed on 29 October 2017.

TROPICOS, 2017. Tropicos.org. <www.tropicos.org>. Missouri Botanical Garden, Saint Louis, Missouri, USA. Accessed on 20 October 2017.

Uddin, M.Z. and Hassan, M.A. 2004. Flora of Rema–Kalenga Wildlife Sanctuary. IUCN Bangladesh Country Office, Dhaka, Bangladesh, pp. 1–120.

Uddin, M.Z. and Hassan, M.A. 2010. Angiosperm diversity of Lawachara National Park (Bangladesh): A preliminary assessment. Bangladesh J. Plant Taxon. **17**(1): 9–22.

van Valkenburg, J.L.C.H. and Bunyapraphatsara. N. (Eds) 2002. Plant Resources of South–East Asia, No. **12**(2). Medicinal and Poisonous Plants 2. Prosea Foundation, Bogor, Indonesia, 782 pp.

Wu, Z.Y. and Raven, P.H. (Eds) 1994-2001. Flora of China, Vols. **8, 15-18, 24**. Missouri Botanical Garden Press, St. Louis, USA.

Wu, Z.Y., Raven, P.H. and Hong, D.Y. (Eds) 1999-2013. Flora of China, Vols. **2-7, 9-14, 19-23, 25**. Missouri Botanical Garden Press, St. Louis.

SPECIES DELINEATION OF THE GENUS *DIPLAZIUM* SWARTZ (ATHYRIACEAE) USING LEAF ARCHITECTURE CHARACTERS

JENNIFER M. CONDA[1] AND INOCENCIO E. BUOT, JR[2]

Department of Science and Technology-Forest Products Research and Development Institute, Los Baños, Laguna, Philippines

Keywords: Leaf architecture; Taxonomic marker; Cladodromous; Reticulodromous; Craspedodromous; Cophenetic correlation.

Abstract

The present study was conducted to delineate *Diplazium* Swartz species based on leaf architecture. Using PAleontological STatistics (PAST), a cluster and Principal Component Analysis of leaf architecture characters of 27 selected *Diplazium* species at the Philippine National Herbarium (PNH) was done. The dendogram (cophenetic correlation = 0.8436) and principal component analysis supported the four clusters of *Diplazium* using leaf architecture characters. At Gower distance of 0.25, *Diplazium* species were categorized as: Cluster 1 (Cladodromous – short stalked, stout and massive 1° vein); Cluster 2 (Reticulodromous – long stalked, moderate 1° vein); Cluster 3 (Craspedodromous – long stalked, stout to massive 1° vein); and Cluster 4 (Craspedodromous – short stalked, stout to massive 1° vein). The unifying characters were apex shape, base symmetry and 1° vein category, while the significant differentiating characters were 2° vein angle of divergence and variation in the 2° vein angle of divergence, 3° vein category, 3° vein angle of divergence, variation in 3° vein angle of divergence, 3° vein spacing and lobation. The successful delineation of *Diplazium* species proved that leaf architecture can be a good taxonomic marker and could be an alternative way of identifying species in the absence of sori.

Introduction

Diplazium Swartz consists of about 400 species distributed mainly in the tropics and sparingly in temperate forest (Kramer *et al.*, 1990). Copeland (1947) enumerated 62 *Diplazium* species in the Philippines. Meanwhile, 49 species of *Diplazium* were listed in Co Digital Flora of the Philippines (http://www. philippineplants.org/Families/Pteridophytes.html). Among genera under Athyriaceae, *Diplazium* species were always included in ethnobotanical studies (Rai *et al.*, 2005; Kumari *et al.*, 2011; Sujarwo *et al.*, 2014) as sources of food, medicine and decorative materials (Vasudeva, 1999). In Asian and Filipino dishes, *Diplazium esculentum* is served as salad, dietary staple, base for spicy condiments and vegetable (Kayang, 2007). As medicine, *Diplazium* species were noted for their antibacterial (Amit *et al.*, 2011), phytochemicals (Sivaraman *et al.*, 2011), antimicrobial and cytotoxic (Akler *et al.*, 2014), analgesic (Chawla *et al.*, 2015), and antioxidant properties (Pradhan *et al.*, 2015).

Despite of the well-studied uses of genus *Diplazium* their taxonomic classification and identification is still controversial among taxonomists and pteridologists. Some of the problems in accurate identification of the genus included insufficient data (Kramer *et al.*, 1990) and continuous changes in taxonomic classification and morphological variations through apparently intermediate

---

[1]Corresponding author. Email: jhen_0421@yahoo.com; jennifermconda@gmail.com
[2]Institute of Biological Sciences, College of Arts and Sciences, University of the Philippines Los Baños, Los Baños, Laguna, Philippines.

forms, which are commonly regarded as putative hybrids (Takamiya *et al.*, 1999). The chance of misidentification is higher especially during field surveys and actual identification because *Diplazium* species are morphologically similar to their sisters *Athrium* and *Deparia* (Kato, 1977) and to some members of Woodsiaceae and Polypodiaceae to which *Diplazium* was formerly circumscribed (Smith *et al.*, 2006). Unconscious identification of *Diplazium* might lead to collection of wrong specimens, thus cannot satisfied the intent use and worst can be hazardous to human health or even cause death. The lack of knowledge or information when collecting for medicinal purposes, toxin-containing plants can result in misidentification with grave consequences (Voncina *et al.*, 2014).

Thus, several classification system and scholarly works were done to differentiate, delineate and investigate the phylogenetic relationship of *Diplazium* species. These include DNA sequencing (Wei *et al.*, 2013), spore morphology (Praptosuwiryo *et al.*, 2007), stelar anatomy (Praptosuwiryo and Darnaedi, 2014), and cytology and reproduction (Takamiya *et al.*, 1999). Takhtajan (1996) pointed out that molecular methods are not necessarily a universal remedy in elucidating the evolution of a certain taxon because molecular characters are also subjected to evolutionary convergence, parallelism and reversal besides random changes in DNA sequence. Further, molecular studies are expensive and not feasible in low cost-funded projects and inefficient in field surveys where actual identification is necessary.

One taxonomic tool useful in differentiating angiosperm taxa and also considered in ferns is leaf architecture, which is defined as the placement and form of elements constituting the outward expression of leaf structure, including venation pattern, marginal configuration, leaf shape, and gland position (Hickey, 1973). Pacheco and Moran (1999) resurrected *Callipteris* in their revision of the Neotropical species, and found diagnostic characters such as anastomosing veins and petiole/rachis scales with bifid-toothed margins. Recent studies on fern leaf architecture were done in the genus *Ophioglossum* (Magrini and Scoppola, 2010) and *Lygodium* (Shinta *et al.*, 2012). Leaf architecture of fern species such as *Blechnum binervatum, Ctenitis falciculata, Magalastrum connexum, Microgramma squamulosa* and *Serpocaulon catharinae* were studied by Larcher *et al.*, (2013).

Though, leaf plasticity had been an issue on the use of leaf architecture as important taxonomic marker it proved its usefulness in differentiating angiosperm. As vascular plants with distinct venation pattern, ferns are expected to have similar stability in terms of venation pattern. In fact, fern stipes are reinforced by a very stiff sclerenchyma consisting of dead cells with non-extensible rigid cell walls (Leroux, 2012) providing support and preserving the leaf architecture (Larcher *et al.*, 2013). In addition, ferns have persisted through their evolutionary history and represent highly successful forms in both past and present (Pittermann, 2010). Therefore, this study aims to delineate some *Diplazium* species of the Philippines using leaf architecture characters.

Materials and Methods

The leaf architecture characters of 27 *Diplazium* species at the Philippine National Herbarium (PNH) were summarized in Table 1 (leaf morphology) and Table 2 (venation pattern). The morphological leaf characters and venation pattern (Conda and Buot, 2017) were used to determine the species delineation of the genus *Diplazium* through Cluster and Principal Component Analysis of Paleontological Statistics (PAST). The distance measure and clustering method used were Gower and Unweighted Pair-Group Method of Arithmetic Mean (UPGMA), respectively.

For data analysis, 21 characters were selected for each species and each character was assigned to a corresponding legend as follows: LO1-6 for leaf organization, BlCl1-7 for blade class, Sh1-3 for shape, ApSh1-2 for apex shape, BaSh1-4 for base shape, BaAn1-3 for base angle, BaSy1-2 for base symmetry, Mar1-3 for margin, St1-2 for stalk, Lob1-4 for lobation, PVC1 for 1° vein category, PVS1-4 for 1° vein size, SVC1-3 for 2° vein category, SAD1-6 for 2° vein angle of divergence, SVAD1-4 for 2° vein, variation in angle of divergence, SVS1-3 for 2° vein spacing, TVC1-3 for 3° vein category, TAD1-6 for 3° vein angle of divergence, TVAD1-3 for 3° vein, variation in angle of divergence, TVS1-3 for 3° vein spacing, and AR1-2 for areole.

Results and Discussion

Leaf architecture characters of 27 *Diplazium* species are presented in Table 1. These characters varied especially in terms of L:W ratio, blade class, base angle and lobation. This interspecific variation illustrated that these characters could be good indicators of identification. The dendrogram (Fig. 1) with cophenetic correlation of 0.8436 and principal component analysis (Fig. 2) consistently separated *Diplazium* species into four clusters. At Gower distance of 0.25, *Diplazium* species were grouped into 4 clusters namely, Cluster 1 (Cladodromous - short stalked, stout and massive 1° vein); Cluster 2 (Reticulodromous - long stalked, moderate 1° vein); Cluster 3 (Craspedodromous - long stalked, stout and massive 1° vein) and Cluster 4 (Craspedodromous - short stalked, stout to massive 1° vein).

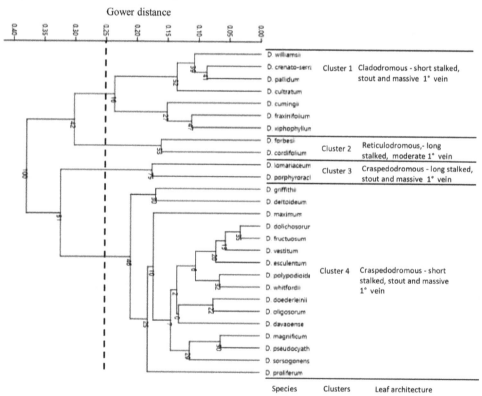

Fig. 1. Dendrogram of the 27 *Diplazium* species constructed by Unweighted Pair-Group of Arithmetic Mean (UPGMA) clustering and Bower using the Paleontological Statistics software. With cophenetic correlation of 0.8436 and gower distance of 0.25, four cluster were identified: Cluster 1 (Cladodromous - short-stalked, stout and massive 1° vein); Cluster 2 (Reticulodromous - long stalked, moderate 1° vein); Cluster 3 (Craspedodromous - long stalked, stout and massive 1° vein) and Cluster 4 (Craspedodromous - short stalked, stout to massive 1° vein).

Table 1. Leaf architecture characters of 27 selected *Diplazium* Swartz species (Athyriaceae): General leaf morphology.

Species	Leaf organization	Shape	Apex shape	L : W ratio	Blade class	Base shape	Base angle	Base symmetry	Margin	Stalk	Lobation
Diplazium cordifolium Bl.	simple-pinnate	lanceolate	acute	2.7-3.4:1	mesophyll	cordate	WO	asymmetrical	entire	LS	unlobed
D. crenato-serratum T. Moore	pinnate	lanceolate	acute	3.4-4.6:1	microphyll	truncate	WO	asymmetrical	serrate	SS	shallow
D. cultratum C. Presl.	pinnate	lanceolate	acute	1.4-5:1	microphyll	truncate	WO	asymmetrical	entire	SS	unlobed
D. cumingii C. Chr.	Pinnate	lanceolate	acute	3.6-9:1	Mesophyll	cuneate	A	asymmetrical	entire	SS	unlobed
D. davaoense Copel	tripinnate	lanceolate	acute	5.2-5.5:1	notophyll	truncate	O	asymmetrical	crenate	SS	shallow
D. deltoideum C. Presl	bipinnatifid	oblong	acute	3.3-5:1	notophyll	truncate	O	asymmetrical	serrate	SS	deep
D. dolichosorum Copel	tripinnate	lanceolate	acute	3.2-3.9:1	microphyll	truncate	O	asymmetrical	serrate	SS	shallow
D. doederleinii (Luerss.) Makino	tripinnatifid	lanceolate	acute	3.7-3.9:1	microphyll	truncate	O	asymmetrical	entire	SS	moderate
D. esculentum (Retz). Sw.	bipinnate-tripinnate	lanceolate	acute	4.1-5:1	microphyll	truncate	O	asymmetrical	serrate	SS	shallow
D. forbesii C. Chr.	pinnate	lanceolate	acute	3.4-7:1	mesophyll	rounded	O	asymmetrical	entire	LS	unlobed
D. fraxinifolium C. Presl.	pinnate	elliptic	acute	4.6-6.1:1	mesophyll	cuneate	A	asymmetrical	entire	SS	unlobed
D. fructuosum Copel	tripinnate	lanceolate	acute	4.7-7.8:1	microphyll	truncate	O	asymmetrical	serrate	SS	moderate
D. griffithii T. Moore	tripinnatifid	oblong	acute	3.1-4.1:1	microphyll	truncate	WO	asymmetrical	serrate	SS	deep
D. lomariaceum (C. Chr.) M.G. Price	pinnatifid	elliptic	acute	10.5-12.4:1	mesophyll	cuneate	A	asymmetrical	entire	LS	deep
D. magnificum (Copel) M.G. Price	tripinnatifid	lanceolate	acute	4-4.9:1	microphyll	truncate	O	asymmetrical	serrate	SS	deep
D. maximum (D. Don) C. Chr.	tripinnate	lanceolate	acute	3.9-5:1	mesophyll	truncate	O	asymmetrical	crenate	SS	shallow
D. oligosorum Copel	tripinnatifid	lanceolate	acute	3.9-4.7:1	microphyll	truncate	WO	asymmetrical	entire	SS	moderate
D. pallidum T. Moore	pinnate	lanceolate	acute	6.3-10.9:1	microphyll	rounded	O	asymmetrical	serrate	SS	shallow
D. polypodioides Blume	tripinnatifid	lanceolate	acute	3.9-5.1:1	microphyll	truncate	O	asymmetrical	serrate	SS	deep
D. porphyrorachis Diers.	pinnatifid	elliptic	acute	5.3-6.9:1	mesophyll	cuneate	O	asymmetrical	entire	LS	deep
D. proliferum (Lam.) Thou.	pinnate	lanceolate	acute	2.8-4.4:1	notophyll	truncate	O	asymmetrical	serrate	SS	shallow
D. pseudocyatheifoleum Rosent	tripinnatifid	lanceolate	acute	3.6-4.9:1	microphyll	rounded	O	asymmetrical	entire	SS	deep
D. sorsogonense C. Presl	bipinnatifid	lanceolate	acute	4.6-7:1	microphyll	truncate	O	asymmetrical	serrate	SS	deep
D. vestitum C. Presl.	tripinnate	lanceolate	acute	3.3-6:1	microphyll	truncate	O	asymmetrical	serrate	SS	shallow
D. whitfordii Copel	bipinnatifid	lanceolate	acute	2.7-3:1	nanophyll	truncate	O	asymmetrical	serrate	SS	deep
D. williamsii Copel	pinnate	lanceolate	acute	2.8-3.3:1	nanophyll	truncate	O	asymmetrical	serrate	SS	shallow
D. xiphophyllum C. Chr.	pinnate	lanceolate	acute	4.8-6.3:1	mesophyll	rounded	O	asymmetrical	entire	SS	unlobed

WO = Wide obtuse, A = Acute, O = Obtuse; LS = Long stalked, SS = Short stalked.

Table 2. Leaf architecture characters of 27 selected *Diplazium* Swartz species (Athyriaceae): Venation Characters.

Species	Primary Vein		Secondary Vein				Tertiary Vein			
	Category	Size	Category	AD	VAD	Spacing	Category	AD	VAD	Spacing
D. cordifolium	Pinnate	moderate	Reticulodromous	moderate	regular	irregular	none	none	none	none
D. crenato-serratum	pinnate	massive	Cladodromous	narrow	upper vein more acute than lower	uniform	none	none	none	none
D. cultratum	pinnate	massive	Cladodromous	wide	upper vein more acute than lower	increasing toward the base	none	none	none	none
D. cumingii	pinnate	stout	Cladodromous	moderate	nearly uniform	uniform	none	none	none	none
D. davaoense	pinnate	massive	Craspedodromous	moderate	upper vein more acute than lower	increasing toward the base	Free end in sinuses	narrow	upper vein more acute than lower	increasing toward the base
D. deltoideum	pinnate	stout	Craspedodromous	right	nearly uniform	irregular	free and forked touching margin	wide	upper 3° vein more acute than lower	upper 3° vein more acute than lower
D. dolichosorum	pinnate	massive	Craspedodromous	moderate	upper 2° vein more acute than lower	increasing toward the base	free end in sinuses	narrow	upper 3° vein more acute than lower	increasing toward the base
D. doederleinii	pinnate	massive	Craspedodromous	wide	upper 2° vein more acute than lower	increasing toward the base	free end in sinuses	narrow	upper 3° vein more acute than lower	increasing toward the base
D. esculentum	pinnate	massive	Craspedodromous	wide	varies irregularly	increasing toward the base	forming commissural vein	narrow	upper 3° vein more acute than lower	increasing toward the base
D. forbesii	pinnate	moderate	Reticulodromous	moderate	varies irregularly	irregular	none	none	none	none
D. fraxinifolium	pinnate	stout	Cladodromous	moderate	upper 2° vein more obtuse than lower	uniform	none	none	none	none
D. fructuosum	pinnate	massive	Craspedodromous	wide	upper 2° vein more acute than lower	increasing toward the base	free end in sinuses	narrow	upper 3° vein more acute than lower	increasing toward the base
D. griffithii	pinnate	massive	Craspedodromous	wide	upper 2° vein more obtuse than lower	increasing toward the base	free and forked touching margin	narrow	upper 3° vein more acute than lower	increasing toward the base
D. lomariaceum	pinnate	massive	Craspedodromous	right	nearly uniform	irregular	free and forked touching margin	moderate	varies irregularly	irregular

Table 2 Contd.

Species	Primary Vein		Secondary Vein					Tertiary Vein			
	Category	Size	Category	AD	VAD	Spacing	Category	AD	VAD	Spacing
D. magnificum	pinnate	massive	Craspedodromous	right	upper 2° vein more acute than lower	increasing toward the base	free end in sinuses	moderate	uniform	increasing toward the base
D. maximum	pinnate	stout	Craspedodromous	wide	upper 2° vein more acute than lower	irregular	free end in sinuses	moderate	uniform	increasing toward the base
D. oligosorum	pinnate	stout	Craspedodromous	moderate	upper 2° vein more acute than lower	increasing toward the base	free end in sinuses	moderate	upper 3° vein more acute than lower	increasing toward the base
D. pallidum	pinnate	massive	Cladodromous	wide	nearly uniform	uniform	none	none	none	none
D. polypodioides	pinnate	stout	Craspedodromous	wide	upper 2° vein more acute than lower	uniform	free end in sinuses	narrow	upper 3° vein more acute than lower	increasing toward the base
D. porphyrorachis	pinnate	stout	Craspedodromous	wide	upper 2° vein more obtuse than lower	uniform	free and forked touching margin	moderate	upper 3° vein more acute than lower	uniform
D. proliferum	pinnate	stout	Craspedodromous	moderate	upper 2° vein more obtuse than lower	increasing toward the base	forming commissural vein	narrow	upper 3° vein more acute than lower	uniform
D. pseudo-cyatheifoleum	pinnate	massive	Craspedodromous	right	upper 2° vein more acute than lower	increasing toward the base	free end in sinuses	moderate	uniform	increasing toward the base
D. sorsogonense	pinnate	massive	Craspedodromous	right	upper 2° vein more acute than lower	increasing toward the base	free end in sinuses	moderate	uniform	increasing toward the base
D. vestitum	pinnate	massive	Craspedodromous	wide	nearly uniform	uniform	free end in sinuses	narrow	upper 3° vein more acute than lower	increasing toward the base
D. whitfordii	pinnate	massive	Craspedodromous	moderate	upper 2° vein more acute than lower	uniform	free end in sinuses	narrow	upper 3° vein more acute than lower	increasing toward the base
D. williamsii	pinnate	massive	Cladodromous	moderate	upper 2° vein more acute than lower	irregular	none	none	none	none
D. xiphophyllum	pinnate	stout	Cladodromous	moderate	nearly uniform	uniform	none	none	none	none

AD = Angle of Divergence, VAD = Variation in Angle of Divergence

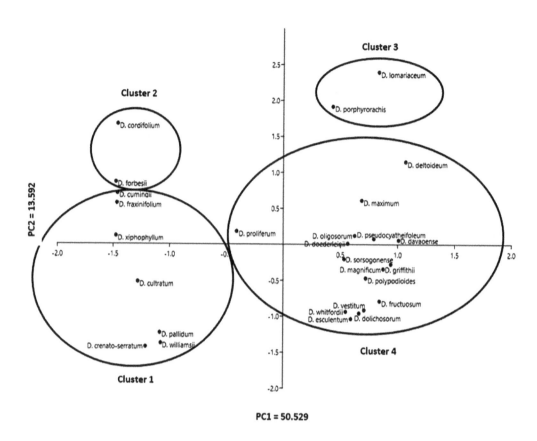

Fig. 2. Principal Component Analysis of 27 *Diplazium* species using PAleotological STatistics (PAST) software. Four clusters were classified: Cluster 1 (Cladodromous – short stalked, stout and massive 1° vein); Cluster 2 (Reticulodromous - long stalked, moderate 1° vein); Cluster 3 (Craspedodromous - long stalked, stout and massive 1° vein) and Cluster 4 (Craspedodromous - short stalked, stout to massive 1° vein).

Cluster 1, the Cladodromous - short stalked, stout to massive 1° vein: It includes *D. williamsii* Copel, *D. crenato-serratum* T. Moore, *D. pallidum* T. Moore, *D. cultratum* C. Presl., *D. cumingii* C. Chr. and *D. xiphophyllum* C. Chr. Sample line drawings of species under Cluster 1 (Figs. 3a-3c) were lifted from Conda and Buot (2017). Based on the illustrations, common leaf architecture characters were: pinnate leaf arrangement, lanceolate shape, acute apex, symmetrical base, entire and serrate margin, unlobed to shallow lobation, short stalked, pinnate 1° vein, stout to massive 1° vein size, cladodromous 2° vein category and absence of areole. In this cluster there is one outlier, *D. fraxinifolium*, which is reticulodromous.

Cluster 2, the Reticulodromous, long stalked - moderate 1° vein: Cluster 2 includes *D. cordifolium* Bl. (Fig. 3d) and *D. forbesii* C. Chr. (Fig. 3e). These species showed pinnate leaf arrangement, lanceolate leaf shape, acute apex, asymmetrical base, entire margin, long-stalked, unlobed blade, mesophyllous blade class, pinnate 1° vein, reticulodromous 2° vein, moderate 2° vein angle of divergence and presence of areole. The two species differed in variation in 2° vein angle of divergence. The former exhibits nearly uniform 2° vein angle of divergence while irregular in the latter. This cluster was found consistent with the classification of *Diplazium* species using stelar anatomy of stipe (Praptosuwiryo and Darnaedi, 2014) and spore morphology specifically perine ornamentation (Praptosuwiryo *et al.*, 2007).

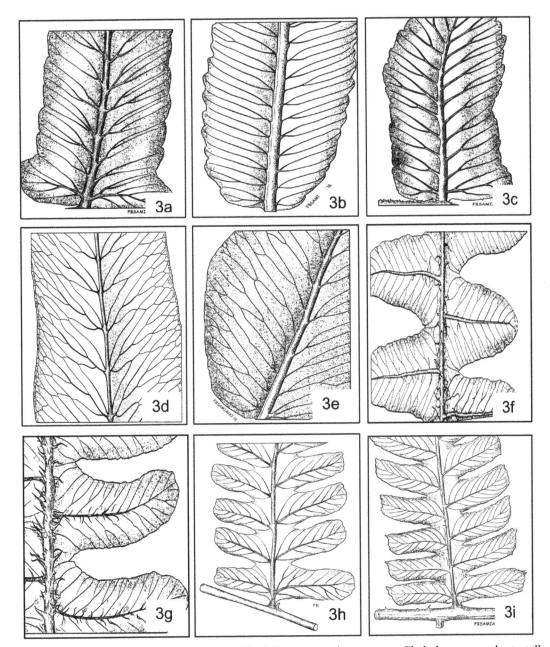

Fig. 3. Line drawings of *Diplazium* species with different venation pattern. Cladodromous - short stalked, stout to massive 1° vein venation pattern: *D. crenato-serratum* T. Moore (3a), *D. pallidum* T. Moore (3b) and *D. cutratum* C. Presl (3c). Reticulodromous - short stalked, moderate 1° vein: *D. cordifolium* Blume (3d) and *D. forbesii* C. Chr. (3e). Craspedodromous - long stalked, stout and massive 1° vein: *D. lomariaceum* (C. Chr.) M.G. Price (3f) and *D. porphyrorachis* Diers. (3g). Craspedodromous - short stalked, stout to massive 1° vein: *D. oligosorum* Copel (3h) and *D. polypodioides* Blume (3i).

Cluster 3, the Craspedodromous - long stalked, stout and massive 1° vein: This cluster consists of *D. lomariaceum* (C. Chr.) M.G. Price (Fig. 3f) and *D. porphyrorachis* Diers (Fig. 3g). They exhibit pinnatifid lamina, elliptic shape, acute apex, cuneate and asymmetrical base, entire margin, long stalked, deeply lobed, mesophyllous blade class, pinnate 1° vein, craspedodromous 2° vein, right 2° vein angle of divergence, free and forked touching margin 3° vein, moderate 3°

vein angle of divergence and absence of areoles. *D. porphyrorachis* differs by having stout 1° vein size, uniform 2° and 3° vein spacing and upper 3° vein more acute than lower variation in 3° vein angle of divergence. While *D. lomariaceum* showed a massive 1° vein size, irregular 2° and 3° vein spacing and irregular variation in 3° vein angle of divergence. This group was strongly supported using spore morphology (Praptosuwiryo et al., 2007).

Cluster 4, the Craspedodromous - short stalked, stout to massive 1° vein: It includes majority of *Diplazium* species (16 individuals) namely, *D. griffithii* T. Moore, *D. deltoideum* (C. Presl.), *D. maximum* (D. Don) C. Chr., *D. proliferum* (Lam.) Thours., *D. esculentum* (Retz.) Sw., *D. oligosorum* (Copel), *D. sorsogonense* (C. Presl.) C. Presl., *D. magnifium* (Copel) M.G. Price, *D. pseudocyatheifolium* Rosent., *D. doederleinii* (Luerss.) Makino, *D. whitfordii* Copel, *D. polypodioides* (Blume), *D. vestitum* C. Presl., *D. fructuosum* (Copel), *D. dolichosorum* (Copel) and *D. davaoense* (Copel). Sample drawings (Figs. 3h & 3i) from Conda and Buot (2017) were incorporated to emphasize the common leaf architecture characters namely, lanceolate to rarely oblong leaf shape, acute apex, truncate base, obtuse to wide obtuse base angle, asymmetrical base, pinnate 1° vein, stout to massive 1° vein size, upper 2° vein more acute than lower variation in angle of divergence and absence of areoles. At Gower distance of 0.19, *D. griffithii* and *D. deltoideum,* having oblong pinnule, was separated from the lanceolate group. Among the lanceolate group, only *D. esculentum* and *D. proliferum* possessed 3° vein forming commissural vein, while the rest have free end in sinuses 3° vein. This cluster coincides mostly with the work of Wei *et al.* (2013) using DNA sequencing of *Diplazium* from different geographical areas. Most species in this study fell under clade IV, subclade E (*Diplazium* species with short branches connecting deeper nodes and long branches leading to tip – occuring in Southeast Asia and adjoining regions) of Wei *et al.,* (2013) phylogram. The analysis of the leaf architecture characters of *D. davaoense, D. esculentum* and *D. doederlenii* (cluster 4) revealed similarities with subclade H (Wei *et al.,* 2013) possibly because these species are Asiatic in nature.

Leaf architecture, particularly the venation pattern, is a good taxonomic tool in delineating *Diplazium* species. Consistency in groupings with spore morphology, stelar anatomy and DNA sequencing proved leaf architecture's usefulness in the classification system for *Diplazium* species. The dendrogram (cophenetic coefficient = 0.8436) and principal component analyses highly supported the four clusters of *Diplazium* using leaf architecture characters, *viz.* Cluster 1 (Cladodromous - short stalked, stout and massive 1° vein); Cluster 2 (Reticulodromous - long stalked, moderate 1° vein); Cluster 3 (Craspedodromous - long stalked, stout and massive 1° vein) and Cluster 4 (Craspedodromous - short stalked, stout to massive 1° vein). The unifying characters in the genus are apex shape, base symmetry and 1° vein category, whereas 2° vein angle of divergence and variation in 2° vein angle of divergence, 3° vein category, 3° vein angle of divergence, variation in 3° vein angle of divergence, 3° vein spacing and lobation are the differentiating features. This study has proved that identification of sterile specimen is now feasible with leaf architecture.

Acknowledgments

We would like to thank the Department of Science and Technology and Forest Products Research and Development Institute for providing scholarship and allowing the senior author to pursue graduate degree studies at the University of the Philippines, Los Baños and the Philippine National Museum, particularly the Philippine National Herbairum, for access to their botanical collections. We also extend our deepest gratitude to Dr. Tito Evangelista, John Rey Callado, Emerita R. Barile and Froilan B. Samiano for the assistance during the course of the study.

References

Akler, S., Hossain, M.M., Ara, I. and Akhtar, P. 2014. Investigation of *in vitro* antioxidant, antimicrobial and cytotoxic activity of *Diplazium esculentum* (Rets.) Sw. International J. Adv. Pharm. Biol. & Chem. **3**(3): 723–733.

Amit, S., Sunil, K. and Arvind, N. 2011. Antibacterial activity of *Diplazium esculentum* Retz.) Sw. Phcog J. **3**(21): 77–79.

Chawla, S., Ram, V., Semwal, V.A. and Singh, R. 2015. Analgesic activity of medicinally important leaf of *Diplazium esculentum*. Afr. J. Pharm. Pharmacol. **9**(25): 628–632.

Co Digital Flora of the Philippines. http://www.philippineplants.org/Families/Pteridophytes.html.> Retrieved on 17 March 2018.

Conda, J.M. and Buot Jr., I.E. 2017. Leaf architecture of selected Philippine *Diplazium* Swartz species (Athyriaceae). THNHMJ. **11**(2): 57–76.

Copeland, E.B. 1947. Genera Filicum. Waltham MA: Chronica Botanica Company.

Hickey, L.J. 1973. Classification of the architecture of dicotyledonous leaves. Am. J. Bot. **60**(1): 17–33.

Kato, M. 1977. Classification of *Athyrium* and allied genera of Japan. Bot. Mag. (Tokyo) **90**: 23–40.

Kayang, H. 2007. Tribal knowledge on wild edible plants of Meghalaya, Northeast India. Indian J. Trad. Knowledge **6**: 177–181.

Kramer, K.U., Holttum, R.E., Moran, R.C. and Smith, A.R. 1990. *Dryopteridaceae. In*: Kramer K.U. and Green, P.S. (Eds), Pteridophytes and Gymnosperms. Berlin: Springer-Verlag, pp. 101–144.

Kumari, P., Otaghvari, A.M., Govindapyari, H., Bahuguna, Y.M. and Uniyal, P.I. 2011. Some ethno-medicinally important pteridophytes of India. Int. J. Med. Arom. Plants **1**(1): 18–22.

Larcher, L., Boeger, M.R.T. and Silveira, T.I. 2013. Leaf architecture of terrestrial and epiphytic ferns from an Araucaria forest in southern Brazil. Botany **91**: 768–773.

Leroux, O. 2012. Collenchyma: a versatile mechanical tissue with dynamic cell walls. Ann Bot. **110**(6): 1083–1098.

Magrini, S. and Scoppola, A. 2010. Geometric morphometrics as a tool to resolve taxonomic problems: the case of *Ophioglossum* species (ferns). *In:* Nimis, P.L. and Lebbe, R.V. (Eds), Tools for identifying biodiversity: progress and problems, pp. 251–256.

Pacheco, L. and Moran, R.C. 1999. Monograph of the Neotropical species of *Callipteris* with anastomosing veins (Woodsiaceae). Brittonia **51**: 343–388.

Pittermann, J. 2010. The evolution of water transport in plants: an intergrated approach. Geobiology **8**: 112–139.

Pradhan, S., Manivannan, S. and Tamang, J.P. 2015. Proximate, mineral composition and antioxidant properties of some wild leafy vegetables. J. Sci. Ind. Res. **74**: 155–159.

Praptosuwiryo, T.N., Kato, M. and Darnaedi, D. 2007. Specific delimitation and relationship among species of *Diplazium* based on spore morphology. Floribunda **3**(3): 57–84.

Praptosuwiryo, T.N. and Darnaedi, D. 2014. The stellar anatomy of stipe and its taxonomic significance in *Diplazium* (Athyriaceae). Floribunda **4**(8): 195–201.

Rai, A.K., Sharma, R.M. and Tamang, J.P. 2005. Food value of common edible plants of Sikkim. J. Hill Res. **18**(2): 99–103.

Shinta, R.N., Arbain, A. and Syamsuardi, D. 2012. The morphometrics study of climbing ferns (*Lygodium*) in West Sumatra. J. Bio. J. Biol. Universitas Andalas **1**(1): 45–53.

Sivaraman, M., Johnson, N. and Babu, A. 2011. Phytochemical studies on selected species of *Diplazium* from Tirunelveli Hills, Western Ghats, South India. Int. J. Basic & Appl. Biol. **5**(3&4): 241–247.

Smith, A.R., Pryer, K.M., Schuettpelz, E., Korall, P., Schneider, H. and Wolf, P. 2006. A classification for extant ferns. Taxon **55**(3): 705–731.

Sujarwo, W., Lugrayasa, I.N. and Caneva, G. 2014. Ethnobotanical study of edible ferns used in Bali, Indonesia. APJSAFE **2**(2): 1–4.

Takhtajan, A. 1996. Diversity and Classification of Flowering Plants. New York: Columbia University Press, pp. 231–234.

Takamiya, M., Takaoka, C. and Ohta, N. 1999. Cytological and reproductive studies on Japanese *Diplazium* (Woodsiaceae: Pteridophyta): apomictic reproduction in *Diplazium* with evergreen bi- to tri-pinnate leaves. J. Plant Res. **112**: 419–436.

Vasudeva, S.M. 1999. Economic importance of Pteridophytes. Indian Fern J. **16**(1-2): 130–152.

Voncina, M., Baricevic, D. and Brvar, M. 2014. Adverse effect and intoxications related to medicinal/harmful plants. Acta Agri. Slov. **103**(2): 263–270.

Wei, R., Schneider, H. and Zhang, X.C. 2013. Towards a new circumscription of the twinsorus-fern genus *Diplazium* (Athyriaceae): a molecular phylogeny with morphological implications and infrageneric taxonomy. Taxon **62**(3): 441–457.

TAXONOMIC REVISION OF THE GENUS *CRINUM* L. (LILIACEAE) OF BANGLADESH

Sumona Afroz, M. Oliur Rahman[1] and Md. Abul Hassan

Department of Botany, University of Dhaka, Dhaka 1000, Bangladesh

Keywords: Crinum L.; Taxonomy; Revision; Amaryllidaceae; Bangladesh.

Abstract

The genus *Crinum* L. represented by eight species in Bangladesh is revised. The species occurring in Bangladesh are *Crinum amabile* Donn, *C. amoenum* Roxb., *C. asiaticum* L., *C. defixum* Ker-Gawl., *C. jagus* (Thomps.) Dandy, *C. latifolium* L., *C. pratense* Herb. and *C. stenophyllum* Baker. Each species is described with updated nomenclature, important synonyms, English and Bangla names, phenology, specimens examined, chromosome number, habitat, distribution, economic value and mode of propagation. A dichotomous bracketed key to the species and illustrations are also provided.

Introduction

The classification of the lilioid monocots has long been problematic (Chase *et al.*, 2009). Some authors treated all lilioid monocots including the genus *Crinum* L. in one family, Liliaceae *s.l.* (Cronquist, 1981). Though the genus *Crinum* L. was formerly included in the family Liliaceae, the Angiosperm Phylogeny Group (APG) reevaluated the taxonomic position of this genus and placed it in the family Amaryllidaceae (APG III, 2009). Linnaeus established the genus *Crinum* in 1737 recognising four species, *viz. Crinum latifolium, C. asiaticum, C. americanum* and *C. africanum* (Nordal, 1977). The pantropical genus *Crinum* L. consists of about 112 species distributed in tropical Africa, America, Asia and Australia (Govaerts *et al.*, 2012). The genus is most diverse in Africa, particularly sub-Saharan Africa. Biogeographical analyses place the origin of *Crinum* in southern Africa (Meerow *et al.*, 2003; Kwembeya *et al.*, 2007).

Crinum are perennial herbs with globose to ovoid subterranean bulbs. Herbert (1837) divided the genus into two sections on the basis of the degree to which the tepals are patent. Baker (1881) provided detailed insight into the genus *Crinum* and divided the genus into three subgenera based on floral characters, *viz., Stenaster, Platyaster* and *Codonocrinum*. The actinomorphic flowers with linear petals were placed in the subgenus *Stenaster*; actinomorphic flowers and lanceolate petals were included in the subgenus *Platyaster,* while the subgenus *Codonocrinum* is characterized by funnel-form, zygomorphic flowers and curved tubes. Later, Baker (1898) submerged *Platyaster* into subgenus *Stenaster*, which must be named subgenus *Crinum* as it contains the type species, *C. americanum* L. (Meerow *et al.*, 2003). In order to resolve the mix-ups in nomenclature in *Crinum* several systematic studies have been carried out (Herbert, 1820; Baker, 1888, 1896; Hooker, 1892; Uphof, 1942; Verdoorn, 1973; Dassanayake, 2000). Though identification of *Crinum* species is straight forward, yet there is species complexity in many cases. In the recent past, many species of *Crinum* were placed under some other genera especially under *Amaryllis*, while many species belonging to other genera were transferred to *Crinum* (Hannibal and Williams, 1998). These snags were mainly due to inadequate research and misinterpretation or misidentifications of the plant specimens (Hannibal and Williams, 1998). Recently, Yakandawala and Samarakoon (2006) made an attempt to solve the taxonomic ambiguity on species limits of *C. latifolium* and *C. zeylanicum*.

---

[1]Corresponding author. Email: prof.oliurrahman@gmail.com; oliur.bot@du.ac.bd

Members of *Crinum* are important for their ornamental, economical and medicinal values. Leaf extract is used for treatment for vomiting and for ear-aches. The bulbs are crushed and applied onto piles and abscesses to cause suppuration. In addition, the roasted bulbs are used as a rubefacient in rheumatism (Jayaweera, 1981). *C. asiaticum* possesses antimicrobial activities (Win, 2011). Phytochemical analysis has recently yielded a vast array of compounds, including more than 150 different alkaloids in the genus *Crinum* (Fennell and van Staden, 2001).

In Bangladesh, *Crinum* L. appears to be the largest genus in the family Liliaceae represented by eight species including both wild and cultivated. In the Indian sub-continent Hooker (1892) was the pioneer on the genus *Crinum* L. who recognised 19 species from this area of which 7 species were treated as doubtful or imperfectly known. Of these, four species were reported from the area of current Bangladesh. Later, Prain (1903) listed four *Crinum* species from the area of present Bangladesh. A very few cytological investigations on some *Crinum* species occurring in Bangladesh were made over last two decades. Alam *et al*. (1998) made a karyotype analysis in *C. pretense* and *C. defixum* with differential banding patterns. Later, Ahmed *et al*. (2004) studied flurescent banding in *C. latifolium* L., *C. asiaticum* L. and *C. amoenum* Roxb. Those studied were concentrated with orcein, CMA and DAPI rather than taxonomy of those species. Recently, Hassan (2007), and Afroz and Hassan (2008) documented six *Crinum* species occurring in Bangladesh with inadequate taxonomic description. There has been no detailed taxonomic studies on this genus in Bangladesh. Therefore, the present study aims to revise the genus *Crinum* L. in Bangladesh.

Materials and Methods

Plant samples of different *Crinum* L. species were collected from different parts of the country and planted in the Dhaka University Botanical Garden for further study. The collected plant specimens were critically studied and examined which were supplemented by the herbarium specimens housed at the Dhaka University Salarkhan Herbarium (DUSH) and Bangladesh National Herbarium (DACB). Identification of the *Crinum* species were confirmed in consultation with standard literature (Hooker, 1892; Karthikeyan *et al*., 1989; Raven and Zhengyi, 2000; Utech, 2002; Hassan, 2007) and matching with authentically identified herbarium specimens deposited in DUSH and DACB. Updated nomenclature is determined consulting The Plant List (2013), a working list of all plant species. Each species is described with updated nomenclature, important synonyms, English and Bangla names, flowering and fruiting period, specimens examined, chromosome number, habitat, distribution, economic value, and mode of propagation. A dichotomous bracketed key to the species and illustrations are also provided. The voucher specimens are deposited at DUSH.

Results

Taxonomic treatment

Genus Crinum L.,

Gen. Pl. ed. 1: 97 (1737); Sp. Pl.: 291 (1753); Benth. & Hook. f., Gen. Pl. 3: 726 (1883); Bak., Handb. Amaryll. : 74 (1888); Fl. Cap. 6: 198 (1896); Fl. Trop. Afr. 7: 373 (1898); Phill., Gen. ed. 2: 203 (1951); Uphof in Herbertia 9: 63 (1942); Traub, the Genera of Amaryllidaceae : 60 (1963). *Crinopsis* Herb., Amaryll. : 270 (1837). *Erigona* Salisb., Gen. Pl. Fragm. : 115 (1866). *Liriamus* Rafin., Fl. Tell. 4: 23 (1836). *Scadianus* Rafin., Atl. Journ. : 164 (1833). *Taenais* Salisb., Gen. Pl. Fragm. : 115 (1856). *Tanghekolli* Adans. Fam. 2: 57 (1763).

Perennial herbs with tunicated bulbs, usually produced at the apex into a short or long false stem. Leaves long, lorate or ensiform, spirally arranged, sessile, with smooth or scabrous edges.

Peduncle compressed, solid. Flowers large, fragrant, umbellate, short-pedicelled or sessile, spathes 2, lanceolate, scarious; bracteoles many, linear. Perianth funnel-shaped or almost salver-shaped, tube long, straight or incurved, perianth segments 6, linear-lanceolate or narrowly oblong, red to white, often striped, streaked, or overlaid with red abaxially. Stamens 6, adnate to the throat of the perianth tube; filaments free, filiform, declinate or diverging; anthers linear or oblong-linear, dorsifixed. Carpels 3, syncarpous. Ovary inferior, 3-celled, ovules few in each locule, biseriate; style long, filiform, more or less declinate; stigma small, sub-capitate. Fruit a capsule, sub-globose or obovoid, membranous or coriaceous, bursting irregularly. Seeds few, large, green, rounded or irregularly compressed.

Key to the species of *Crinum* L. occurring in Bangladesh

1.	Perianth lobes linear	2
-	Perianth lobes oblong or lanceolate	5
2.	Umbels more than 15-flowered	3
-	Umbels up to 15-flowered	4
3.	Scape purplish, shorter than the leaves	*C. amabile*
-	Scape green, longer than the leaves	*C. asiaticum*
4.	Bulbs with a fusiform, stoloniferous base	*C. defixum*
-	Bulbs not stoloniferous	*C. stenophyllum*
5.	Perianth tube erect; stamens spreading	6
-	Perianth tube upcurved; stamens declinate	7
6.	Leaves acuminate, scabrous; perianth lobes shorter than the tube	*C. amoenum*
-	Leaves obtuse or sub-acute; perianth lobes longer than the tube	*C. pratense*
7.	Leaf margin scabrous; perianth vertically reddish on the back	*C. latifolium*
-	Leaf margin smooth; perianth white	*C. jagus*

Crinum amabile Donn, Hort. Cantabring. ed. 6: 82 (1811). *Crinum augustum* Roxb., Fl. Ind. 2: 136 (1832). **(Figs 1 & 7A-C).**

English names: Purple Spider Lily, Pink Crinum lily, Giant Spider Lily, Tiger Lily.

Bangla name: *Sukhdarshan*.

A perennial herb with a large tunicated bulb, bulb c. 40 × 12 cm with long stem; roots c. 15 cm long. Leaves long, c. 60-170 × 7-20 cm, lorate, entire, acute, glabrous, green in colour. Scape solid, 60-130 cm long, purplish, 20-50 flowered umbel, green, glabrous, arise from the side of the stem. Flowers large, actinomorphic, bisexual, epigynous, purple, fragrant at night, pedicellate, pedicel c. 3.7 cm long. Spathes 2, 15-25 × 7.0-12.5 cm, lanceolate, purplish-green or purple, bracteoles many, linear, c. 10.2 × 0.5 cm, white in colour. Perianth segments 6, c. 17 × 3 cm, purple, lower parts forming a long, slightly curved tube, tube c. 13 cm long, purple. Stamens 6, adnate to the throat of the perianth tube; filaments filiform, c. 9 cm long, purplish; anthers linear, 1.5-2.5 cm long, dorsifixed, yellow. Carpels 3, syncarpous; ovary inferior, c. 1.8 cm long, 3-celled, purple; style single, filiform, c. 22 cm long; stigma sub-capitate; placentation axile. Fruit not formed.

Flowering: Almost throughout the year.

Specimens examined: **Dhaka**: Dhaka University Botanical Garden, 15.11.2006, Sumona 3 (DUSH); Cantonment, Shaheed Anwar Girls College campus, 15.11.2006, 06.10.2016, Sumona 105 (DUSH).

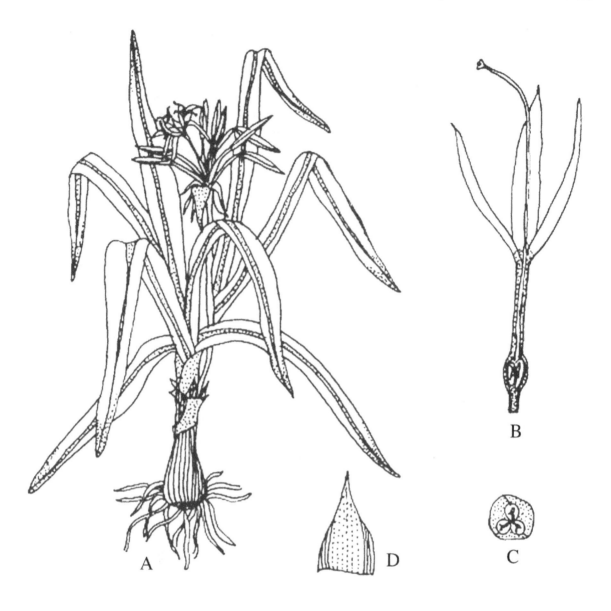

Fig. 1. *Crinum amabile* Donn: A. Habit (×0.1); B. L.S. of a flower (×0.2); C. T.S. of ovary (×2); D. Bract (× 0.1).

Chromosome number: 2n = 33 (Ahmed *et al.*, 2004).
Habitat: Cultivated in gardens.
Distribution: South Africa, Tropical regions of Asia. In Bangladesh, the species is cultivated in some private institutions and roadsides.
Economic value: Ornamental.
Propagation: By bulb separation.

Crinum amoenum Roxb., Hort. Beng. : 23 (1814); Roxb., Fl. Ind. 2: 127 (1832); Hook. f., Fl. Brit. Ind. 6: 282 (1892); Prain, Beng. Pl. 2: 798 (1903); Hassan, Encycl. Flora & Fauna of Bangladesh 11: 340 (2007). *Crinum himalense* Royle, Ill. Bot. Himal. Mts. (1839); *Crinum verecundum* Carey *ex* M. Roem., Fam. Nat. Syn. Monogr. : 75 (1847). **(Figs 2 & 7D-F)**.

English names: Himalayan Crinum, Tiger Lily.

Bangla name: *Gang Kachu*.

A bulbous perennial herb, bulb globose, 5.0-7.5 cm in diameter. Leaves 45-60×2.5-4.0 cm, bright-green, sub-erect, ensiform, tapering from the base to the tip, acuminate, margin sub-scabrous. Scape 30-60 cm long, rather slender, sub-cylindric, greenish-purple. Inflorescence of 6-12 flowered umbels; spathes 2, c. 5 cm long, lanceolate; bracteoles many. Flowers sub-sessile. Perianth tube green, 7.5-10.0 cm long, lobes 5.0-7.5 cm long, linear-lanceolate, longer than the filaments, white. Stamens 6; filaments red, c. 6 cm long, shorter than the perianth lobes; anthers oblong, dorsifixed. Carpels 3; ovary 3-celled, inferior, c. 1.6 cm long; placentation axile. Fruit a capsule. Seeds 1-5, irregularly round.

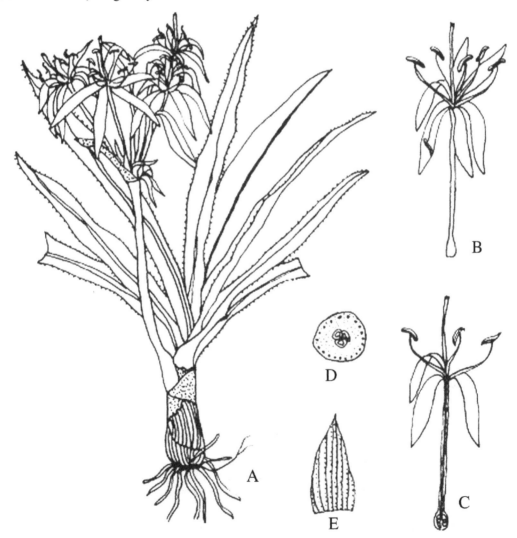

Fig. 2. *Crinum amoenum* Roxb.: A. Habit (×0.2); B. Flower (×0.3); C. L.S. of a flower (×0.3); D. T.S. of ovary (×3); E. Bract (×0.5).

Flowering and fruiting: May–August.

Specimens examined: **Dhaka**: Baldha Garden, 26.05.2007, Sumona 38 (DUSH); Dokkhin Middle Faidabad, 24.05.2007, Sumona 36 (DUSH). **Patuakhali**: Galachipa, Rangabali, 23.03.2006, M. Sultana 1208 (DUSH); Patuakhali Sadar, Laukathi, 15.05.2006, M. Sultana 1268 (DUSH). **Chittagong**: Chunati, Goalmara, 28.06.1997, Rahman *et al.* 663B (HCU). **Cox's Bazar**: Teknaf, Upazila Sadar, 25.05.2014, Sumona 88 (DUSH).

Chromosome number: $2n = 18, 22$ (Kumar and Subramaniam, 1986).

Habitat: In forests, plain lands and gardens.

Distribution: Tropical Himalayas, India (Sikkim and Khasia Hills), Nepal and Myanmar. In Bangladesh, it is distributed in Dhaka, Patuakhali, Sylhet, Cox's Bazar and Chittagong districts.

Economic value: Ornamental.

Propagation: By seeds and sucker formation.

Crinum asiaticum L., Sp. Pl.: 292 (1753); Hook. f., Fl. Brit. Ind. 6: 280 (1892); Prain, Beng. Pl. 2: 797 (1903); Utech, Fl. North Am. 26: 279 (2002); Hassan, Encycl. Flora & Fauna of Bangladesh 11: 340 (2007). *Amaryllis carnosa* Herb. Ham. *ex* Hook. f., Fl. Brit. Ind. 6: 280 (1892). *Crinum albiflorum* Noronha, Verh. Batav. Genootsch. Kunst. 5(Art. 4): 12 (1790). *Crinum angustifolium* Herb. *ex* Steud., Nomencl. Bot. ed. 2, 1: 438 1(840). *Crinum bancanum* Kurz, Tijdschr. Nederl. Ind. 27: 231 (1864). *Crinum bracteatum* Willd., Sp. Pl., ed. 4. 2(1): 47 (1799). *Crinum hornemannianum* M. Roem., Fam. Nat. Syn. Monogr. : 71 (1847). *Crinum macrocarpum* Carey *ex* Kunth, Enum. Pl. 5: 553 (1850). *Crinum plicatum* Livings. *ex* Hook., Bot. Mag. 56: t. 2908 (1829). *Crinum rumphii* Merr., Interpr. Rumph. Herb. Amboin. : 141 (1917). *Crinum sumatranum* Roxb., Fl. Ind. 2: 131 (1832). *Crinum umbellatum* Carey *ex* Herb., Bot. Mag. 47: sub t. 2121, p. 7 (1820). *Crinum woolliamsii* L.S. Hannibal, Herbert. 43(1): 14 (1987). *Crinum toxicarium* Roxb., Fl. Ind. 2: 134 (1832). **(Figs 3 & 7G-I)**.

English names: Poison Bulb, Giant Crinum Lily, Crinum Lily.

Bangla names: *Bara Kanur, Nagdal, Kachori, Sukhdarshan, Gaerhonar-pata*.

A perennial herb with a large tunicated bulb. Leaves long, 36-48×3-5 cm, lorate, margin entire, acute, wavy, glabrous, green in colour. Scape solid, 15-50 flowered umbels, green, glabrous. Flowers large, actinomorphic, bisexual, epigynous, white, fragrant at night, pedicellate; pedicel c. 3.3 cm long. Bracts 2, c. 6.5×3.2 cm, ovate-lanceolate, acute, greenish-white, bracteoles many, linear, white in colour. Perianth segments 6, c. 8×1 cm, white, lower parts forming a long, straight tube, tube erect, greenish, c. 7.5 cm long, equalling the linear lobes, lobes revolute. Stamens 6, adnate to the throat of the perianth tube; filaments filiform, c. 4.6 cm long, purplish in upper half and white in lower half; anthers linear, 1.5-2.5 cm long, dorsifixed, yellow. Carpels 3, syncarpous, green; ovary inferior, c. 1.5 cm long, 3-celled, placentation axile; style single, filiform; stigma sub-capitate. Fruit a capsule, c. 3.0×1.5 cm, sub-globose, beaked, green, bursting irregularly. Seeds round, concave.

Flowering and fruiting: March–November.

Specimens examined: **Dhaka**: Dhaka University Botanical Garden, 08.08.2007, Sumona 43 (DUSH); *ibid*, 01.07.1968, Mozahar 155; 05.09.1994, M.M. Khan 89; Uttara, Sector No. 8, 12.07.2007, Sumona 41 (DUSH). **Jhalakathi**: Chankati, 03.03.1987, Huq & Mia 6667 (DACB). **Khulna**: Sundarban, Manderbaria, 21.08.2002, S. Nasir Uddin N-1386(1); Sundarban, Kotka, 24.08.2010, Sumona 65; Kotka, 21.09.2011 Sumona 71 (DACB). **Mymensingh**: Bhaluka, 03.07.2001, M.S. Hossain 229; Ishwarganj, 05.07.2001, M.S. Hossain 261 (DACB). **Patuakhali**: Kalapara, Nilganj, 11.03.1999, M. Sultana 320 (DUSH); Patuakhali Sadar, Lohalia, 14.05.2005,

M. Sultana 714 (DUSH); Kalapara, Gongamoti, 07.01.2006, M. Sultana 935 (DUSH); Kalapara, 08.08.2013, Sumona 81 (DUSH).

Chromosome number: 2n = 22 (Kumar and Subramaniam, 1986).

Habitat: Homesteads, coastal areas, and also cultivated in gardens.

Distribution: Throughout the tropical parts of India, Sri Lanka and Nepal. In Bangladesh, it is common in the Sundarbans and coastal areas of Chittagong, and also planted in gardens.

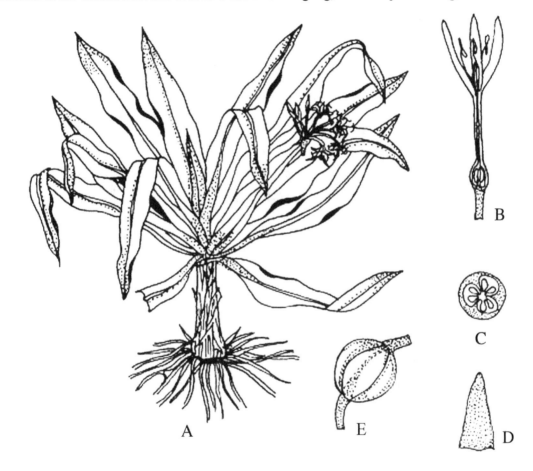

Fig. 3. *Crinum asiaticum* L.: A. Habit (×0.1); B. L.S of a flower (×0.4); C. T.S. of ovary (×5); D. Bract (×0.3); E. Fruit (×1).

Economic value: Widely planted in the gardens for its beautiful flowers. The bulb contains the alkaloids lycorine, crinidine and hamayne (Ghani, 2003). The bitter bulb is tonic, laxative, expectorant, used in biliousness and strangury and other urinary complaints. Fresh root is emetic, nauseant and diaphoretic. Seeds are purgative, diuretic, emmenagogue and tonic. Leaves are expectorant, applied to skin diseases and to reduce inflammation (Sinha, 1996). Tuber is useful in bronchitis and diseases of the chest and lungs, gonorrhoea, night blindness and defective vision, disease of the spleen, urinary conceretions, lumbago, anuria, toothache and snake-bite (Kirtikar *et al.*, 1935).

Ethnobotanical information: Leaf juice is used in ear-ache (Yadav and Bhandoria, 2013).

Propagation: By bulbs and seeds.

Crinum defixum Ker-Gawl., Quart. Journ. Sci. 3: 105 (1817). Hook. f., Fl. Brit. Ind. 6: 281 (1892); Prain, Beng. Pl. 2: 798 (1903); Cooke, Fl. Pres. Bomb. 2: 749 (1908); Haines, Bot. Bih. Or.: 1108 (1924); Hassan, Encycl. Flora & Fauna of Bangladesh 11: 341 (2007). *Crinum asiaticum* Roxb., Hort. Beng. : 23 (1814). *Crinum viviparum* (Lamk.) R. Ansari & V.J. Nair, J. Econ. Taxon. Bot. 11(1): 205 (1988). **(Figs 4 & 7J)**.

English names: Poison Bulb, Crinum Lily.

Bangla name: *Sukhdarshan*.

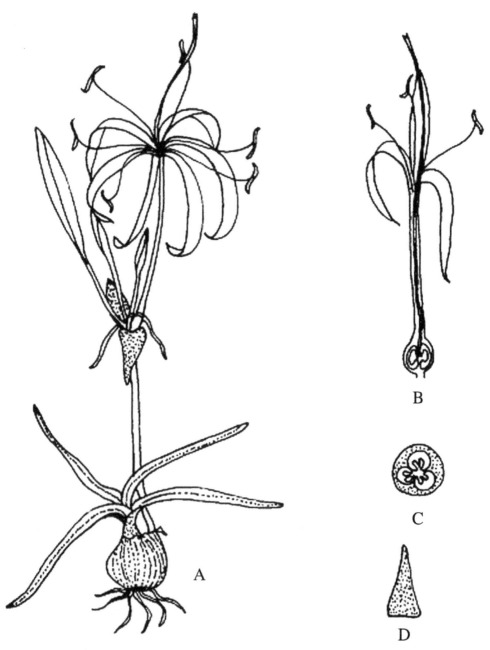

Fig. 4. *Crinum defixum* Ker-Gawl.: A. Habit (×0.2); B. L.S. of a flower (×0.4); C. T.S of ovary (×0.3); D. Bract (×0.2).

Very stout bulbous herb, bulb with a fusiform stoloniferous base, neck cylindric. Leaves 30-80×2-3 cm, linear or linear-lanceolate, concave, smooth, entire, obtuse. Scape 35-50 cm long, usually shorter than the leaves, compressed, smooth; spathe 2-leaved, bracteoles filiform. Flowers in umbels, umbel usually 6-15 flowered, bisexual, large, shortly pedicellate. Perianth white, tube cylindric, 6.0-7.5 cm long, segments 6, linear, nearly as long as the tube. Stamens 6, adnate to the throat of the perianth tube, spreading, recurved; filaments white or pink, shorter than the perianth lobes; anthers oblong, brown, versatile. Carpels 3, syncarpous; ovary inferior, 3-celled; style erect, exserted; stigma simple. Fruit a capsule, ellipsoid, c 2.5 cm long, 1-2 seeded. Seeds large, rugose.

Flowering and fruiting: May–August.

Specimens examined: **Dhaka**: Dhaka University Botanical Garden (originally collected from Char Kukri Mukri), 05.07.2017, Sumona 110 (DUSH); Savar: Jahangirnagar University campus, 30.04.2015, Sumona 94 (DUSH). **Patuakhali**: Bhupal, Kalaiya, 13.03.1973, M. S. Khan K-2843 (DACB); Patuakhali Sadar, Lohalia, 18.11.2004, M. Sultana 462 (DUSH); Mirzaganj, Subidkhali, 20.11.2004, M. Sultana 565 (DUSH); Galachipa, Basbunia, 01.03.2005, M. Sultana 619 (DUSH); Galachipa, Panpotti, 18.12.2010, M. Sultana 1860 (DUSH).

Chromosome number: 2n = 22 (Alam *et al.*, 1998); 50, 60 (Kumar and Subramaniam, 1986).

Habitat: Swampy river banks and gardens where it is commonly cultivated.

Distribution: Throughout tropical India and Sri Lanka. In Bangladesh, it is well represented in forests and many gardens.

Economic value: Commonly cultivated in the gardens for its beautiful large fragrant flowers. Bulb is nauseous, emollient, emetic and diaphoretic. The plant is toxic to cattle (Sinha, 1996). Bulb and stolon are administered in the treatment of burns and carbuncle. In otitis a few drops of juice of leaves are instilled into the ear. In Rema Kalenga area of Moulvi Bazar district bulbs are used for the treatment of stomach complaints of cow (Yusuf *et al.*, 2009).

Propagation: By bulbs.

Crinum jagus (Thomps.) Dandy, Journ. Bot. Lond. 77: 64 (1939). *Amaryllis jagus* Thomps., Bot. Displ. : t. 6 (1798); *Crinum giganteum* Andr., Bot. Rep. : t. 169 (1810). **(Figs 5 & 7K)**.

English name: Giant Crinum.

Bangla name: *Sukhdarshan*.

A bulbous perennial herb, bulb globose, 12.5-15.0 cm in diameter with c. 7 cm long neck. Leaves many, 60-90×7-12 cm, lorate or lanceolate, margin entire, wavy, acute or obtuse. Scape 30-90 cm long, green; spathes 2, greenish-white, ovate-lanceolate, c. 9.7×5.9 cm, obtuse; bracteoles 4-8, linear-lanceolate, c. 8.0×0.7 cm, greenish-white. Inflorescence of 4-8 flowered umbels, short-pedicelled or sessile. Perianth segments 6, c. 11.5×4.0 cm, ovate-lanceolate, fragrant, white, lobes as long as or shorter than the tube, tube c. 19 cm long, green. Stamens 6; filaments adnate to the throat of the perianth tube, 6-8 cm long, shorter than the perianth lobes, curved, white; anthers oblong, c. 1.5×0.2 cm, dorsifixed, versatile, spiral after bursting. Carpels 3, syncarpous; ovary 3-celled, inferior, c. 2.5×1.5 cm; placentation axile; style with stigma c. 9.5 cm long, green. Fruit a sub-globose capsule. Seeds not found.

Flowering and fruiting: April–July.

Specimens examined: **Dhaka**: Dhaka University campus, Science Library, 03.05.2007, Sumona 25 (DUSH); Near Charukala Institute, 26.05.2007, Sumona 39 (DUSH); Dhaka University Btanical Garden (originally collected from Char Kukri Mukri), 10.05.2017, Sumona 106 (DUSH).

Chromosome number: 2n = 33 (Kumar and Subramaniam, 1986).

Habitat: Soil rich in organic matter.

Distribution: Native to tropical Africa. Found in Sri Lanka, India, Myanmar, and Malaysia. In Bangladesh, it is found to be grown in different gardens.

Economic value: Cultivated in the gardens for its large beautiful flowers.

Ethnobotanical information: Crushed and roasted bulbs are used in rheumatism. Leaf juice is used in ear-ache (Sinha, 1996).

Propagation: By bulbs.

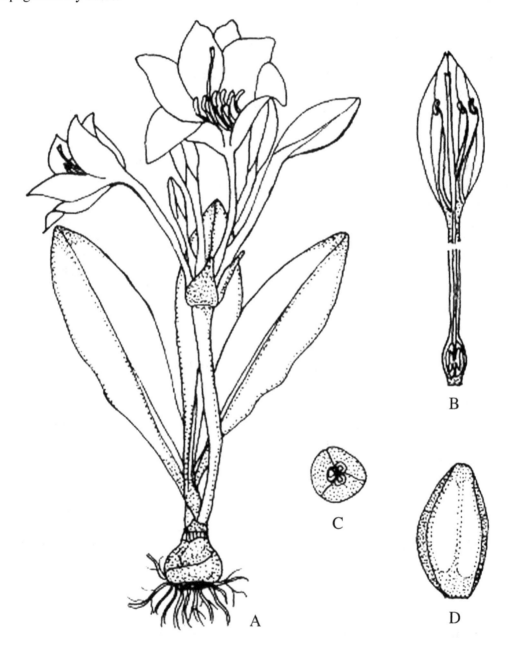

Fig. 5. *Crinum jagus* (Thomps.) Dandy: A. Habit (×0.1); B. L.S. of a flower (×0.1); C. T.S. of ovary (×2); D. Bract (×0.1).

Crinum latifolium L., Sp. Pl.: 291 (1753); Hook. f., Fl. Brit. Ind. 6: 283 (1892); Prain, Beng. Pl. 2: 798 (1903); Raven and Zhengyi, Fl. China 24: 265 (2000); Hassan, Encycl. Flora & Fauna of Bangladesh 11: 341 (2007). *Crinum ornatum* Herb., Amaryll. : 262 (1837). *Crinum moluccanum* Roxb., Fl. Ind. 2: 140 (1859). *Crinum zeylanicum* L., Syst. ed. 12 (1767). **(Figs 6 & 7L)**.

English name: Pink Striped Trumpet Lily.

Bangla name: *Sukhdarshan*.

A bulbous perennial herb, bulb globose, 12.5-15.0 cm in diameter with a short neck. Leaves many, 60-90×7-12 cm, lorate, margin sub-scabrid. Scape 60-90 cm long, greenish-purple or yellowish-green; spathes 2, reddish-green or purple, lanceolate. Inflorescence of 6-12 flowered umbels, short-pedicelled. Perianth segments 6, c. 12.2×3.0 cm, perianth tube curved, c. 7 cm long, lobes 7-15 cm long, as long as or shorter than the tube, elliptic-oblong or elliptic-lanceolate, fragrant, white, more or less streaked or tinged with red towards the centre, sometimes red-purple, nearly all over the back. Stamens 6, declinate; filaments adnate to the throat of the perianth tube, 6-8 cm long, shorter than the perianth lobes; anthers oblong, 1.3-2.0 cm long, grey, dorsifixed, versatile. Carpels 3, syncarpous; ovary inferior, 3-celled, c. 1 cm long; placentation axile. Fruit a sub-globose capsule, c. 4.5×3.0 cm, pinkish-maroon.

Flowering and fruiting: May–September.

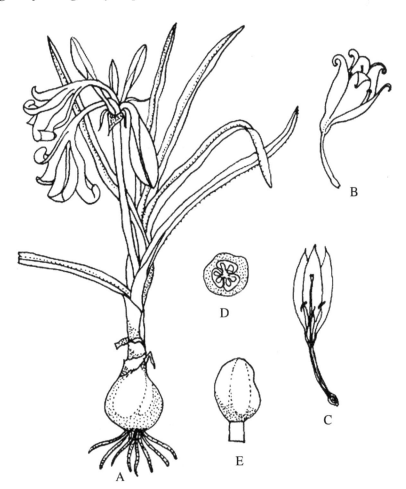

Fig. 6. *Crinum latifolium* L.: A. Habit (×0.1); B. Flower (×0.1); C. L.S. of a flower (×0.1); D. T.S. of ovary (×2); E. Fruit (×0.4).

Specimens examined: **Dhaka**: Dhaka University Botanic Garden, 28.04.2007, Sumona 24 (DUSH); Dhaka University Campus, Science Library, 19.09.2007, Sumona 44 (DUSH); *ibid*. 20.08.2012, Sumona 74 (DUSH).

Chromosome number: 2n = 22, 33 (Kumar and Subramaniam, 1986).

Habitat: Soil rich in organic matter.

Distribution: Native to tropical Asia. Distributed throughout Sri Lanka, India and Myanmar, also in Malaysia and Africa. In Bangladesh, it is cultivated in different gardens.

Economic value: The bulbs are extremely acidic. In India, when roasted, they are used as rubifacient, or crushed on piles and abscesses to cause suppuration. Leaf juice is used for ear-ache (van Valkenburg and Bunyapraphatsara, 2002). Crushed and roasted bulbs are used in rheumatism (Sinha, 1996).

Ethnobotanical information: In some parts of India bulbs are used in traditional medicine (Kehimkar, 2000).

Propagation: By bulbs.

Crinum pratense Herb., Amaryll.: 256 (1837). Hook. f., Fl. Brit. Ind. 6: 282 (1892); Prain, Beng. Pl. 2: 798 (1903); Cooke, Fl. Pres. Bomb. : 750 (1908); Hassan, Encycl. Flora & Fauna of Bangladesh 11: 342 (2007). *Crinum longifolium* Roxb., Fl. Ind. 2: 130 (1832). *Crinum lorifolium* Roxb. *ex* Ker-Gawl., J. Sci. Arts 3(5): 110 (1817).

Bangla names: *Sukhdarshan, Bon Peyaj*.

A bulbous perennial herb, bulb ovoid or spherical, 10-13 cm in diameter, neck 5-7 cm across. Leaves 45-90 cm long, linear, channelled, sub-erect or declinate, entire, obtuse. Scape c. 30 cm or more long, compressed, decumbent; spathe 5.0-7.5 cm long, deltoid-lanceolate. Flowers in umbels, white, fragrant, shortly pedicellate, bisexual, epigynous. Perianth tube 7.5-10.0 cm long, perianth lobes lanceolate. Stamens 6, adnate to the throat of the perianth tube; filaments filiform, red; anthers oblong, dorsifixed, bursting longitudinally. Carpels 3, syncarpous; ovary inferior, 3-celled; style single; stigma simple. Fruit a capsule.

Flowering and fruiting: May–August.

Specimen examidned: **Dhaka**: Dhaka University Botanic Garden (originally collected from Chanbari beat of Rema-Kalenga Wildlife Sanctuary in Habiganj), 01.06.2000, Zashim Uddin 835 (DACB).

Chromosome number: 2n = 22 (Alam *et al.*, 1998).

Habitat: Plain lands, also on the bank of channel (Uddin and Hassan, 2004).

Distribution: Plains of India and Myanmar. In Bangladesh, it is found both in wild and planted in household gardens.

Economic value: Used as an ornamental herb.

Propagation: By bulbs.

Crinum stenophyllum Baker, Gard. Chron. 1: 786 (1881); Handb. Amaryl. : 75 (1888); Hook. f., Fl. Brit. Ind. 6: 281 (1892); Hassan, Encycl. Flora & Fauna of Bangladesh 11: 342 (2007).

Herbs. Leaves 90×0.6-1.0 cm, linear, flaccid. Scape very slender, 2-edged. Inflorescence umbel, 4-6 flowered. Spathe c. 5 cm long, lanceolate. Pedicel c. 0.6 cm long. Perianth tube 7-10 cm long, very slender, lobes half as long or longer.

Specimen examined: No specimen was examined because of unavailability in nature and in any herbarium of Bangladesh.

Distribution: India, Bangladesh and Myanmar.

Notes: J.D. Hooker reported this species from Sylhet district in 1892. Since then there has been no further report of its occurrence from anywhere Bangladesh and no specimen available at any herbarium of Bangladesh. Hence, the species is presumed to be extinct in Bangladesh.

Fig. 7. Photographs of *Crinum* L. species: A-C. *Crinum amabile* Donn; D-F. *C. amoenum* Roxb.; G-I. *C. asiaticum* L.; J. *C. defixum* Ker-Gawl.; K. *C. jagus* (Thomps.) Dandy; L. *C. latifolium* L.

Acknowledgement

The first author is thankful to Prime Minister's Office for financial support under 'Prime Minister's Research and Higher Education Supporting Fund' for carrying out the study.

References

APG III. 2009. An update of the Angiosperm Phylogeny Group classification for the orders and families of flowering plants: APG III. Bot. J. Linn. Soc. **161**: 105–121.

Afroz, S. and Hassan, M.A. 2008. Systematic studies in the family Liliaceae from Bangladesh. Bangladesh J. Plant Taxon. **15**(2): 115–128.

Ahmed, L., Begum, R., Noor, S.S., Zaman, M.A. and Alam, S.S. 2004. Reversible flurescent chromosome banding in three *Crinum* spp. (Amaryllidaceae). Cytologia **69**(1): 69–74.

Alam, S.S., Sharmin, S.A., Sarker, R.H. and Zaman, M.A. 1998. Karyotype analysis with differential bandings in *Crinum pratense* and *C. defixum*. Cytologia **63**: 223–227.

Baker, J.G. 1881. Synopsis of known species of *Crinum*. I. Gard. Chron. **15**: 763.

Baker, J.G. 1888. Handbook of the Amaryllidaceae. London: George Bell and Sons, London, pp. 74–95.

Baker, J.G. 1896. Amaryllidaceae. Flora Capensis. L. Reeve and Co. Ltd., Kent, England. **6**: 171–246.

Baker, J.G. 1898. Amaryllidaceae. *In*: Thiselton-Dyer, W.T. (Ed.), Flora of Tropical Africa, Vol. **7**. London: L. Reeve, pp. 376–413.

Chase, M.W., Reveal, J.L. and Fay, M.F. 2009. A subfamilial classification for the expanded asparagalean families Amaryllidaceae, Asparagaceae and Xanthorrhoeaceae Bot. J. Linn. Soc. **161**: 132–136.

Cronquist, A. 1981. An Integrated System of Classification of Flowering Plants. Columbia University Press, New York.

Dassanayake, M.D. 2000. Amaryllidaceae. *In*: Dassanayake, M.D. and Clayton, W.D. (Eds), A Revised Handbook to the Flora of Ceylon. Oxford and IBH Publishing Co. Pvt. Ltd., New Delhi & Calcutta. **14**: 18–20.

Fennell, C.W. and van Staden, J. 2001. *Crinum* species in traditional and modern medicine. J. Ethnopharmacol. **78**: 15–26.

Ghani, A. 2003. Medicinal Plants of Bangladesh. Asiatic Society of Bangladesh, Dhaka, pp. 1-460.

Govaerts, R., Snijman, D.A., Marcucci, R., Silverstone-Sopkin, P.A. and Brullo, S. 2012. World checklist of Amaryllidaceae. Facilitated by the Royal Botanic Gardens, Kew. <http://apps.kew.org/wcsp>.

Hannibal, L.S. and Williams, A. 1998. The Genus *Crinum* (Amaryllidaceae). <http://www.crinum.org>. Retrieved on 15 July 2018

Hassan, M.A. 2007. Liliaceae. *In*: Siddique, K.U., Islam, M.A., Ahmed, Z.U., Begum, Z.N.T., Hassan, M.A., Khondker, M, Rahman, M.M., Kabir, S.M.H., Ahmad, A.T.A., Rahman, A.K.A. and Haque, E.U. (Eds), Encyclopedia of Flora and Fauna of Bangladesh. Vol. **11**. Angiosperms: Monocotyledons (Agavaceae-Najadaceae). Asiatic Society of Bangladesh, Dhaka, pp. 339–343.

Herbert, W.M. 1820. Specierum Enumeratio *ex* pt. *Crinum broussonetii*. Botanical Magazine 47: 4–8.

Herbert, W. 1837. Amaryllidaceae. James Ridgway & Sons, Piccadilly, London, pp. 242–275.

Hooker, J.D. 1892. The Flora of British India. Vol. **6**. L. Reeve and Co. Ltd., Kent, England, pp. 280–284.

Jayaweera, D.M.A. 1981. Medicinal Plants used in Ceylon. Part I. National Science Council of Sri Lanka, Colombo, pp. 1–61.

Karthikeyan, S., Jain, S.K., Nayar, M.P. and Sanjappa, M. 1989. Flora of India. Ser. **4**. Botanical Survey of India, Pune, India, pp. 1–102.

Kehimkar, I. 2000. Common Indian Wild Flowers. Bombay Natural History Society. Oxford University Press, Calcutta, India, pp. 1–141.

Kirtikar, K.R., Basu, B.D. and An, I.C.S. 1935 (reprint ed. 1994). Indian Medicinal Plants (Second edition). Vol. **4**. Bishen Singh Mahendra Pal Singh, Dehra Dun, India, pp. 2465–2528.

Kumar, V. and Subramaniam, B. 1986. Chromosome Atlas of Flowering Plants of the Indian Subcontinent. Vol. **II**, Botanical Survey of India, Calcutta, India, pp. 1–464.

Kwembeya, E.G., Bjora, C.S., Stedje, B. and Nordal, I. 2007. Phylogenetic relationship in the genus *Crinum* (Amaryllidaceae) with emphasis on tropical African species: evidence from *trn*L-F and nuclear ITS DNA sequence data. Taxon **56**: 801–810.

Meerow, A.W., Lehmiller, D.J. and Clayton, J.R. 2003. Phylogeny and biogeography of *Crinum* L. (Amaryllidaceae) inferred from nuclear and limited plastid non-coding DNA sequences. Bot. J. Linn. Soc. **141**: 349–363.

Nordal, I. 1977. Revision of the East African taxa of the genus *Crinum* (Amaryllidaceae). Norwegian J. Bot. **24**: 179–194.

Prain, D. 1903. Bengal Plants, Vol. **2**. Botanical Survey of India, Calcutta, pp. 797–798.

Raven, P. and Zhengyi, W. (Eds) 2000. Flora of China, Vol. **24**. Flagillariaceae through Marantaceae. Science Press, Beijing, and Missouri Botanical Garden Press, St. Louis, pp. 1–431.

Sinha, S.C. 1996. Medicinal Plants of Manipur. Mass and Sinha, Manipur Cultural Integration Conference Palace Compound, Imphal, India, pp. 1–238.

The Plant List 2013. Version 1.1. Published on the Internet; http://www.theplantlist.org/ (Accessed on 30 July 2018).

Uddin, M.Z. and Hassan, M.A. 2004. Flora of Rema-Kalenga Wildlife Sanctuary. IUCN-Bangladesh Country Office, Dhaka, Bangladesh, pp. 1–122.

Uphof, J.C.T. 1942. A review of the species of *Crinum*. Herbertia **9**: 63–84.

Utech, F.H. 2002. Flora of North America: North of Mexico. Vol. **26**. Flora of North America Editorial Committee (Eds), Oxford University Press, New York, USA, pp. 1–752.

van Valkenburg, J.L.C.H. and Bunyapraphatsara, N. (Eds). 2002. Plant Resources of South-East Asia. No. **12**. Medicinal and Poisonous Plants 2. Backhuys Publishers, Leiden, the Netherlands, pp. 1–782.

Verdoorn, C. 1973. The Genus *Crinum* in Southern Africa. Bolhalia **11**(1&2): 27–52.

Win, Z.Z. 2011. Phytochemical investigation and antimicrobial activities of the leaves of *Crinum asiaticum* L. Univ. Res. J. **4**(1): 123–137.

Yadav, S.S. and Bhandoria, M.S. 2013. Ethnobotanical exploration in Mahendergarh district of Haryana (India). J. Med. Plants Res. **7**(18): 1263–1671.

Yakandawala, D.M.D. and Samarakoon, T.M. 2006. An empirical study on the taxonomy of *Crinum zeylanicum* (L.) L. and *Crinum latifolium* L. (Amaryllidaceae) occurring in Sri Lanka. Cey. J. Sci. (Biol. Sci.) **35**(1): 53–72.

Yusuf, M., Chowdhury, J.U., Haque, M.N. and Begum, J. 2009. Medicinal Plants of Bangladesh. Bangladesh Council of Scientific and Industrial Research, Chittagong, Bangladesh, pp. 1–794.

PERMISSIONS

All chapters in this book were first published in BJPT, by Bangladesh Association of Plant Taxonomists; hereby published with permission under the Creative Commons Attribution License or equivalent. Every chapter published in this book has been scrutinized by our experts. Their significance has been extensively debated. The topics covered herein carry significant findings which will fuel the growth of the discipline. They may even be implemented as practical applications or may be referred to as a beginning point for another development.

The contributors of this book come from diverse backgrounds, making this book a truly international effort. This book will bring forth new frontiers with its revolutionizing research information and detailed analysis of the nascent developments around the world.

We would like to thank all the contributing authors for lending their expertise to make the book truly unique. They have played a crucial role in the development of this book. Without their invaluable contributions this book wouldn't have been possible. They have made vital efforts to compile up to date information on the varied aspects of this subject to make this book a valuable addition to the collection of many professionals and students.

This book was conceptualized with the vision of imparting up-to-date information and advanced data in this field. To ensure the same, a matchless editorial board was set up. Every individual on the board went through rigorous rounds of assessment to prove their worth. After which they invested a large part of their time researching and compiling the most relevant data for our readers.

The editorial board has been involved in producing this book since its inception. They have spent rigorous hours researching and exploring the diverse topics which have resulted in the successful publishing of this book. They have passed on their knowledge of decades through this book. To expedite this challenging task, the publisher supported the team at every step. A small team of assistant editors was also appointed to further simplify the editing procedure and attain best results for the readers.

Apart from the editorial board, the designing team has also invested a significant amount of their time in understanding the subject and creating the most relevant covers. They scrutinized every image to scout for the most suitable representation of the subject and create an appropriate cover for the book.

The publishing team has been an ardent support to the editorial, designing and production team. Their endless efforts to recruit the best for this project, has resulted in the accomplishment of this book. They are a veteran in the field of academics and their pool of knowledge is as vast as their experience in printing. Their expertise and guidance has proved useful at every step. Their uncompromising quality standards have made this book an exceptional effort. Their encouragement from time to time has been an inspiration for everyone.

The publisher and the editorial board hope that this book will prove to be a valuable piece of knowledge for researchers, students, practitioners and scholars across the globe.

LIST OF CONTRIBUTORS

Gül Tarimcilar, Özer Yilmaz, Ruziye Daşkin and Gönül Kaynak
Department of Biology, Faculty of Arts and Science, Uludag University, 16059 Görükle Bursa, Turkey

Abeer Al-Andal and Suliman Alruman
Department of Biology, College of Science, King Khalid University, Abha, Kingdom of Saudi Arabia

Mahmoud Moustafa
Research Center for Advanced Materials Science (RCAMS), King Khalid mUniversity, Abha, Saudi Arabia
Department of Botany, Faculty of Science, South Valley University, Qena, Egypt

Jess H. Jumawan and Inocencio E. Buot, Jr
Institute of Biological Sciences, University of the Philippines, Los Baños, College, Laguna, Philippines

W. Jiang, Z. M. Tao and Z. G. Wu
College of Life science, Zhejiang Sci-Tech University, Hangzhou 310018, China

N. Mantri
Zhejiang Institute of Subtropical Crops, Zhejiang Academy of Agricultural Sciences Wenzhou 325005, China

H. F. Lu and Z. S. Liang
School of Applied Sciences, Health Innovations Research Institute, RMIT University, Melbourne 3000, Victoria, Australia

Ebadi-Nahari Mostafa
Department of Biology, Faculty of Science, Azarbaijan Shahid Madani University, Tabriz, Iran

Nikzat-Siahkolaee Sedigheh and Eftekharian Rosa
Faculty of Biological Sciences, Shahid Beheshti University, Tehran, Iran

Krishna Kumar Rawat
CSIR-National Botanical Research Institute, Lucknow, India

Afroz Alam
Department of Bioscience and Biotechnology, Banasthali University, Rajasthan, India

Praveen Kumar Verma
Forest Research Institute, Dehra Dun, India

Dhafer Ahmed Alzahrani
Department of Biological Sciences, Faculty of Science, King Abdulaziz University, Jeddah, Saudi Arabia

Enas Jameel Albokhari
Department of Biological Sciences, Faculty of Applied Sciences, Umm Al-Qura University, Makkah, Saudi Arabia

Kenan Yazici
Biology Department, Faculty of Science, Karadeniz Technical University, 61080, Trabzon, Turkey

André Aptroot
ABL Herbarium G.v.d. Veenstraat 107 NL-3762 XK Soest, The Netherlands

L. Rasingam, J. Swamy and S. Nagaraju
Botanical Survey of India, Deccan Regional Centre, Plot. No. 366/1, Attapur, Hyderguda Post, Hyderabad-500048, Telangana, India

Keum Seon Jeong and Jae Kwon Shin
Division of Forest Biodiversity, Korea National Arboretum, Pocheon, Gyeonggi-do 487-821, Korea

Masayuki Maki
Division of Ecology and Evolutionary Biology, Graduate School of Life Sciences, Tohoku University, Aoba, Sendai 980-8578, Japan

Jae-Hong Pak
Research Institute for Dok-do and Ulleung-do Island, Kyungpook National University, Daegu 702-701, Korea

Rajeev Kumar Singh
Botanical Survey of India (BSI), Southern Regional Centre (SRC), TNAU Campus, Lawley Road, Coimbatore 641 003, Tamil Nadu, India

Sumona Afroz, M. Oliur Rahman and Md. Abul Hassan
Department of Botany, University of Dhaka, Dhaka 1000, Bangladesh

S. Bandyopadhyay and P. P. Ghoshal
Central National Herbarium, Botanical Survey of India, Botanic Garden, Howrah 711 103, West Bengal, India

Kakali Sen
Department of Botany, University of Kalyani, Pin-741235, Kalyani, Nadia, West Bengal, India

Radhanath Mukhopadhyay
CAS, Department of Botany, University of Burdwan, Pin-713104, Burdwan, West Bengal, India

M. Oliur Rahman and Md. Abul Hassan
Department of Botany, University of Dhaka, Dhaka 1000, Bangladesh

M. Enamur Rashid and M. Atiqur Rahman
Department of Botany, University of Chittagong, Chittagong-4331, Bangladesh

A. K. M. Kamrul Haque and Saleh Ahammad Khan
Department of Botany, Jahangirnagar University, Savar, Dhaka 1342, Bangladesh

Sarder Nasir Uddin and Shayla Sharmin Shetu
Bangladesh National Herbarium, Zoo Road, Mirpur-1, Dhaka 1216, Bangladesh

Jennifer M. Conda
Department of Science and Technology-Forest Products Research and Development Institute, Los Baños, Laguna, Philippines

Inocencio E. Buot
Institute of Biological Sciences, College of Arts and Sciences, University of the Philippines Los Baños, Los Baños, Laguna, Philippines

Sumona Afroz, M. Oliur Rahman and Md. Abul Hassan
Department of Botany, University of Dhaka, Dhaka 1000, Bangladesh

Index

A
Alchornea, 155, 188
Aleuritopteris, 144-147, 149-153
Amaryllis, 125-127, 211, 216, 219
Anacardiaceae, 11, 189
Apocynaceae, 21, 28-29, 101, 162, 175-176, 190
Araceae, 135, 155, 158, 179-180, 194
Asclepiadaceae, 29, 96, 98, 101, 177, 190
Aspicilia, 89-91, 94-95
Athyriaceae, 200, 209-210

B
Bauhinia, 136-143
Bryophytes, 63

C
C. Rhytidophylla, 33, 38, 42
Caesalpinioideae, 136, 142-143
Camellia, 30-33, 37-40, 42-45
Cephalaria, 46-47
Cheilanthes, 144-146, 148-153
Cheilanthoid Ferns, 144, 146, 149, 153
Chukrasia, 12, 20
Codonocrinum, 211
Craspedodromous, 200, 202, 206-208
Crinum, 133, 197, 211-225
Cryptocarya, 155
Cyperaceae, 179-180, 195

D
Diplazium, 200-202, 206-210
Dipsacus, 46-47, 53
Doryopteris, 144-146, 149, 152

E
Ericaceae, 162, 172, 177-178
Eriodontes, 1
Euphorbiaceae, 155, 157-158, 179-180, 188

F
Fissidens, 54, 57-59, 62

G
Galium, 102-105, 107-108, 110, 156

H
H. Buotii, 21-22, 25-26, 28
H. Mindorensis, 21-22, 25-26, 28
Halconensis, 21-22, 25-26, 28
Hallieriana Elmer, 136-137
Hoya R. Br., 21, 29
Hylaea, 102, 104
Hymeneliaceae, 89
Hyophila, 54, 59-60, 62

K
Knautia, 46

L
L. Nepetifolia, 112, 116
Lamiaceae, 1, 9-10, 112, 121-122, 191
Lecanora Subcarnea, 89-92, 94
Lens Spp, 12
Leucas, 112-122, 191
Liliaceae, 123, 133-134, 155-156, 158, 160, 197, 211-212, 224
Liliopsida, 179-180, 194
Lomelosia, 46, 52-53

M
M. Dumetorum, 2
M. Pulegium, 1-2, 4-5, 8
M. Spicata, 1-6, 8-9
M. Spicata Subsp. Tomentosa, 1, 3-4, 6, 8-9
M. Suaveolens, 1-5, 7-8
M. × Piperita, 1-2, 4-5, 8
M. × Rotundifolia, 1-4, 7-9
M. × Villoso-nervata, 1-4, 7-9
Magnoliopsida, 179-180, 182
Mentha, 1-4, 7-10

N
Notholaena, 144-146, 149

O
Oleaceae, 162, 174-175, 191

P
Pellaea, 144-146, 149, 152
Phanera, 136-143
Physcomitrium, 54, 60, 62
Polyploidisation, 1
Pterocephalus, 46-47, 49, 51-53

R
Reticulodromous, 200, 202, 206-208
Rubiaceae, 102, 110-111, 155-160, 162-169, 179-180, 192

S
Scabiosa, 46-47, 52-53
Schinus Molle, 11, 20
Sect. Furfuracea, 30-31, 33, 37-39, 42-43
Styraceae, 162, 174

Subgenus Platyaster, 211
Symplocaceae, 162, 174, 186

T
Tetraena, 64-83, 86-87
Thelidium, 89, 91-95
Theopsis, 30-33, 37-39, 42
Trachygalium, 102, 105, 108-109
Tubulosae, 1
Tylophora, 96-101

V
Vaccinaceae, 162

Z
Z. Atamasco, 123-125, 128-133
Zephyranthes, 123-127, 129-130, 132-135
Ziziphus Spina-christi, 12, 20
Zygophyllaceae, 64-66, 86-88
Zygophyllum, 64-69, 71, 73, 76-78, 80-81, 83-88